STATA TIME-SERIES
REFERENCE MANUAL
RELEASE 8

A Stata Press Publication
STATA CORPORATION
College Station, Texas

Stata Press, 4905 Lakeway Drive, College Station, Texas 77845

The suggested citation for this software is

StataCorp. 2003. *Stata Statistical Software: Release 8.0*. College Station, TX: Stata Corporation.

Table of Contents

Cross-Referencing the Documentation

When reading this manual, you will find references to other Stata manuals. For example,

[U] **29 Overview of Stata estimation commands**

[R] **regress**

[P] **matrix define**

The first is a reference to Chapter 29, *Overview of Stata estimation commands* in the *Stata User's Guide*, the second is a reference to the `regress` entry in the *Base Reference Manual*, and the third is a reference to the `matrix define` entry in the *Programming Reference Manual*.

All of the manuals in the Stata Documentation have a shorthand notation, such as [U] for the *User's Guide* and [R] for the *Base Reference Manual*.

The complete list of shorthand notations and manuals is as follows:

[GSM]	*Getting Started with Stata for Macintosh*
[GSU]	*Getting Started with Stata for Unix*
[GSW]	*Getting Started with Stata for Windows*
[U]	*Stata User's Guide*
[R]	*Stata Base Reference Manual*
[G]	*Stata Graphics Reference Manual*
[P]	*Stata Programming Reference Manual*
[CL]	*Stata Cluster Analysis Reference Manual*
[XT]	*Stata Cross-Sectional Time-Series Reference Manual*
[SVY]	*Stata Survey Data Reference Manual*
[ST]	*Stata Survival Analysis & Epidemiological Tables Reference Manual*
[TS]	*Stata Time-Series Reference Manual*

Detailed information about each of these manuals may be found online at

`http://www.stata-press.com/manuals/`

Title

> **intro** — Introduction to time-series manual

Description

This entry describes the *Stata Time-Series Reference Manual.*

Remarks

This manual documents the time-series commands, and is referred to as [TS] in cross-references. Following this entry, [TS] **time series** provides an overview of the ts commands.

This manual is arranged alphabetically. If you are new to Stata's time-series commands, we recommend that you read the following sections first:

> [TS] **time series**   Introduction to time-series commands
> [TS] **tsset**   Declare dataset to be time-series data

Stata is continually being updated. Stata users are always writing new commands, as well. To find out about the latest time-series features, type search time series after installing the latest official updates; see [R] **update**.

What's new

Stata 8 has a number of new time-series commands. This section is intended for previous Stata users. If you are new to Stata, you may as well skip it.

1. var fits vector autoregressive models.

2. svar fits structural vector autoregressive models.

3. varbasic fits simple vector autoregressive models and graphs the impulse–response functions or the forecast-error variance decompositions.

4. varfcast produces dynamic forecasts of the dependent variables in a previously fitted var or svar.

5. vargranger performs Granger causality tests after var or svar.

6. varirf estimates the impulse–response functions, the cumulative impulse–response functions, the orthogonalized impulse functions, and the forecast error decomposition after var. In addition, varirf estimates the structural impulse–response functions and the structural forecast-error variance decompositions after svar.

7. varlmar computes Lagrange multiplier test statistics for residual autocorrelation after var or svar.

8. varnorm computes a series of test statistics against the null hypothesis that the disturbances are normally distributed after estimation with var. For each equation, and for all equations jointly, three statistics are computed: a skewness statistic, a kurtosis statistic, and the Jarque–Bera statistic.

9. varsoc reports a series of lag-order selection statistics.

10. varstable checks the eigenvalue stability condition after fitting a VAR or SVAR.

11. varwle performs a Wald test for each equation in a VAR, and all equations jointly, of the hypothesis that all the endogenous variables of a given lag are jointly zero.

1

12. There are six new time-series smoothers:

tssmooth ma	weighted or unweighted moving average
tssmooth exp	single exponential smoother
tssmooth dexp	double exponential smoother
tssmooth hwinters	Holt–Winters nonseasonal smoother
tssmooth shwinters	Holt–Winters seasonal smoother
tssmooth nl	nonlinear smoother

13. tsappend appends observations to a time-series dataset. tsappend uses the information set by tsset, automatically fills in the time variable, and fills in the panel variable if the panel variable was set.

14. dfgls performs the Dickey–Fuller GLS test for a unit root on a time series. dfgls reports the test statistic based on models with 1 to *maxlags* lags of the first-differenced variable in the augmented Dickey–Fuller regression.

15. arima now allows estimation and prediction using samples that are wholly contained within one panel of a panel dataset (a dataset that has been tsset with both panel and time identifiers). This allows arima to be used in loops over panels. Previously, unless the condition option was specified, arima refused to fit any models on panel data.

    Also, arima may now be used with by.

16. archlm computes a Lagrange multiplier test for autoregressive conditional heteroskedasticity (ARCH) effects in the residuals after regress.

17. bgodfrey computes the Breusch–Godfrey Lagrange multiplier (LM) test for serial correlation in the error distribution after regress.

18. durbina computes Durbin's alternative statistic to test for serial correlation in the disturbances after regress when some of the regressors are not strictly exogenous.

For a complete list of all the new features in Stata 8, see [U] **1.3 What's new**.

Also See

Complementary: [U] **1.3 What's new**

Background: [R] **intro**

Title

> **time series** — Introduction to time-series commands

Description

The *Time-Series Reference Manual* organizes the commands alphabetically, which makes it easy to find individual command entries if you know the name of the command. The overview that is presented below organizes and presents the commands conceptually; that is, according to the similarities in the functions that they perform.

The commands under the heading **Data management tools and time-series operators** help you prepare your data for further analysis. The commands under the heading **Univariate time-series** are grouped together because they are either estimators or filters that are designed for univariate time-series, or they are pre-estimation or post-estimation commands that are conceptually related to one or more univariate time-series estimators. The commands under the heading **Multivariate time-series** are similarly grouped together because they are either estimators designed for multivariate time-series, or they are pre-estimation or post-estimation commands conceptually related to one or more multivariate time-series estimators. Some further grouping is also possible within these three broad categories, and the commands in these subcategories have likewise been grouped together because of their common, conceptually related functions.

Data management tools and time-series operators

tsset	Declare dataset to be time-series data
tsfill	Fill in missing times with missing observations
tsappend	Add observations to a time-series dataset
tsreport	Report time-series aspects of dataset or estimation sample
tsrevar	Time-series operator programming command

Univariate time-series

Estimators

arima	Autoregressive integrated moving average models
arch	Autoregressive conditional heteroskedasticity (ARCH) family of estimators
newey	Regression with Newey–West standard errors
prais	Prais–Winsten regression and Cochrane–Orcutt regression

Diagnostic tools

corrgram	Correlogram
xcorr	Cross-correlogram for bivariate time series
cumsp	Cumulative spectral distribution
pergram	Periodogram
dfgls	Perform DF-GLS unit-root test
dfuller	Augmented Dickey–Fuller test for a unit root
pperron	Phillips–Perron test for unit roots
dwstat	Durbin–Watson d statistic
durbina	Durbin's alternative test for serial correlation
bgodfrey	Breusch–Godfrey test for higher-order serial correlation
archlm	Engle's LM test for the presence of autoregressive conditional heteroskedasticity
wntestb	Bartlett's periodogram-based test for white noise
wntestq	Portmanteau (Q) test for white noise

Time-series smoothers and filters

tssmooth ma	Moving-average filter
tssmooth dexponential	Double exponential smoothing
tssmooth exponential	Exponential smoothing
tssmooth hwinters	Holt–Winters nonseasonal smoothing
tssmooth shwinters	Holt–Winters seasonal smoothing
tssmooth nl	Nonlinear filter

Multivariate time-series

Estimators

var	Vector autoregression models
svar	Structural vector autoregression models
varbasic	Fit a simple VAR and graph impulse–response functions

Forecasting, inference, and interpretation

varirf create	Obtain impulse–response functions and forecast-error variance decompositions
varfcast compute	Compute dynamic forecasts of dependent variables after var or svar
vargranger	Perform pairwise Granger causality tests after var or svar

Diagnostic tools

varlmar	Obtain LM statistics for residual autocorrelation after var or svar
varnorm	Test for normally distributed disturbances after var or svar
varsoc	Obtain lag-order selection statistics for a set of VARs
varstable	Check stability condition of var or svar estimates
varwle	Obtain Wald lag exclusion statistics after var or svar

Graphs and tables

varfcast graph	Graph forecasts of dependent variables after var or svar
varirf graph	Graph impulse–response functions and FEVDs
varirf cgraph	Make combined graphs of impulse–response functions and FEVDs
varirf ograph	Graph overlaid impulse–response functions and FEVDs
varirf table	Create tables of impulse–response functions and FEVDs
varirf ctable	Make combined tables of impulse–response functions and FEVDs

Results management tools

varfcast clear	Drop variables containing previous forecasts from varfcast
varirf add	Add VARIRF results from one VARIRF file to another
varirf describe	Describe a VARIRF file
varirf dir	List the VARIRF files in a directory
varirf drop	Drop VARIRF results from the active VARIRF file
varirf erase	Erase a VARIRF file
varirf rename	Rename a VARIRF result in a VARIRF file
varirf set	Set active VARIRF file

Remarks

Data management tools and time-series operators

Since time-series estimators are, by definition, a function of the temporal ordering of the observations in the estimation sample, Stata's time-series commands require the data to be sorted and indexed by time, using the `tsset` command, before they can be used. `tsset` is simply a way for you to tell Stata which variable in your dataset represents time; `tsset` then sorts and indexes the data appropriately for use with the time-series commands. Once your dataset has been `tsset`, you may use Stata's time-series operators in any data manipulation or programming you might undertake using that particular dataset, and you may use these operators when specifying the syntax for most time-series commands. Stata has time-series operators for representing the lags, leads, differences, and seasonal differences of a variable. The time-series operators are documented in [TS] **tsset**.

`tsset` may also be used to declare your dataset to contain cross-sectional time-series data, often referred to as panel data. When you use `tsset` to declare your dataset to contain panel data, you specify a variable that identifies the panels as well as identifying the time variable. Once your dataset has been `tsset` as panel data, the time-series operators work appropriately for such data.

`tsfill`, which is documented in [TS] **tsset**, can be used after `tsset` to fill in missing times with missing observations. `tsset` will tell you if there are any gaps in your data, and `tsreport` will provide additional details as to where the gaps are. `tsappend` adds observations to a time-series dataset using the information set by `tsset`. This can be particularly useful when you wish to predict out-of-sample after fitting a model with a time-series estimator. `tsrevar` is a programmer's command that provides a way to use *varlist*s that contain time-series operators with commands that do not otherwise support time-series operators.

Univariate time-series

Estimators

The four univariate time-series estimators currently available in Stata are `arima`, `arch`, `newey`, and `prais`. The latter two, `prais` and `newey`, are really just extensions to ordinary linear regression. When fitting a linear regression on time-series data via ordinary least squares, if the disturbances are autocorrelated, the parameter estimates are consistent, but the estimated standard errors tend to be biased downward. As a result, a number of estimators were developed to deal with this problem. One strategy is to use OLS for estimating the regression parameters, but to use a different estimator for the variances; one that is consistent in the presence of autocorrelated disturbances. The Newey–West estimator that is implemented in `newey` is such an estimator. An alternative strategy is to attempt to model the dynamics of the disturbances. The estimators found in `prais`, `arima`, and `arch` are based upon such a strategy.

`prais` implements two such estimators: the Prais–Winsten and the Cochrane–Orcutt GLS estimators. These estimators are generalized least-squares estimators, but they are fairly restrictive in that they only permit first-order autocorrelation in the disturbances. While they have certain pedagogical value and are of historic interest, they are somewhat obsolete. Faster computers with greater random access memory (RAM) capacity have made it possible to implement Full Information Maximum Likelihood (FIML) estimators like Stata's `arima` command. These estimators permit much greater flexibility when modeling the disturbances and are unquestionably more efficient estimators.

Stata's `arima` command provides the means to fit linear models with autoregressive moving-average (ARMA) disturbances, or in the absence of linear predictors, autoregressive integrated moving-average

(ARIMA) models. This means that regardless of whether you think your data are best represented in terms of a distributed lag model, a transfer function model, a stochastic difference equation, or you simply wish to employ a Box–Jenkins type filter to your data, the model can be fit using arima. arch, a conditional maximum likelihood estimator, has similar modeling capabilities for the mean of the time series, but, in addition, can model autoregressive conditional heteroskedasticity in the disturbances, with a wide variety of possible specifications for the variance equation.

Time-series smoothers and filters

In addition to the estimators mentioned above, Stata also provides six time-series filters or smoothers. Included are a simple uniformly weighted moving-average filter with unit weights, a weighted moving-average filter where the user is allowed to specify the weights, single and double exponential smoothers, Holt–Winters seasonal and nonseasonal smoothers, and a nonlinear smoother.

Most of these smoothers were originally developed as *ad hoc* procedures and are used as a means of reducing the noise in a time series (smoothing) or for forecasting. While there are limitations regarding the general applicability of smoothers such as these for signal extraction, they have all been found to be optimal for some underlying modern time-series model.

Diagnostic tools

Stata's time-series commands also include a number of pre-estimation and post-estimation diagnostic commands. corrgram provides estimates of the autocorrelation function and partial autocorrelation function of a univariate time-series, as well as Q statistics. These functions and statistics are often used to determine the appropriate model specification prior to fitting ARIMA models. corrgram can also be used in combination with wntestb and wntestq to examine the residuals after fitting a model to determine if there is evidence of model misspecification. Stata's time-series commands also include the commands pergram and cumsp, which provide the log standardized periodogram and the cumulative sample spectral distribution, respectively, for those time-series analysts who prefer to work in the frequency domain rather than in the time domain.

xcorr estimates the cross-correlogram for bivariate time-series, and can similarly be used both as a pre-estimation tool and as a post-estimation tool. For example, the cross-correlogram can be used prior to fitting a transfer function model to produce initial estimates of the impulse–response function. This estimate can then be used to determine the optimal lag length of the input series to include in the model specification. It can also be used as a post-estimation tool after fitting a transfer function. The cross-correlogram between the residual from a transfer function model and the pre-whitened input series of the model can be examined to determine if there is evidence of model misspecification.

When fitting ARMA or ARIMA models, it is necessary for the dependent variable being modeled to be covariance-stationary (ARMA models) or for the order of integration to be known (ARIMA models). Stata has three commands that can be used to test for the presence of a unit root in a time-series variable. dfuller performs the augmented Dickey–Fuller test, pperron performs the Phillips–Perron test, and dfgls performs a modified Dickey–Fuller test.

The remaining diagnostic tools for univariate time-series are for use after fitting a linear model via OLS with Stata's regress command. They are documented collectively in [TS] **regression diagnostics**. They include dwstat, durbina, bgodfrey, and archlm. dwstat computes the Durbin–Watson d statistic to test for the presence of first-order autocorrelation in the OLS residuals. durbina likewise performs a test for the presence of autocorrelation in the residuals. By comparison, however, Durbin's alternative test is more general and easier to use than the Durbin–Watson test. With durbina, you can test for higher orders of autocorrelation, the assumption that the covariates in the model are strictly exogenous is relaxed, and there is no need to consult tables to compute rejection regions as you must with the Durbin–Watson test. bgodfrey computes the Breusch–Godfrey test for autocorrelation in the residuals, and while the computations are different, the test in `bgodfrey` is asymptotically equivalent to the test in `durbina`. Finally, `archlm` is available to perform Engle's LM test for the presence of autoregressive conditional heteroskedasticity.

Multivariate time series

At present, the multivariate time-series commands all revolve around fitting vector autoregressions. [TS] **var intro** presents an introduction and overview of the commands with an outline very similar to the one presented at the beginning of this entry. The interested reader should turn there to read about these commands.

References

Hamilton, J. D. 1994. *Time Series Analysis*. Princeton: Princeton University Press.

Lütkepohl, H. 1993. *Introduction to Multiple Time Series Analysis*. 2d ed. New York: Springer.

Pisati, M. 2001. sg162: Tools for spatial data analysis. *Stata Technical Bulletin* 60: 21–37. Reprinted in *Stata Technical Bulletin Reprints*, vol. 10, pp. 277–298.

Stock, J. H. and M. W. Watson. 2001. Vector autoregressions. *Journal of Economic Perspectives* 15(4): 101–115.

Also See

Complementary:	[U] **1.3 What's new**
Background:	[R] **intro**

Title

> **arch** — Autoregressive conditional heteroskedasticity (ARCH) family of estimators

Syntax

arch *depvar* [*varlist*] [*weight*] [if *exp*] [in *range*] [, arch(*numlist*) garch(*numlist*)

saarch(*numlist*) tarch(*numlist*) aarch(*numlist*) narch(*numlist*) narchk(*numlist*)

abarch(*numlist*) atarch(*numlist*) sdgarch(*numlist*) earch(*numlist*) egarch(*numlist*)

parch(*numlist*) tparch(*numlist*) aparch(*numlist*) nparch(*numlist*) nparchk(*numlist*)

pgarch(*numlist*) het(*varlist*) archm archmlags(*numlist*) archmexp(*exp*)

ar(*numlist*) ma(*numlist*) arima($\#_p$,$\#_d$,$\#_q$) noconstant constraints(*numlist*)

hessian opg robust score(*newvarlist* | *stub**) arch0(*cond_method*)

arma0(*cond_method*) condobs(#) savespace detail level(#)

maximize_options from(*initial_values*) gtolerance(#)

bhhh dfp bfgs nr bhhhbfgs(#,#) bhhhdfp(#,#)]

To fit an ARCH($\#_m$) model, type

    . arch *depvar* ... , arch(1/$\#_m$)

To fit a GARCH($\#_m$,$\#_k$) model, type

    . arch *depvar* ... , arch(1/$\#_m$) garch(1/$\#_k$)

Fitting of other models is possible.

You must tsset your data before using arch; see [TS] **tsset**.

depvar and *varlist* may contain time-series operators; see [U] **14.4.3 Time-series varlists**.

iweights are allowed; see [U] **14.1.6 weight**.

arch shares the features of all estimation commands; see [U] **23 Estimation and post-estimation commands**.

Details of syntax

The basic model arch fits is

$$y_t = \mathbf{x}_t \boldsymbol{\beta} + \epsilon_t$$
$$\mathrm{Var}(\epsilon_t) = \sigma_t^2 = \gamma_0 + A(\boldsymbol{\sigma}, \boldsymbol{\epsilon}) + B(\boldsymbol{\sigma}, \boldsymbol{\epsilon})^2 \tag{1}$$

The y_t equation may optionally include ARCH-in-mean and/or ARMA terms:

$$y_t = \mathbf{x}_t \boldsymbol{\beta} + \sum_i \psi_i g(\sigma_{t-i}^2) + \mathrm{ARMA}(p, q) + \epsilon_t$$

If no options are specified, $A() = B() = 0$ and the model collapses to linear regression. The following options add to $A()$ (α, γ, and κ represent parameters to be estimated):

8

Option	Terms added to $A()$
arch()	$A() = A()+\alpha_{1,1}\epsilon_{t-1}^2 + \alpha_{1,2}\epsilon_{t-2}^2 + \cdots$
garch()	$A() = A()+\alpha_{2,1}\sigma_{t-1}^2 + \alpha_{2,2}\sigma_{t-2}^2 + \cdots$
saarch()	$A() = A()+\alpha_{3,1}\epsilon_{t-1} + \alpha_{3,2}\epsilon_{t-2} + \cdots$
tarch()	$A() = A()+\alpha_{4,1}\epsilon_{t-1}^2(\epsilon_{t-1} > 0) + \alpha_{4,2}\epsilon_{t-2}^2(\epsilon_{t-2} > 0) + \cdots$
aarch()	$A() = A()+\alpha_{5,1}(|\epsilon_{t-1}| + \gamma_{5,1}\epsilon_{t-1})^2 + \alpha_{5,2}(|\epsilon_{t-2}| + \gamma_{5,2}\epsilon_{t-2})^2 + \cdots$
narch()	$A() = A()+\alpha_{6,1}(\epsilon_{t-1} - \kappa_{6,1})^2 + \alpha_{6,2}(\epsilon_{t-2} - \kappa_{6,2})^2 + \cdots$
narchk()	$A() = A()+\alpha_{7,1}(\epsilon_{t-1} - \kappa_7)^2 + \alpha_{7,2}(\epsilon_{t-2} - \kappa_7)^2 + \cdots$

The following options add to $B()$:

Option	Terms added to $B()$
abarch()	$B() = B()+\alpha_{8,1}|\epsilon_{t-1}| + \alpha_{8,2}|\epsilon_{t-2}| + \cdots$
atarch()	$B() = B()+\alpha_{9,1}|\epsilon_{t-1}|(\epsilon_{t-1} > 0) + \alpha_{9,2}|\epsilon_{t-2}|(\epsilon_{t-2} > 0) + \cdots$
sdgarch()	$B() = B()+\alpha_{10,1}\sigma_{t-1} + \alpha_{10,2}\sigma_{t-2} + \cdots$

Each of the options requires a *numlist* argument (see [U] **14.1.8 numlist**). The *numlist* determines the lagged terms included. For instance, arch(1) specifies $\alpha_{1,1}\epsilon_{t-1}^2$, arch(2) specifies $\alpha_{1,2}\epsilon_{t-2}^2$, arch(1,2) specifies $\alpha_{1,1}\epsilon_{t-1}^2 + \alpha_{1,2}\epsilon_{t-2}^2$, arch(1/3) specifies $\alpha_{1,1}\epsilon_{t-1}^2 + \alpha_{1,2}\epsilon_{t-2}^2 + \alpha_{1,3}\epsilon_{t-3}^2$, etc.

If options earch() and/or egarch() are specified, the basic model fitted is

$$y_t = \mathbf{x}_t\boldsymbol{\beta} + \sum_i \psi_i g(\sigma_{t-i}^2) + \text{ARMA}(p, q) + \epsilon_t \tag{2}$$

$$\ln \text{Var}(\epsilon_t) = \ln \sigma_t^2 = \gamma_0 + C(\ln\boldsymbol{\sigma}, \mathbf{z}) + A(\boldsymbol{\sigma}, \boldsymbol{\epsilon}) + B(\boldsymbol{\sigma}, \boldsymbol{\epsilon})^2$$

where $z_t = \epsilon_t/\sigma_t$. $A()$ and $B()$ are given as above, but note that $A()$ and $B()$ now add to $\ln \sigma_t^2$ rather than σ_t^2. (The options corresponding to $A()$ and $B()$ are rarely specified in this case.) $C()$ is given by

Option	Terms added to $C()$
earch()	$C() = C() +\alpha_{11,1}z_{t-1} + \gamma_{11,1}(|z_{t-1}| - \sqrt{2/\pi})$ $+\alpha_{11,2}z_{t-2} + \gamma_{11,2}(|z_{t-2}| - \sqrt{2/\pi}) + \cdots$
egarch()	$C() = C() +\alpha_{12,1}\ln\sigma_{t-1}^2 + \alpha_{12,2}\ln\sigma_{t-2}^2 + \cdots$

Alternatively, if options parch(), tparch(), aparch(), nparch(), nparchk(), and/or pgarch() are specified, the basic model fitted is

$$y_t = \mathbf{x}_t\boldsymbol{\beta} + \sum_i \psi_i g(\sigma_{t-i}^2) + \text{ARMA}(p, q) + \epsilon_t \tag{3}$$

$$\{\text{Var}(\epsilon_t)\}^{\varphi/2} = \sigma_t^\varphi = \gamma_0 + D(\boldsymbol{\sigma}, \boldsymbol{\epsilon}) + A(\boldsymbol{\sigma}, \boldsymbol{\epsilon}) + B(\boldsymbol{\sigma}, \boldsymbol{\epsilon})^2$$

where φ is a parameter to be estimated. $A()$ and $B()$ are given as above, but note that $A()$ and $B()$ now add to σ_t^φ. (The options corresponding to $A()$ and $B()$ are rarely specified in this case.) $D()$ is given by

Option	Terms added to $D()$
`parch()`	$D() = D() + \alpha_{13,1}\epsilon_{t-1}^\varphi + \alpha_{13,2}\epsilon_{t-2}^\varphi + \cdots$
`tparch()`	$D() = D() + \alpha_{14,1}\epsilon_{t-1}^\varphi(\epsilon_{t-1} > 0) + \alpha_{14,2}\epsilon_{t-2}^\varphi(\epsilon_{t-2} > 0) + \cdots$
`aparch()`	$D() = D() + \alpha_{15,1}(|\epsilon_{t-1}| + \gamma_{15,1}\epsilon_{t-1})^\varphi + \alpha_{15,2}(|\epsilon_{t-2}| + \gamma_{15,2}\epsilon_{t-2})^\varphi + \cdots$
`nparch()`	$D() = D() + \alpha_{16,1}|\epsilon_{t-1} - \kappa_{16,1}|^\varphi + \alpha_{16,2}|\epsilon_{t-2} - \kappa_{16,2}|^\varphi + \cdots$
`nparchk()`	$D() = D() + \alpha_{17,1}|\epsilon_{t-1} - \kappa_{17}|^\varphi + \alpha_{17,2}|\epsilon_{t-2} - \kappa_{17}|^\varphi + \cdots$
`pgarch()`	$D() = D() + \alpha_{18,1}\sigma_{t-1}^\varphi + \alpha_{18,2}\sigma_{t-2}^\varphi + \cdots$

Commonly fitted models

Common term	Options to specify
ARCH (Engle 1982)	`arch()`
GARCH (Bollerslev 1986)	`arch() garch()`
ARCH-in-mean (Engle et al. 1987)	`archm arch()` $[$`garch()`$]$
GARCH with ARMA terms	`arch() garch() ar() ma()`
EGARCH (Nelson 1991)	`earch() egarch()`
TARCH, threshold ARCH (Zakoian 1990)	`abarch() atarch() sdgarch()`
GJR, form of threshold ARCH (Glosten et al. 1993)	`arch() tarch()` $[$`garch()`$]$
SAARCH, simple asymmetric ARCH (Engle 1990)	`arch() saarch()` $[$`garch()`$]$
PARCH, power ARCH (Higgins and Bera 1992)	`parch()` $[$`pgarch()`$]$
NARCH, nonlinear ARCH	`narch()` $[$`garch()`$]$
NARCHK, nonlinear ARCH with a single shift	`narchk()` $[$`garch()`$]$
A-PARCH, asymmetric power ARCH (Ding et al. 1993)	`aparch()` $[$`pgarch()`$]$
NPARCH, nonlinear power ARCH	`nparch()` $[$`pgarch()`$]$

In all cases, you type

$$\texttt{arch } depvar \; [indepvars] \text{, } options$$

where you obtain the options from the table above. Each option requires that you specify a *numlist* as its argument; the *numlist* specifies the lags to be included. For the vast majority of ARCH models, that value will be 1. For instance, to fit the classic first-order GARCH model on `cpi`, you would type

    . arch cpi, arch(1) garch(1)

If you wanted to fit a first-order GARCH model of `cpi` on `wage`, you would type

    . arch cpi wage, arch(1) garch(1)

If, for any of the options, you want first- and second-order terms, specify *optionname*(1/2). Specifying `garch(1) arch(1/2)` would fit a GARCH model with first- and second-order ARCH terms. If you specified simply `arch(2)`, only the lag 2 term would be included.

Reading arch output

The regression table reported by `arch` will appear as

op.depvar		Coef.	Std. Err.	z	P>|z|	[95% Conf. Interval]
depvar						
x1		# ...				
x2						
	L1	# ...				
	L2	# ...				
_cons		# ...				
ARCHM						
sigma2		# ...				
ARMA						
ar						
	L1	# ...				
ma						
	L1	# ...				
HET						
z1		# ...				
z2						
	L1	# ...				
	L2	# ...				
ARCH						
arch						
	L1	# ...				
garch						
	L1	# ...				
aparch						
	L1	# ...				
etc.						
_cons		# ...				
POWER						
power		# ...				

Dividing lines separate "equations".

The first one, two, or three equations report the mean model

$$y_t = \mathbf{x}_t\boldsymbol{\beta} + \sum_i \psi_i g(\sigma_{t-i}^2) + \text{ARMA}(p,q) + \epsilon_t$$

The first equation reports $\boldsymbol{\beta}$, and the equation will be named [*depvar*]. (Say you fitted a model on `d.cpi`; then the first equation would be named [cpi].) In Stata, the coefficient on x1 in the above example could be referred to as [*depvar*]_b[x1]. The coefficient on the lag 2 value of x2 would be referred to as [*depvar*]_b[L2.x2]. Such notation would be used, for instance, in a subsequent `test` command; see [R] **test**.

The [ARCHM] equation reports the ψ coefficient(s) if your model includes ARCH-in-mean terms; see *Options for specifying ARCH-in-mean terms*. Most ARCH-in-mean models include only a contemporaneous variance term, so the term $\sum_i \psi_i g(\sigma_{t-i}^2)$ becomes $\psi\sigma_t^2$. The coefficient ψ will be, in Stata/SE, [ARCHM]_b[sigma2]. If your model includes lags of σ_t^2, the additional coefficients will be [ARCHM]_b[L1.sigma2], and so on. If you specify a transformation $g()$ (option archmexp()), the coefficients will be [ARCHM]_b[sigma2ex], [ARCHM]_b[L1.sigma2ex], and so on. sigma2ex refers to $g(\sigma_t^2)$, the transformed value of the conditional variance.

The [ARMA] equation reports the ARMA coefficients if your model includes them; see *Options for specifying ARIMA terms*. This equation includes one or two "variables" named ar and ma. In subsequent test statements, one could refer to the coefficient on the first lag of the autoregressive term by typing [ARMA]_b[L1.ar] or simply [ARMA]_b[L.ar] because the L operator is assumed to be lag 1 if you do not specify otherwise. The second lag on the moving-average term, if there were one, could be referred to by typing [ARMA]_b[L2.ma].

The last one, two, or three equations report the variance model.

The [HET] equation reports the multiplicative heteroskedasticity if the model includes such; see *Other options affecting specification of variance*. When you fit such a model, you specify the variables (and their lags) determining the multiplicative heteroskedasticity and, after estimation, their coefficients are simply [HET]_b[*op.varname*].

The [ARCH] equation reports the ARCH, GARCH, etc. terms by referring to "variables" arch, garch, and so on. For instance, if you specified arch(1) garch(1) when you fitted the model, the conditional variance is given by $\sigma_t^2 = \gamma_0 + \alpha_{1,1}\epsilon_{t-1}^2 + \alpha_{2,1}\sigma_{t-1}^2$. The coefficients would be named [ARCH]_b[_cons] (γ_0), [ARCH]_b[L.arch] ($\alpha_{1,1}$), and [ARCH]_b[L.garch] ($\alpha_{2,1}$).

The [POWER] equation appears only if you are fitting a variance model in the form of (3) above; the estimated φ is the coefficient [POWER]_b[power].

The naming convention for estimated ARCH, GARCH, etc., parameters is (definitions for parameters α_i, γ_i, and κ_i can be found in the tables for $A()$, $B()$, $C()$, and $D()$ above)

Option	1st parameter	2nd parameter	common parameter
arch()	$\alpha_1 = $ [ARCH]_b[arch]		
garch()	$\alpha_2 = $ [ARCH]_b[garch]		
saarch()	$\alpha_3 = $ [ARCH]_b[saarch]		
tarch()	$\alpha_4 = $ [ARCH]_b[tarch]		
aarch()	$\alpha_5 = $ [ARCH]_b[aarch]	$\gamma_5 = $ [ARCH]_b[aarch_e]	
narch()	$\alpha_6 = $ [ARCH]_b[narch]	$\kappa_6 = $ [ARCH]_b[narch_k]	
narchk()	$\alpha_7 = $ [ARCH]_b[narch]	$\kappa_7 = $ [ARCH]_b[narch_k]	
abarch()	$\alpha_8 = $ [ARCH]_b[abarch]		
atarch()	$\alpha_9 = $ [ARCH]_b[atarch]		
sdgarch()	$\alpha_{10} = $ [ARCH]_b[sdgarch]		
earch()	$\alpha_{11} = $ [ARCH]_b[earch]	$\gamma_{11} = $ [ARCH]_b[earch_a]	
egarch()	$\alpha_{12} = $ [ARCH]_b[egarch]		
parch()	$\alpha_{13} = $ [ARCH]_b[parch]		$\varphi = $ [POWER]_b[power]
tparch()	$\alpha_{14} = $ [ARCH]_b[tparch]		$\varphi = $ [POWER]_b[power]
aparch()	$\alpha_{15} = $ [ARCH]_b[aparch]	$\gamma_{15} = $ [ARCH]_b[aparch_e]	$\varphi = $ [POWER]_b[power]
nparch()	$\alpha_{16} = $ [ARCH]_b[nparch]	$\kappa_{16} = $ [ARCH]_b[nparch_k]	$\varphi = $ [POWER]_b[power]
nparchk()	$\alpha_{17} = $ [ARCH]_b[nparch]	$\kappa_{17} = $ [ARCH]_b[nparch_k]	$\varphi = $ [POWER]_b[power]
pgarch()	$\alpha_{18} = $ [ARCH]_b[pgarch]		$\varphi = $ [POWER]_b[power]

Syntax for predict

predict [*type*] *newvarname* [if *exp*] [in *range*] [, *statistic*

t0(*time_constant*) <u>struc</u>tural

<u>d</u>ynamic(*time_constant*) at({*varname$_\epsilon$* | #$_\epsilon$} {*varname$_{\sigma^2}$* | #$_{\sigma^2}$})]

where *statistic* is

xb	predicted values for mean equation—the differenced series; the default
y	predicted values for the mean equation in y—the undifferenced series
variance	predicted values for the conditional variance
<u>het</u>	predicted values of the variance considering only the multiplicative heteroskedasticity
<u>res</u>iduals	residuals or predicted innovations
<u>yres</u>iduals	residuals or predicted innovations in y—the undifferenced series

and *time_constant* is a # or a time literal, such as d(1jan1995) or q(1995q1), etc.; see [U] **27.3 Time-series dates**.

These statistics are available both in and out of sample; type predict ... if e(sample) ... if wanted only for the estimation sample.

Description

arch fits models of autoregressive conditional heteroskedasticity (ARCH) using conditional maximum likelihood. In addition to ARCH terms, models may include multiplicative heteroskedasticity.

Concerning the regression equation itself, models may also contain ARCH-in-mean and/or ARMA terms.

Options

Options for specifying terms appearing in A()

arch(*numlist*) specifies the ARCH terms (lags of ϵ_t^2).

Specify arch(1) to include first-order terms, arch(1/2) to specify first- and second-order terms, arch(1/3) to specify first-, second-, and third-order terms, etc. Terms may be omitted. Specify arch(1/3 5) to specify terms with lags 1, 2, 3, and 5. All the options work like this.

arch() may not be specified with aarch(), narch(), narchk(), nparchk(), or nparch(), as this would result in collinear terms.

garch(*numlist*) specifies the GARCH terms (lags of σ_t^2).

saarch(*numlist*) specifies the simple asymmetric ARCH terms. Adding these terms is one way to make the standard ARCH and GARCH models respond asymmetrically to positive and negative innovations. Specifying saarch() with arch() and garch() corresponds to the SAARCH model of Engle (1990).

saarch() may not be specified with narch(), narchk(), nparchk(), or nparch(), as this would result in collinear terms.

tarch(*numlist*) specifies the threshold ARCH terms. Adding these is another way to make the standard ARCH and GARCH models respond asymmetrically to positive and negative innovations. Specifying tarch() with arch() and garch() corresponds to one form of the GJR model (Glosten, Jagannathan, and Runkle 1993).

Note that tarch() may not be specified with tparch() or aarch(), as this would result in collinear terms.

aarch(*numlist*) specifies the lags of the two-parameter term $\alpha_i(|\epsilon_t| + \gamma_i \epsilon_t)^2$. This term provides the same underlying form of asymmetry as including arch() and tarch(); it is just expressed in a different way.

aarch() may not be specified with arch() or tarch() as this would result in collinear terms.

narch(*numlist*) specifies the lags of the two-parameter term $\alpha_i(\epsilon_t - \kappa_i)^2$. This term allows the minimum conditional variance to occur at a value of lagged innovations other than zero. For any given term specified at lag L, the minimum contribution to conditional variance of that lag occurs when $\epsilon_{t-L}^2 = \kappa_L$—the squared innovations at that lag are equal to the estimated constant κ_L.

narch() may not be specified with arch(), saarch(), narchk(), nparchk(), or nparch(), as this would result in collinear terms.

narchk(*numlist*) specifies the lags of the two-parameter term $\alpha_i(\epsilon_t - \kappa)^2$; note that this is a variation on narch() with κ held constant for all lags.

narchk() may not be specified with arch(), saarch(), narch(), nparchk(), or nparch(), as this would result in collinear terms.

Options specifying terms appearing in B()

abarch(*numlist*) specifies lags of the term $|\epsilon_t|$.

atarch(*numlist*) specifies lags of $|\epsilon_t|(\epsilon_t > 0)$, where $(\epsilon_t > 0)$ represents the indicator function returning 1 when true and 0 when false. Like the TARCH terms, these ATARCH terms allow the effect of unanticipated innovations to be asymmetric about zero.

sdgarch(*numlist*) specifies lags of σ_t. Combining atarch(), abarch(), and sdgarch() produces the model by Zakoian (1990) that the author called the TARCH model. The acronym TARCH, however, is often used to refer to any model using thresholding to obtain asymmetry.

Options for terms appearing in C()

earch(*numlist*) specifies lags of the two-parameter term $\alpha z_t + \gamma(|z_t| - \sqrt{2/\pi})$. These terms represent the influence of news—lagged innovations—in Nelson's (1991) EGARCH model. For these terms, $z_t = \epsilon_t/\sigma_t$, and arch assumes $z_t \sim N(0,1)$. Nelson (1991) derived the general form of an EGARCH model for any assumed distribution and performed estimation assuming a Generalized Error Distribution (GED). See Hamilton (1994) for a derivation where z_t is assumed normal. The z_t terms can be parameterized in at least two equivalent ways. arch uses Nelson's (1991) original parameterization; see Hamilton (1994) for an equivalent alternative.

egarch(*numlist*) specifies lags of $\ln(\sigma_t^2)$.

Options for terms appearing in D()

Note: The model is parameterized in terms of $h(\epsilon_t)^\varphi$ and σ_t^φ. A single φ is estimated even when more than one option is specified.

parch(*numlist*) specifies lags of $|\epsilon_t|^\varphi$. parch() combined with pgarch() corresponds to the class of nonlinear models of conditional variance suggested by Higgins and Bera (1992).

tparch(*numlist*) specifies lags of $(\epsilon_t > 0)|\epsilon_t|^\varphi$, where $(\epsilon_t > 0)$ represents the indicator function returning 1 when true and 0 when false. As with tarch(), tparch() specifies terms that allow for a differential impact of "good" (positive innovations) and "bad" (negative innovations) news for lags specified by *numlist*.

Note that tparch() may not be specified with tarch(), as this would result in collinear terms.

aparch(*numlist*) specifies lags of the two-parameter term $\alpha(|\epsilon_t| + \gamma\epsilon_t)^\varphi$. This asymmetric power ARCH model, A-PARCH, was proposed by Ding et al. (1993), and corresponds to a Box–Cox function in the lagged innovations. The authors fitted the original A-PARCH model on over 16,000 daily observations of the Standard and Poor's 500, and not without good reason. As the number of parameters and the flexibility of the specification increase, larger amounts of data are required to estimate the parameters of the conditional heteroskedasticity. See Ding et al. (1993) for a discussion of how 7 popular ARCH models nest within the A-PARCH model.

Note that when γ goes to 1, the full term goes to zero for many observations, and this point can be numerically unstable.

nparch(*numlist*) specifies lags of the two-parameter term $\alpha|\epsilon_t - \kappa_i|^\varphi$.

nparch() may not be specified with arch(), saarch(), narch(), narchk(), or nparchk(), as this would result in collinear terms.

nparchk(*numlist*) specifies lags of the two-parameter term $\alpha|\epsilon_t - \kappa|^\varphi$; note that this is a variation on nparch() with κ held constant for all lags. This is the direct analog of narchk() except for the power of φ. nparchk() corresponds to an extended form of the model of Higgins and Bera (1992) as presented by Bollerslev et al. (1994). nparchk() would typically be combined with the option pgarch().

nparchk() may not be specified with arch(), saarch(), narch(), narchk(), or nparch() as this would result in collinear terms.

pgarch(*numlist*) specifies lags of σ_t^φ.

Other options affecting specification of variance

het(*varlist*) specifies that *varlist* be included in the specification of the conditional variance. *varlist* may contain time-series operators. This varlist enters the variance specification collectively as multiplicative heteroskedasticity; see Judge et al. (1985). If het() is not specified, the model will not contain multiplicative heteroskedasticity.

Assume the conditional variance is thought to depend on variables x and w while also having an ARCH(1) component. We request this specification by using the options het(x w) arch(1), and this corresponds to the conditional-variance model

$$\sigma_t^2 = \exp(\lambda_0 + \lambda_1 x_t + \lambda_2 w_t) + \alpha\epsilon_{t-1}^2$$

Multiplicative heteroskedasticity enters differently with an EGARCH model because the variance is already specified in logs. For the options het(x w) earch(1) egarch(1), the variance model is

$$\ln(\sigma_t^2) = \lambda_0 + \lambda_1 x_t + \lambda_2 w_t + \alpha z_{t-1} + \gamma(|z_{t-1}| - \sqrt{2/\pi}) + \delta\ln(\sigma_{t-1}^2)$$

Options for specifying ARCH-in-mean terms

archm specifies that an ARCH-in-mean term be included in the specification of the mean equation. This term allows the expected value of *depvar* to depend on the conditional variance. ARCH-in-mean is most commonly used in evaluating financial time series when a theory supports a trade-off between asset riskiness and asset return. By default, no ARCH-in-mean terms are included in the model.

archm specifies that the contemporaneous expected conditional variance be included in the mean equation. For example, typing

        . arch y x, archm arch(1)

specifies the model

$$y_t = \beta_0 + \beta_1 x_t + \psi \sigma_t^2 + \epsilon_t$$
$$\sigma_t^2 = \gamma_0 + \gamma \epsilon_{t-1}^2$$

archmlags(*numlist*) is an expansion of archm and specifies that lags of the conditional variance σ_t^2 be included in the mean equation. To specify a contemporaneous and once-lagged variance, either specify archm archmlags(1) or specify archmlags(0/1).

archmexp(*exp*) specifies the transformation in *exp* be applied to any ARCH-in-mean terms in the model. The expression should contain an X wherever a value of the conditional variance is to enter the expression. This option can be used to produce the commonly used ARCH-in-mean of the conditional standard deviation. Using the example from archm, typing

        . arch y x, archm arch(1) archmexp(sqrt(X))

specifies the mean equation $y_t = \beta_0 + \beta_1 x_t + \psi \sigma_t + \epsilon_t$. Alternatively, typing

        . arch y x, archm arch(1) archmexp(1/sqrt(X))

specifies $y_t = \beta_0 + \beta_1 x_t + \psi/\sigma_t + \epsilon_t$.

Options for specifying ARIMA terms

ar(*numlist*) specifies the autoregressive terms to be included in the model. These are the autoregressive terms of the structural model disturbance. For example, ar(1/3) specifies that lags of 1, 2, and 3 of the structural disturbance are to be included in the model. ar(1,4) specifies that lags 1 and 4 are to be included, possibly to account for quarterly effects.

If the model does not contain any regressors, these terms can also be considered autoregressive terms for the dependent variable; see [TS] **arima**.

ma(*numlist*) specifies the moving average terms to be included in the model. These are the terms for the lagged innovations, or white-noise disturbances.

arima($\#_p$,$\#_d$,$\#_q$) is an alternate, shorthand notation for specifying models that are autoregressive in the dependent variable. The dependent variable and any independent variables are differenced $\#_d$ times, 1 through $\#_p$ lags of autocorrelations are included, and 1 through $\#_q$ lags of moving averages are included. For example, the specification

        . arch y, arima(2,1,3)

is equivalent to

        . arch D.y, ar(1/2) ma(1/3)

The former is easier to write for "classic" ARIMA models of the mean equation, but it is not nearly as expressive as the latter. If gaps in the AR or MA lags are to be modeled, or if different operators are to be applied to independent variables, the latter syntax will generally be required.

Other options affecting the mean and/or variance specifications

noconstant suppresses the constant term (intercept) in the equation for the conditional mean.

constraints(*numlist*) specifies the constraint numbers of the linear constraints to be applied during estimation. The default is to perform unconstrained estimation. Constraints are specified using the constraint command; see [R] **constraint** (also see [R] **reg3** for the use of constraint in multiple-equation contexts).

Options affecting the estimated standard errors

hessian and opg specify how standard errors are to be calculated. The default is opg unless one of the options bfgs, dfp, or nr is specified, in which case, the default is hessian.

hessian specifies that the standard errors and coefficient covariance matrix be estimated from the full Hessian—the matrix of negative second derivatives of the log-likelihood function. These are the estimates produced by most of Stata's maximum likelihood estimators.

opg specifies that the standard errors and coefficient covariance matrix be estimated using the outer product of the coefficient gradients with respect to the observation likelihoods.

hessian and opg provide asymptotically equivalent estimates of the standard errors and covariance matrix, and there is no theoretical justification for preferring either estimate.

If you obtain your standard errors from the Hessian because you either specify hessian or use an optimization method that implies hessian, be aware that the part of the calculation that occurs after convergence can take a while. Evaluating the second derivatives numerically to estimate the covariance matrix is an $O(k^2/2)$ process, where k is the number of parameters. If the model contains 5 parameters, producing the covariance matrix at the final step will take about 12.5 times longer than a single iteration in finding the maximum. If you have 10 parameters, it will take about 50 times longer. (This is assuming that you did not use method nr. Method nr requires the longer time at every iteration.)

robust specifies that the Huber/White/sandwich estimator of variance is to be used in place of the traditional calculation; see [U] **23.14 Obtaining robust variance estimates**.

For ARCH models, the robust or quasi-maximum likelihood estimates (QMLE) of variance are robust to symmetric nonnormality in the disturbances. The robust variance estimates are not generally robust to functional misspecification of the mean equation; see Bollerslev and Wooldridge (1992).

Note that the robust variance estimates computed by arch are based on the full Huber/White formulation as discussed in [P] **_robust**. In fact, many software packages report robust estimates that set some terms to their expectations of zero (Bollerslev and Wooldridge 1992), which saves them from having to calculate second derivatives of the log-likelihood function.

score(*newvarlist* | *stub**) creates a new variable for each parameter in the model. Each new variable contains the derivative of the model log-likelihood with respect to the parameter for each observation in the estimation sample: $\partial L_t / \partial \beta_k$, where L_t is the log likelihood for observation t and β_k is the kth parameter in the model.

If score(*newvarlist*) is specified, the *newvarlist* must contain a new variable for each parameter in the model. If score(*stub**) is specified, variables named *stub#* are created for each parameter in the model. The *newvarlist* is filled, or the #'s in *stub#* are created, in the order in which the estimated parameters are reported in the estimation results table.

Unlike scores for most other models, the scores from arch are individual gradients of the log likelihood with respect to the variables, not with respect to $x_t\beta$. Since the ARCH model is inherently nonlinear, the scores with respect to $x_t\beta$ could not be used to reconstruct the gradients for the individual parameters.

Options affecting conditioning (priming) values

arch0(*cond_method*) is a rarely used option to specify how the conditioning (presample or priming) values for σ_t^2 and ϵ_t^2 are to be computed. In the presample period, it is assumed that $\sigma_t^2 = \epsilon_t^2$ and that this value is constant. If arch0() is not specified, the priming values are computed as the expected unconditional variance given the current estimates of the β coefficients and any ARMA parameters.

arch0(xb) is the default. It specifies that the priming values are the expected unconditional variance of the model, which is $\sum_1^T \widehat{\epsilon}_t^2 / T$, where $\widehat{\epsilon}_t$ is computed from the mean equation and any ARMA terms.

arch0(xb0) specifies that the priming values are the estimated variance of the residuals from an OLS estimate of the mean equation.

arch0(xbwt) specifies that the priming values are the weighted sum of the $\widehat{\epsilon}_t^2$ from the current conditional mean equation (and ARMA terms) that places more weight on estimates of ϵ_t^2 at the beginning of the sample.

arch0(xb0wt) specifies that the priming values are the weighted sum of the $\widehat{\epsilon}_t^2$ from an OLS estimate of the mean equation (and ARMA terms) that places more weight on estimates of ϵ_t^2 at the beginning of the sample.

arch0(zero) specifies that the priming values are 0. Unlike the priming values for ARIMA models, 0 is generally not a consistent estimate of the presample conditional variance or squared innovations.

arch0(#) specifies that $\sigma_t^2 = \epsilon_t^2 = $ # for any specified nonnegative #. Thus, arch0(0) is equivalent of arch0(zero).

arma0(*cond_method*) is a rarely used option to specify how the ϵ_t values are initialized at the beginning of the sample for the ARMA component of the model, if it has such a component. This option has an effect only when AR or MA terms are included in the model (options ar(), ma(), or arima() specified).

arma0(zero) is the default. This specifies that all priming values of ϵ_t are to be taken to be 0. This fits the model over the entire requested sample and takes ϵ_t to be its expected value of 0 for all lags required by the ARMA terms; see Judge et al. (1985).

arma0(p), arma0(q), and arma0(pq) specify that the estimation begin after priming the recursions for a certain number of observations. p specifies that estimation begin after the pth observation in the sample, where p is the maximum AR lag in the model; q specifies that estimation begin after the qth observation in the sample, where q is the maximum MA lag in the model; and pq specifies that estimation begin after the $(p + q)$th observation in the sample.

During the priming period, the recursions necessary to generate predicted disturbances are performed, but results are used only for the purpose of initializing pre-estimation values of ϵ_t. Understand the definition of pre-estimation: say you fit a model in 10/100. If the model is specified with ar(1,2), then pre-estimation refers to observations 10 and 11.

The ARCH terms σ_t^2 and ϵ_t^2 are also updated over these observations. Any required lags of ϵ_t prior to the priming period are taken to be their expected value of 0, while ϵ_t^2 and σ_t^2 take the values specified in arch0()

arma0(#) specifies that the presample values of ϵ_t are to be taken as # for all lags required by the ARMA terms. Thus, arma0(0) is equivalent to arma0(zero).

condobs(#) is a rarely used option to specify a fixed number of conditioning observations at the start of the sample. Over these priming observations, the recursions necessary to generate predicted disturbances are performed, but only for the purpose of initializing pre-estimation values of ϵ_t, ϵ_t^2, and σ_t^2. Any required lags of ϵ_t prior to the initialization period are taken to be their expected value of 0 (or the value specified in arma0()), and required values of ϵ_t^2 and σ_t^2 assume the values specified by arch0(). condobs() can be used if conditioning observations are desired for the lags in the ARCH terms of the model. If arma() is also specified, the maximum of the number of conditioning observations required by arma() and condobs(#) is used.

Other options exclusive of optimization options

savespace specifies that memory use be conserved by retaining only those variables required for estimation. The original dataset is restored after estimation. This option is rarely used, and should be specified only if there is insufficient space to fit a model without the option. Note that arch requires considerably more temporary storage during estimation than most estimation commands in Stata.

detail specifies that a detailed list of any gaps in the series be reported. These include gaps due to missing observations or missing data for the dependent variable or independent variables.

level(#) specifies the confidence level, in percent, for confidence intervals of the coefficients.

Options for controlling maximization

maximize_options control the maximization process; see [R] **maximize**. These options are often more important for ARCH models than other maximum likelihood models because of convergence problems associated with ARCH models—ARCH model likelihoods are notoriously difficult to maximize.

Several alternate optimization methods such as Berndt–Hall–Hall–Hausman (BHHH) and Broyden–Fletcher–Goldfarb–Shanno (BFGS) are provided for arch models. Since each method attacks the optimization differently, some problems can be successfully optimized by an alternate method when one method fails.

The default optimization method for arch is a hybrid method combining BHHH and BFGS iterations. This combination has been found operationally to provide good convergence properties on difficult likelihoods. However, sometimes a likelihood is particularly deceptive to one or both of these methods.

from(*initial_values*) allows specifying the initial values of the coefficients. ARCH models may be sensitive to initial values, and may have coefficient values that correspond to local maxima. The default starting values are obtained via a series of regressions producing results that, based on asymptotic theory, are consistent for the β and ARMA parameters, and are, we believe, reasonable for the rest. Nevertheless, these values will sometimes prove to be infeasible in that the likelihood function cannot be evaluated at the initial values arch first chooses. In such cases, the estimation is restarted with ARCH and ARMA parameters initialized to zero. It is possible, but unlikely, that even these values will be infeasible, and that you will have to supply initial values yourself.

The standard syntax for from() accepts a matrix, a list of values, or coefficient name value pairs; see [R] **maximize**. In addition, arch allows the following:

from(archb0) specifies that the starting value for all the ARCH/GARCH/... parameters in the conditional-variance equation be set to 0.

from(armab0) specifies that the starting value for all ARMA parameters in the model be set to 0.

from(archb0 armab0) specifies that the starting value for all ARCH/GARCH/... and ARMA parameters be set to 0.

gtolerance(#) specifies the threshold for the relative size of the gradient; see [R] **maximize**. The default for arch is gtolerance(.05).

gtolerance(999) may be specified to disable the gradient criterion. If the optimizer becomes stuck with repeated "(backed up)" messages, it is likely that the gradient still contains substantial values, but an uphill direction cannot be found for the likelihood. With this option, results can often be obtained, but it is unclear whether the global maximum likelihood has been found.

When the maximization is not going well, it is also possible to set the maximum number of iterations, see [R] **maximize**, to the point where the optimizer appears to be stuck and to inspect the estimation results at that point.

bhhh, dfp, bfgs, nr, bhhhbfgs(), and bhhhdfp() specify how the likelihood function is to be maximized. bhhhbfgs(5,10) is the default.

bhhh specifies that the Berndt–Hall–Hall–Hausman (BHHH, Berndt et al. 1974) method be used. While it is difficult to make general statements about convergence properties of nonlinear optimization methods, BHHH tends to do well in areas far from the maximum, but does not have quadratic convergence in areas near the maximum.

dfp specifies that the Davidon–Fletcher–Powell (DFP) method be used; see Press et al. (1992). As currently implemented, dfp requires substantially less temporary storage space than the other methods (with the exception of bfgs), and this may be an advantage for models with many parameters.

bfgs specifies that the Broyden–Fletcher–Goldfarb–Shanno (BFGS) method be used; see Press et al. (1992). BFGS optimization is similar to DFP with second-order terms included when updating the Hessian. bfgs, like dfp, requires little memory.

nr specifies that Stata's modified Newton–Raphson method be used. Since all derivatives for arch are taken numerically, this method can be slow for models with many parameters. However, its choice of direction is computed quite differently from DFP, BFGS, and BHHH, and so nr is sometimes successful when the other methods have difficulty. (When you specify nr, arch automatically specifies the maximizer's difficult option for you; see [R] **maximize**.)

bhhhbfgs(#₁,#₂) specifies BHHH and BFGS be combined. $#_1$ designates the number of BHHH steps; $#_2$, the number of BFGS steps. Optimization alternates between these sets of BHHH and BFGS steps until convergence is achieved. The default optimization method is bhhhbfgs(5, 10).

bhhhdfp(#₁,#₂) specifies that BHHH and DFP be combined. $#_1$ designates the number of BHHH steps; $#_2$, the number of DFP steps. The optimization alternates between these sets of BHHH and DFP steps until convergence is achieved.

Options for predict

Six statistics can be computed by using predict after arch: the predictions of the mean equation (option xb, the default), the undifferenced predictions of the mean equation (option y), the predictions of the conditional variance (option variance), the predictions of the multiplicative heteroskedasticity component of variance (option het), the predictions of residuals or innovations (option residuals), and the predictions of residuals or innovations in terms of y (option yresiduals). Given the dynamic nature of ARCH models and that the dependent variable might be differenced, there are alternate ways of computing each statistic. We can use all the data on the dependent variable available right up to the time of each prediction (the default, which is often called a one-step prediction), or we can use the data up to a particular time, after which the predicted value of the dependent variable is used recursively to make subsequent predictions (option dynamic()). Either way, we can consider or ignore the ARMA disturbance component (the component is considered by default, and is ignored if you specify option `structural`). We might also be interested in predictions at certain fixed points where we specify the prior values of ϵ_t and σ_t^2 (option `at()`).

`xb` (the default) calculates the predictions from the mean equation. If $D.depvar$ is the dependent variable, these predictions are of $D.depvar$ and not $depvar$ itself.

`y` specifies that predictions of $depvar$ are to be made even if the model was specified in terms of, say, $D.depvar$.

`variance` calculates predictions of the conditional variance $\widehat{\sigma}_t^2$.

`het` calculates predictions of the multiplicative heteroskedasticity component of variance.

`residuals` calculates the residuals. If no other options are specified, these are the predicted innovations ϵ_t; i.e., they include any ARMA component. If option `structural` is specified, these are the residuals from the mean equation ignoring any ARMA terms; see `structural` below. The residuals are always from the estimated equation, which may have a differenced dependent variable; if $depvar$ is differenced, they are not the residuals of the undifferenced $depvar$.

`yresiduals` calculates the residuals in terms of $depvar$, even if the model was specified in terms of, say, $D.depvar$. As with `residuals`, the `yresiduals` are computed from the model including any ARMA component. If option `structural` is specified, any ARMA component is ignored and `yresiduals` are the residuals from the structural equation; see `structural` below.

`t0`(*time_constant*) specifies the starting point for the recursions to compute the predicted statistics; disturbances are assumed to be 0 for $t < $ `t0()`. The default is to set `t0()` to the minimum t observed in the estimation sample, meaning that observations prior to that are assumed to have disturbances of 0.

> `t0()` is irrelevant if `structural` is specified because in that case, all observations are assumed to have disturbances of 0.

> `t0(5)` would begin recursions at $t = 5$. If your data were quarterly, you might instead type `t0(q(1961q2))` to obtain the same result.

> Note that any ARMA component in the mean equation or GARCH term in the conditional-variance equation makes `arch` recursive and dependent on the starting point of the predictions. This includes one-step-ahead predictions.

`structural` specifies that the calculation is to be made considering the structural component only, ignoring any ARMA terms, and producing the steady-state equilibrium predictions.

`dynamic`(*time_constant*) specifies how lags of y_t in the model are to be handled. If `dynamic()` is not specified, actual values are used everywhere lagged values of y_t appear in the model to produce one-step-ahead forecasts.

> `dynamic`(*time_constant*) produces dynamic (also known as recursive) forecasts. *time_constant* specifies when the forecast is to switch from one-step ahead to dynamic. In dynamic forecasts, references to y evaluate to the prediction of y for all periods at or after *time_constant*; they evaluate to the actual value of y for all prior periods.

> `dynamic(10)` would calculate predictions where any reference to y_t with $t < 10$ evaluates to the actual value of y_t, and any reference to y_t with $t \geq 10$ evaluates to the prediction of y_t. This means that one-step ahead predictions are calculated for $t < 10$, and dynamic predictions thereafter. Depending on the lag structure of the model, the dynamic predictions might still reference some actual values of y_t.

> In addition, you may specify `dynamic(.)` to have `predict` automatically switch from one-step to dynamic predictions at $p + q$, where p is the maximum AR lag and q is the maximum MA lag.

$at(varname_\epsilon \mid \#_\epsilon \; varname_{\sigma^2} \mid \#_{\sigma^2})$ specifies that very static predictions are to be made. $at()$ and $dynamic()$ may not be specified together.

$at()$ specifies two sets of values to be used for ϵ_t and σ_t^2, the dynamic components in the model. These specified values are treated as given. In addition, lagged values of *depvar*, if they occur in the model, are obtained from the real values of the dependent variable. All computations are based on actual data and the given values. The purpose of $at()$ is to allow static evaluation of results for a given set of disturbances. This is useful, for instance, in generating the news response function.

$at()$ requires that you specify two arguments. Each argument can be either a variable name or a number. The first argument supplies the values to be used for ϵ_t; the second supplies the values to be used for σ_t^2. If σ_t^2 plays no role in your model, the second argument may be specified as '.' to indicate missing.

Remarks

The basic premise of ARCH models is that the volatility of a series is not constant through time. Instead, periods of relatively low volatility and periods of relatively high volatility tend to be grouped together. This is a commonly observed characteristic of economic time-series, and is even more pronounced in many frequently-sampled financial series. ARCH models seek to estimate this time-dependent volatility as a function of observed prior volatility. In some cases, the model of volatility is of more interest than the model of the conditional mean. As implemented in `arch`, the volatility model may also include regressors to account for a structural component in the volatility—usually referred to as multiplicative heteroskedasticity.

ARCH models were introduced by Engle (1982) in a study of inflation rates, and there has since been a barrage of proposed parametric and nonparametric specifications of autoregressive conditional heteroskedasticity. Overviews of the literature can found in Bollerslev, Engle, and Nelson (1994) and Bollerslev, Chou, and Kroner (1992). Introductions to basic ARCH models appear in many general econometrics texts, including Davidson and MacKinnon (1993), Greene (2003), Kmenta (1997), Johnston and DiNardo (1997), and Wooldridge (2002). Harvey (1989) and Enders (1995) provide introductions to ARCH in the larger context of econometric time-series modeling, and Hamilton (1994) provides considerably more detail in the same context.

`arch` fits models of autoregressive conditional heteroskedasticity (ARCH, GARCH, etc.) using conditional maximum likelihood. By "conditional", we mean that the likelihood is computed based on an assumed or estimated set of priming values for the squared innovations ϵ_t^2 and variances σ_t^2 prior to the estimation sample; see, for example, Hamilton (1994) or Bollerslev (1986). Sometimes additional conditioning is done on the first a, g, or $a + g$ observations in the sample, where a is the maximum ARCH term lag and g is the maximum GARCH term lag (or the maximum lag(s) from the other ARCH family terms).

The original ARCH model proposed by Engle (1982) modeled the variance of a regression model's disturbances as a linear function of lagged values of the squared regression disturbances. We can write an $ARCH(m)$ model as

$$y_t = \mathbf{x}_t \boldsymbol{\beta} + \epsilon_t \qquad \text{(conditional mean)}$$
$$\sigma_t^2 = \gamma_0 + \gamma_1 \epsilon_{t-1}^2 + \gamma_2 \epsilon_{t-2}^2 + \cdots + \gamma_m \epsilon_{t-m}^2 \qquad \text{(conditional variance)}$$

where

$$\epsilon_t \sim N(0, \sigma_t^2)$$

ϵ_t^2 is the squared residuals (or innovations)

γ_i are the ARCH parameters

The ARCH model has a specification for both the conditional mean and the conditional variance, and the variance is a function of the size of prior unanticipated innovations—ϵ_t^2. This model was generalized by Bollerslev (1986) to include lagged values of the conditional variance—a GARCH model. The GARCH(m, k) model is written as

$$y_t = \mathbf{x}_t \boldsymbol{\beta} + \epsilon_t$$
$$\sigma_t^2 = \gamma_0 + \gamma_1 \epsilon_{t-1}^2 + \gamma_2 \epsilon_{t-2}^2 + \cdots + \gamma_m \epsilon_{t-m}^2 + \delta_1 \sigma_{t-1}^2 + \delta_2 \sigma_{t-2}^2 + \cdots + \delta_k \sigma_{t-k}^2$$

where

$$\gamma_i \text{ are the ARCH parameters}$$
$$\delta_i \text{ are the GARCH parameters}$$

Without proof, we note that the GARCH model of conditional variance can be considered an ARMA process in the squared innovations, although not in the variances as the equations might seem to suggest; see, for example, Hamilton (1994). Specifically, the standard GARCH model implies that the squared innovations result from the process

$$\epsilon_t^2 = \gamma_0 + (\gamma_1 + \delta_1)\epsilon_{t-1}^2 + (\gamma_2 + \delta_2)\epsilon_{t-2}^2 + \cdots + (\gamma_k + \delta_k)\epsilon_{t-k}^2 + w_t - \delta_1 w_{t-1} - \delta_2 w_{t-2} - \delta_3 w_{t-3}$$

where

$$w_t = \epsilon_t^2 - \sigma_t^2$$
$$w_t \text{ is a white-noise process that is fundamental for } \epsilon_t^2$$

One of the primary benefits of the GARCH specification is parsimony in identifying the conditional variance. As with ARIMA models, the ARMA specification in GARCH allows the structure of the conditional variance to be modeled with fewer parameters than with an ARCH specification alone. Empirically, many series with a conditionally heteroskedastic disturbance have been found to be adequately modeled with a GARCH(1,1) specification.

An ARMA process in the disturbances can be easily added to the mean equation. For example, the mean equation can be written with an ARMA$(1, 1)$ disturbance as

$$y_t = \mathbf{x}_t \boldsymbol{\beta} + \rho(y_{t-1} - \mathbf{x}_{t-1}\boldsymbol{\beta}) + \theta \epsilon_{t-1} + \epsilon_t$$

with an obvious generalization to ARMA(p, q) by adding additional terms; see [TS] **arima** for more discussion of this specification. This change affects only the conditional-variance specification in that ϵ_t^2 now results from a different specification of the conditional mean.

Much of the literature on ARCH models has focused on alternate specifications of the variance equation. **arch** allows many of these alternate specifications to be requested using the options `saarch()` through `pgarch()`. In all cases, these options imply that one or more terms be changed or added to the specification of the variance equation.

One of the areas addressed by many of the alternate specifications is asymmetry. Both the ARCH and GARCH specifications imply a symmetric impact of innovations. Whether an innovation ϵ_t^2 is positive or negative makes no difference to the expected variance σ_t^2 in the ensuing periods; only the size of the innovation matters—good news and bad news have the same effect. Many theories, however, suggest that positive and negative innovations should vary in their impact. For risk-averse investors, a large unanticipated drop in the market is more likely to lead to higher volatility than a large unanticipated increase (see Black 1976, Nelson 1991). The options `saarch()`, `tarch()`, `aarch()`, `abarch()`, `earch()`, `aparch()`, and `tparch()` allow various specifications of asymmetric effects.

The options `narch()`, `narchk()`, `nparch()`, and `nparchk()` also imply an asymmetric impact, but of a very specific form. All of the models considered so far have a minimum conditional variance when the lagged innovations are all zero. "No news is good news" when it comes to keeping the conditional variance small. The `narch()`, `narchk()`, `nparch()`, and `nparchk()` options also have a symmetric response to innovations, but they are not centered at zero. The entire news response function (response to innovations) is shifted horizontally such that minimum variance lies at some specific positive or negative value for prior innovations.

ARCH-in-mean models allow the conditional variance of the series to influence the conditional mean. This is particularly convenient for modeling the risk/return relationship in financial series; the riskier an investment, all else equal, the lower its expected return. ARCH-in-mean models modify the specification of the conditional mean equation to be

$$y_t = \mathbf{x_t}\boldsymbol{\beta} + \psi\sigma_t^2 + \epsilon_t \qquad \text{(ARCH-in-mean)}$$

While this linear form in the current conditional variance has dominated the literature, `arch` allows the conditional variance to enter the mean equation through a nonlinear transformation $g()$, and for this transformed term to be included contemporaneously or lagged.

$$y_t = \mathbf{x_t}\boldsymbol{\beta} + \psi_0 g(\sigma_t^2) + \psi_1 g(\sigma_{t-1}^2) + \psi_2 g(\sigma_{t-2}^2) + \cdots + \epsilon_t$$

Square root is the most commonly used $g()$ transformation because researchers want to include a linear term for the conditional standard deviation, but any transform $g()$ is allowed.

▷ Example

We will consider a simple model of the US Wholesale Price Index (WPI) (Enders 1995, 106–110), which we also consider in [TS] **arima**. The data are quarterly over the period 1960q1 through 1990q4.

In [TS] **arima**, we fit a model of the continuously compounded rate of change in the WPI, $\ln(\text{WPI}_t) - \ln(\text{WPI}_{t-1})$. The graph of the differenced series—see [TS] **arima**—clearly shows periods of high volatility and other periods of relative tranquility. This makes the series a good candidate for ARCH modeling. Indeed, price indices have been a common target of ARCH models. Engle (1982) presented the original ARCH formulation in an analysis of UK inflation rates.

First, we fit a constant-only model by OLS and test ARCH effects using Engle's Lagrange multiplier test (`archlm`).

```
. use http://www.stata-press.com/data/r8/wpi1
. regress D.ln_wpi
```

Source	SS	df	MS		Number of obs =	123
					F(0, 122) =	0.00
Model	0	0	.		Prob > F =	.
Residual	.02521709	122	.000206697		R-squared =	0.0000
					Adj R-squared =	0.0000
Total	.02521709	122	.000206697		Root MSE =	.01438

D.ln_wpi	Coef.	Std. Err.	t	P>|t|	[95% Conf. Interval]
_cons	.0108215	.0012963	8.35	0.000	.0082553 .0133878

```
. archlm,lags(1)
```
LM test for autoregressive conditional heteroskedasticity (ARCH)

lags(p)	chi2	df	Prob > chi2
1	8.366	1	0.0038

H0: no ARCH effects   *vs.* H1: ARCH(p) disturbance

Noting that the LM test shows a p-value of 0.0038, which is well below 0.05, the null hypothesis of no ARCH(1) effects is rejected. Thus, we can further estimate the ARCH(1) parameter by specifying `arch(1)` option. You may see [TS] **regression diagnostics** for more information on Engle's LM test.

The first-order generalized ARCH model (GARCH, Bollerslev 1986) is the most commonly used specification for the conditional variance in empirical work, and is typically written GARCH(1, 1). We can estimate a GARCH(1, 1) process for the log-differenced series by typing

```
. arch D.ln_wpi, arch(1) garch(1)

(setting optimization to BHHH)
Iteration 0:   log likelihood =   355.2346
Iteration 1:   log likelihood = 365.64589
 (output omitted )
Iteration 10:  log likelihood = 373.23397

ARCH family regression
```

Sample: 1960q2 to 1990q4

Log likelihood = 373.234

Number of obs = 123
Wald chi2(.) = .
Prob > chi2 = .

D.ln_wpi	Coef.	OPG Std. Err.	z	P>|z|	[95% Conf. Interval]	
ln_wpi						
_cons	.0061167	.0010616	5.76	0.000	.0040361	.0081974
ARCH						
arch						
L1	.4364123	.2437428	1.79	0.073	-.0413147	.9141394
garch						
L1	.4544606	.1866605	2.43	0.015	.0886126	.8203085
_cons	.0000269	.0000122	2.20	0.028	2.97e-06	.0000508

We have estimated the ARCH(1) parameter to be .436 and the GARCH(1) parameter to be .454, so our fitted GARCH(1, 1) model is

$$y_t = .0061 + \epsilon_t$$
$$\sigma_t^2 = .436\, \epsilon_{t-1}^2 + .454\, \sigma_{t-1}^2$$

where $y_t = \ln(\text{wpi}_t) - \ln(\text{wpi}_{t-1})$.

Note that the model Wald test and probability are both reported as missing (.). By convention, Stata reports the model test for the mean equation. In this case, and fairly often for ARCH models, the mean equation consists only of a constant, and there is nothing to test.   ◁

▷ Example

We can retain the GARCH(1, 1) specification for the conditional variance and model the means as an AR(1) and MA(1) process with an additional seasonal MA term at lag 4 by typing

```
. arch D.ln_wpi, ar(1) ma(1 4) arch(1) garch(1)

(setting optimization to BHHH)
Iteration 0:   log likelihood = 380.99952
Iteration 1:   log likelihood = 388.57801
Iteration 2:   log likelihood = 391.34179
Iteration 3:   log likelihood = 396.37029
Iteration 4:   log likelihood = 398.01112
(switching optimization to BFGS)
```

```
Iteration 5:   log likelihood =  398.23657
BFGS stepping has contracted, resetting BFGS Hessian (0)
Iteration 6:   log likelihood =  399.21491
Iteration 7:   log likelihood =  399.21531  (backed up)
  (output omitted )
Iteration 12:  log likelihood =   399.4934  (backed up)
Iteration 13:  log likelihood =  399.49607
Iteration 14:  log likelihood =  399.51241
(switching optimization to BHHH)
Iteration 15:  log likelihood =  399.51441
Iteration 16:  log likelihood =  399.51443
```

ARCH family regression -- ARMA disturbances

Sample: 1960q2 to 1990q4

Log likelihood = 399.5144

Number of obs   =   123
Wald chi2(3)   =   153.60
Prob > chi2   =   0.0000

D.ln_wpi		Coef.	OPG Std. Err.	z	P>|z|	[95% Conf. Interval]	
ln_wpi							
_cons		.0069556	.0039511	1.76	0.078	-.0007884	.0146996
ARMA							
ar							
	L1	.7922546	.1072028	7.39	0.000	.5821409	1.002368
ma							
	L1	-.3417593	.1499399	-2.28	0.023	-.6356361	-.0478824
	L4	.2452439	.1251166	1.96	0.050	.00002	.4904679
ARCH							
arch							
	L1	.2037791	.1243096	1.64	0.101	-.0398632	.4474214
garch							
	L1	.6953231	.1890604	3.68	0.000	.3247715	1.065875
_cons		.0000119	.0000104	1.14	0.253	-8.51e-06	.0000324

To clarify exactly what we have estimated, we could write our model

$$y_t = .007 + .792\,(y_{t-1} - .007) - .342\,\epsilon_{t-1} + .245\,\epsilon_{t-4} + \epsilon_t$$
$$\sigma_t^2 = .204\,\epsilon_{t-1}^2 + .695\,\sigma_{t-1}^2$$

where $y_t = \ln(\texttt{wpi}_t) - \ln(\texttt{wpi}_{t-1})$.

The ARCH(1) coefficient, .204, is not significantly different from zero, but it is clear that collectively, the ARCH(1) and GARCH(1) coefficients are significant. If there is any doubt, you can check the conjecture with test.

```
. test [ARCH]L1.arch [ARCH]L1.garch
 ( 1)  [ARCH]L.arch = 0
 ( 2)  [ARCH]L.garch = 0

        chi2( 2) =     85.10
      Prob > chi2 =     0.0000
```

(Note that for comparison, we fitted the model over the same sample used in the example in [TS] **arima**; Enders fits this GARCH model, but over a slightly different sample.)

◁

❑ Technical Note

The rather ugly iteration log on the prior result is not atypical. Difficulty in converging is common in ARCH models. This is actually a fairly well-behaved likelihood for an ARCH model. The "switching optimization to ... " messages are standard messages from the default optimization method for `arch`. The "backed up" messages are typical of BFGS stepping as the BFGS Hessian is often over-optimistic, particularly during early iterations. These are nothing to be concerned about.

Nevertheless, watch out for the messages "BFGS stepping has contracted, resetting BFGS Hessian" and "backed up". Both can flag problems. Problems, if they arise, will result in an iteration log that goes on and on; Stata will never report convergence, and so will never report final results. The question is: when do you give up and press *Break* and, if you do, what do you do then?

The "BFGS stepping has contracted" message, if it occurs repeatedly (more than, say, five times), often indicates that convergence will never be achieved. Literally, it means that the BFGS algorithm was stuck, and needed to reset its Hessian and take a steepest descent step.

The "backed up" message, if it occurs repeatedly, also indicates problems, but only if the likelihood value is simultaneously not changing. If the message occurs repeatedly but the likelihood value is changing, as it did above, all is going well; it is just going slowly.

If you have convergence problems, you can specify options attempting to assist the current maximization method or try a different method. Or, it simply might be that your model specification and your data lead to a likelihood that is nonconvex in the allowable region and thus cannot be maximized.

Concerning the "backed up" message with no change in the likelihood: You can try resetting the gradient tolerance to a larger value. Specifying option `gtolerance(999)` will disable gradient checking, allowing convergence to be declared more easily. This does not guarantee that convergence will be declared and, even if convergence is declared, it is unclear whether the global maximum likelihood has been found.

You can also try to specify initial values.

Finally, see *Options for controlling maximization*. You can try a different maximization method.

Realize that the ARCH family of models are notorious for convergence difficulties. Unlike most estimators in Stata, it is not uncommon for convergence to require many, many steps, or even to fail. This is particularly true of the explicitly nonlinear terms such as `aarch()`, `narch()`, `aparch()`, or `archm` (ARCH-in-mean), and of any model with several lags in the ARCH terms. There is not always a solution. Alternate maximization methods or possibly different starting values can be tried, but if your data do not support your assumed ARCH structure, convergence simply may not be possible.

ARCH models can be susceptible to irrelevant regressors or unnecessary lags, whether in the specification of the conditional mean or in the conditional variance. In these situations, `arch` will often continue to iterate, making little to no improvement in the likelihood. We view this conservative approach as better than declaring convergence prematurely when the likelihood has not been fully maximized. `arch` is estimating the conditional form of second sample moments, often with flexible functions, and that is asking much of the data.

❑

❑ Technical Note

`if` *exp* and `in` *range* have a somewhat different interpretation with commands accepting time-series operators. The time-series operators are resolved *before* the conditions are tested. We believe that this is what one would expect when typing a command, but it may lead to some confusion. Note the results of the following `list` commands:

```
. list t y l.y in 5/10
```

	t	y	L.y
5.	1961q1	30.8	30.7
6.	1961q2	30.5	30.8
7.	1961q3	30.5	30.5
8.	1961q4	30.6	30.5
9.	1962q1	30.7	30.6
10.	1962q2	30.6	30.7

```
. keep in 5/10
(124 observations deleted)
. list t y l.y
```

	t	y	L.y
1.	1961q1	30.8	.
2.	1961q2	30.5	30.8
3.	1961q3	30.5	30.5
4.	1961q4	30.6	30.5
5.	1962q1	30.7	30.6
6.	1962q2	30.6	30.7

We have one additional lagged observation for y in the first case. l.y was resolved before the in restriction was applied. In the second case, the dataset no longer contains the value of y to compute the first lag. This means that

```
. arch y l.x, arch(1), if twithin(1930, 1990)
```

is not the same as

```
. keep if twithin(1930, 1990)
. arch y l.x, arch(1)
```

❏

▷ Example

Continuing with the WPI data, we might be concerned that the economy as a whole responds differently to unanticipated increases in wholesale prices than it does to unanticipated decreases. Perhaps unanticipated increases lead to cash flow issues that affect inventories and lead to more volatility. We can see if the data supports this supposition by specifying an ARCH model that allows an asymmetric effect of "news"—innovations or unanticipated changes. One of the most popular such models is EGARCH (Nelson 1991). The full first-order EGARCH model for the WPI can be specified as

(Continued on next page)

```
. arch D.ln_wpi, ar(1) ma(1 4) earch(1) egarch(1)

(setting optimization to BHHH)
Iteration 0:   log likelihood =   227.5251
Iteration 1:   log likelihood = 381.69177
 (output omitted)
Iteration 21:  log likelihood = 405.31453

ARCH family regression -- ARMA disturbances
```

Sample: 1960q2 to 1990q4

Log likelihood = 405.3145

			Number of obs	=	123
			Wald chi2(3)	=	156.04
			Prob > chi2	=	0.0000

D.ln_wpi		Coof.	OPG Std. Err.	z	P>|z|	[95% Conf. Interval]	
ln_wpi _cons		.0087343	.0034006	2.57	0.010	.0020692	.0153994
ARMA ar							
	L1	.7692304	.0968317	7.94	0.000	.5794438	.9590169
ma							
	L1	-.3554775	.1265674	-2.81	0.005	-.603545	-.10741
	L4	.241473	.086382	2.80	0.005	.0721674	.4107787
ARCH earch							
	L1	.4064035	.1163528	3.49	0.000	.1783561	.6344509
earch_a							
	L1	.2467611	.1233453	2.00	0.045	.0050088	.4885134
egarch							
	L1	.8417318	.0704089	11.95	0.000	.7037329	.9797307
_cons		-1.488376	.6604486	-2.25	0.024	-2.782831	-.1939203

Our result for the variance is

$$\ln(\sigma_t^2) = -1.49 + .406\, z_{t-1} + .247 \left| z_{t-1} - \sqrt{2/\pi} \right| + .842 \ln(\sigma_{t-1}^2)$$

where $z_t = \epsilon_t/\sigma_t$, which is distributed as $N(0,1)$.

We have a strong indication for a leverage effect. The positive L1.earch coefficient implies that positive innovations (unanticipated price increases) are more destabilizing than negative innovations. The effect appears quite strong (.406), and is substantially larger than the symmetric effect (.247). In fact, the relative scales of the two coefficients imply that the positive leverage completely dominates the symmetric effect.

This can readily be seen if we plot what is often referred to as the news response or news impact function. This curve just shows the resulting conditional variance as a function of unanticipated news, in the form of innovations. That is, it is the conditional variance σ_t^2 as a function of ϵ_t. Thus, we need to evaluate σ_t^2 for various values of ϵ_t—say, −4 to 4—and then graph the result.

predict, at() will calculate σ_t^2 given a set of specified innovations $(\epsilon_t, \epsilon_{t-1}, \ldots)$ and prior conditional variances $(\sigma_{t-1}^2, \sigma_{t-2}^2, \ldots)$. The syntax is

```
. predict newvar, variance at(epsilon sigma2)
```

epsilon and sigma2 are either variables or numbers. The use of sigma2 is a little tricky because you specify values of σ_t^2, and σ_t^2 is what predict is supposed to predict. predict does not simply copy variable *sigma2* into *newvar*. Rather, it uses the lagged values contained in *sigma2* to produce the currently predicted value of σ_t^2; it does this for all t, and those results are stored in *newvar*. (If you are interested in dynamic predictions of σ_t^2, see *Options for predict*.)

We will generate predictions for σ_t^2 assuming the lagged values of σ_t^2 are 1, and we will vary ϵ_t from -4 to 4. First, we will create variable et containing ϵ_t, and then we will create and graph the predictions:

```
. generate et = (_n-64)/15
. predict sigma2, variance at(et 1)
. line sigma2 et in 2/1, m(i) c(l) title(News response function)
```

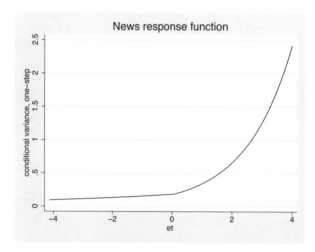

The positive asymmetry does indeed dominate the shape of the news response function. In fact, the response is a monotonically increasing function of news. The form of the response function shows that, for our simple model, only positive unanticipated price increases have the destabilizing effect that we observe as larger conditional variances.

◁

▷ Example

As an example of a frequently sampled, long-run series, consider the daily closing indices of the Dow Jones Industrial Average, variable dowclose. Only data after 1jan1953 are used to avoid the first half of the century when the New York Stock Exchange was open for Saturday trading. The compound return of the series is used as the dependent variable and is graphed below.

(Graph on next page)

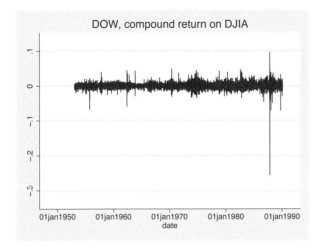

We should probably examine how we formed this difference, because it is somewhat different from most time series.

We formed this difference by referring to D.ln_dow, but only after playing a trick. The series is daily and each observation represents the Dow closing index for the day. Our data included a time variable recorded as a daily date. We wanted, however, to model the log differences in the series, and we wanted the span from Friday to Monday to appear as a single-period difference. That is, the day before Monday is Friday. Because our dataset was tsset with date, the span from Friday to Monday was 3 days. The solution was to create a second variable that sequentially numbered the observations. By tsseting the data with this new variable, we obtained the desired differences.

```
. generate t = _n
. tsset t
```

Now, our data look like

```
. use http://www.stata-press.com/data/r8/dow1
. generate dayofwk = dow(date)
. list date dayofwk t ln_dow D.ln_dow in 1/8
```

	date	dayofwk	t	ln_dow	D.ln_dow
1.	02jan1953	5	1	5.677096	.
2.	05jan1953	1	2	5.682899	.0058026
3.	06jan1953	2	3	5.677439	−.0054603
4.	07jan1953	3	4	5.672636	−.0048032
5.	08jan1953	4	5	5.671259	−.0013762
6.	09jan1953	5	6	5.661223	−.0100365
7.	12jan1953	1	7	5.653191	−.0080323
8.	13jan1953	2	8	5.659134	.0059433

(Continued on next page)

```
. list date dayofwk t ln_dow D.ln_dow in -8/1
```

	date	dayofwk	t	ln_dow	D.ln_dow
9334.	08feb1990	4	9334	7.880188	.0016198
9335.	09feb1990	5	9335	7.881635	.0014472
9336.	12feb1990	1	9336	7.870601	-.011034
9337.	13feb1990	2	9337	7.872665	.0020638
9338.	14feb1990	3	9338	7.872577	-.0000877
9339.	15feb1990	4	9339	7.88213	.009553
9340.	16feb1990	5	9340	7.876863	-.0052676
9341.	20feb1990	2	9341	7.862054	-.0148082

We can see that the difference operator D spans weekends because the currently specified time variable t is not a true date, and has a difference of 1 for all observations. We must leave this contrived time variable in place during estimation, or arch will be convinced that our dataset has gaps. If we were using calendar dates, we would indeed have gaps.

Ding, Granger, and Engle (1993) fitted an A-PARCH model of daily returns of the Standard and Poor's 500 (S&P 500) for 3jan1928 through 30aug1991. We will fit the same model for the Dow data shown above. The model includes an AR(1) term as well as the A-PARCH specification of conditional variance.

```
. arch D.ln_dow, ar(1) aparch(1) pgarch(1)

(setting optimization to BHHH)
Iteration 0:   log likelihood =  31139.553
Iteration 1:   log likelihood =  31350.751
Iteration 2:   log likelihood =  31351.082  (backed up)
 (output omitted )
BFGS stepping has contracted, resetting BFGS Hessian (5)
Iteration 58:  log likelihood =  32273.555

ARCH family regression -- AR disturbances

Sample:  2 to 9341                         Number of obs    =      9340
                                           Wald chi2(1)     =    175.46
Log likelihood =  32273.56                 Prob > chi2      =    0.0000
```

D.ln_dow		Coef.	OPG Std. Err.	z	P>|z|	[95% Conf.	Interval]
ln_dow							
_cons		.0001786	.0000875	2.04	0.041	7.15e-06	.00035
ARMA							
ar							
	L1	.1410944	.0106519	13.25	0.000	.1202171	.1619717
ARCH							
aparch							
	L1	.0626324	.0034307	18.26	0.000	.0559083	.0693565
aparch_e							
	L1	-.3645107	.0378486	-9.63	0.000	-.4386925	-.2903288
pgarch							
	L1	.9299016	.0030998	299.99	0.000	.9238261	.9359771
_cons		7.19e-06	2.53e-06	2.84	0.004	2.23e-06	.0000121
POWER							
power		1.585178	.0629183	25.19	0.000	1.461861	1.708496

Turning first to the iteration log, we note that the final iteration reports the message "backed up". Ending on a "backed up" message for most estimators would be a cause for great concern, but not with arch or, for that matter, arima, as long as you do not specify option gtolerance(). That is because arch and arima, by default, monitor the gradient, and declare convergence only if, in addition to everything else, the gradient is sufficiently small.

The fitted model demonstrates substantial asymmetry, with the large negative L1.aparch_e coefficient indicating that the market responds with much more volatility to unexpected drops in returns, "bad news", than it does to increases in returns, "good news".

◁

▷ Example

Engle's (1982) original model that sparked the interest in ARCH provides an example requiring constraints. Most current ARCH specifications make use of GARCH terms to provide flexible dynamic properties without estimating an excessive number of parameters. The original model was limited to ARCH terms, and to help cope with the collinearity of the terms, a declining lag structure was imposed in the parameters. The conditional variance equation was specified as

$$\sigma_t^2 = \alpha_0 + \alpha(.4\,\epsilon_{t-1} + .3\,\epsilon_{t-2} + .2\,\epsilon_{t-3} + .1\,\epsilon_{t-4})$$
$$= \alpha_0 + .4\,\alpha\epsilon_{t-1} + .3\,\alpha\epsilon_{t-2} + .2\,\alpha\epsilon_{t-3} + .1\,\alpha\epsilon_{t-4}$$

From the earlier arch output, we know how the coefficients will be named. In Stata, the formula is

$$\sigma_t^2 = [\text{ARCH}]\_\text{cons} + .4\,[\text{ARCH}]\text{L1.arch}\,\epsilon_{t-1} + .3\,[\text{ARCH}]\text{L2.arch}\,\epsilon_{t-2}$$
$$+ .2\,[\text{ARCH}]\text{L3.arch}\,\epsilon_{t-3} + .1\,[\text{ARCH}]\text{L4.arch}\,\epsilon_{t-3}$$

We could specify these linear constraints any number of ways, but the following seems fairly intuitive; see [R] **constraint** for syntax.

```
. use http://www.stata-press.com/data/r8/wpi1
. constraint define 1 (3/4)*[ARCH]l1.arch = [ARCH]l2.arch
. constraint define 2 (2/4)*[ARCH]l1.arch = [ARCH]l3.arch
. constraint define 3 (1/4)*[ARCH]l1.arch = [ARCH]l4.arch
```

The original model was fitted on UK inflation; we will again use the WPI data and retain our earlier specification of the mean equation, which differs from Engle's UK inflation model. With our constraints, we type

(Continued on next page)

```
. arch D.ln_wpi, ar(1) ma(1 4) arch(1/4) constraint(1/3)
(setting optimization to BHHH)
Iteration 0:   log likelihood =  396.80192
Iteration 1:   log likelihood =  399.07808
  (output omitted)
Iteration 9:   log likelihood =  399.46243

ARCH family regression -- ARMA disturbances
```

Sample: 1960q2 to 1990q4

				Number of obs	=	123
				Wald chi2(3)	=	123.32
Log likelihood = 399.4624				Prob > chi2	=	0.0000

Constraints:
(1)   .75 [ARCH]L.arch - [ARCH]L2.arch = 0
(2)   .5 [ARCH]L.arch - [ARCH]L3.arch = 0
(3)   .25 [ARCH]L.arch - [ARCH]L4.arch = 0

		Coef.	OPG Std. Err.	z	P>|z|	[95% Conf. Interval]	
ln_wpi							
_cons		.0077204	.0034531	2.24	0.025	.0009525	.0144883
ARMA							
ar							
	L1	.7388168	.1126811	6.56	0.000	.517966	.9596676
ma							
	L1	-.2559691	.1442861	-1.77	0.076	-.5387646	.0268264
	L4	.2528922	.1140185	2.22	0.027	.02942	.4763644
ARCH							
arch							
	L1	.2180138	.0737787	2.95	0.003	.0734101	.3626174
	L2	.1635103	.055334	2.95	0.003	.0550576	.2719631
	L3	.1090069	.0368894	2.95	0.003	.0367051	.1813087
	L4	.0545034	.0184447	2.95	0.003	.0183525	.0906544
_cons		.0000483	7.66e-06	6.30	0.000	.0000333	.0000633

L1.arch, L2.arch, L3.arch, and L4.arch coefficients have the constrained relative sizes. We can also recover the α parameter from the original specification by using lincom.

```
. lincom [ARCH]l1.arch/.4
 ( 1)   2.5 [ARCH]L.arch = 0
```

D.ln_wpi	Coef.	Std. Err.	z	P>|z|	[95% Conf. Interval]	
(1)	.5450344	.1844468	2.95	0.003	.1835253	.9065436

Any of the arch parameters could be used to produce an identical estimate.

◁

(Continued on next page)

Saved Results

arch saves in e():

Scalars

e(N)	number of observations	e(ic)	number of iterations
e(k)	number of variables	e(rank)	rank of e(V)
e(k_eq)	number of equations	e(power)	φ for power arch terms
e(k_dv)	number of dependent variables	e(tmin)	minimum time
e(df_m)	model degrees of freedom	e(tmax)	maximum time
e(ll)	log likelihood	e(N_gaps)	number of gaps
e(rc)	return code	e(archi)	$\sigma_0^2 = \epsilon_0^2$, priming values
e(chi2)	χ^2	e(condobs)	# of conditioning observations
e(p)	significance		

Macros

e(cmd)	arch	e(archm)	ARCH-in-mean lags
e(depvar)	name of dependent variable	e(archmexp)	ARCH-in-mean exp
e(title)	title in estimation output	e(mhet)	1 if multiplicative heteroskedasticity
e(title2)	secondary title in estimation output	e(earch)	lags for EARCH terms
e(eqnames)	names of equations	e(egarch)	lags for EGARCH terms
e(wtype)	weight type	e(aarch)	lags for AARCH terms
e(wexp)	weight expression	e(narch)	lags for NARCH terms
e(vcetype)	covariance estimation method	e(aparch)	lags for APARCH terms
e(user)	name of likelihood-evaluator program	e(nparch)	lags for NPARCH terms
e(opt)	type of optimization	e(saarch)	lags for SAARCH terms
e(chi2type)	Wald; type of model χ^2 test	e(parch)	lags for PAARCH terms
e(tech)	maximization technique	e(tparch)	lags for TPARCH terms
e(ma)	lags for moving average terms	e(abarch)	lags for ABARCH terms
e(ar)	lags for autoregressive terms	e(tarch)	lags for TARCH terms
e(tmins)	formatted minimum time	e(sdgarch)	lags for SDGARCH terms
e(tmaxs)	formatted maximum time	e(pgarch)	lags for PGARCH terms
e(scorevars)	variables containing scores	e(garch)	lags for GARCH terms
e(predict)	program used to implement predict	e(cond)	flag if called by arima
e(arch)	lags for ARCH terms		

Matrices

e(b)	coefficient vector	e(V)	variance–covariance matrix of
e(ilog)	iteration log (up to 20 iterations)		the estimators

Functions

e(sample)	marks estimation sample

Methods and Formulas

arch is implemented as an ado-file using the ml commands; see [R] **ml**. The robust variance computation is performed by _robust; see [P] **_robust**.

The mean equation for the model fitted by arch and with ARMA terms can be written as

$$y_t = \mathbf{x_t}\boldsymbol{\beta} + \sum_{i=1}^{p} \psi_i g(\sigma_{t-i}^2) + \sum_{j=1}^{p} \rho_j \left\{ y_{t-j} - x_{t-j}\boldsymbol{\beta} - \sum_{i=1}^{p} \psi_i g(\sigma_{t-j-i}^2) \right\}$$

$$+ \sum_{k=1}^{q} \theta_k \epsilon_{t-k} + \epsilon_t \qquad\qquad \text{(conditional mean)}$$

where

$\boldsymbol{\beta}$ are the regression parameters

$\boldsymbol{\psi}$ are the ARCH-in-mean parameters

$\boldsymbol{\rho}$ are the autoregression parameters

$\boldsymbol{\theta}$ are the moving average parameters

$g()$ is a general function, see option `archmexp()`

Any or all of the parameters in this full specification of the conditional mean may be zero. For example, the model need not have moving average parameters ($\boldsymbol{\theta} = 0$), or ARCH-in-mean parameters ($\boldsymbol{\psi} = 0$).

The variance equation will be one of

$$\sigma^2 = \gamma_0 + A(\boldsymbol{\sigma}, \boldsymbol{\epsilon}) + B(\boldsymbol{\sigma}, \boldsymbol{\epsilon})^2 \tag{1}$$

$$\ln \sigma_t^2 = \gamma_0 + C(\ln \boldsymbol{\sigma}, \mathbf{z}) + A(\boldsymbol{\sigma}, \boldsymbol{\epsilon}) + B(\boldsymbol{\sigma}, \boldsymbol{\epsilon})^2 \tag{2}$$

$$\sigma_t^\varphi = \gamma_0 + D(\boldsymbol{\sigma}, \boldsymbol{\epsilon}) + A(\boldsymbol{\sigma}, \boldsymbol{\epsilon}) + B(\boldsymbol{\sigma}, \boldsymbol{\epsilon})^2 \tag{3}$$

where $A(\boldsymbol{\sigma}, \boldsymbol{\epsilon})$, $B(\boldsymbol{\sigma}, \boldsymbol{\epsilon})$, $C(\ln \boldsymbol{\sigma}, \mathbf{z})$, and $D(\boldsymbol{\sigma}, \boldsymbol{\epsilon})$ are just linear sums of the appropriate ARCH terms; see *Details of syntax*. Equation (1) is used if no EGARCH or power ARCH terms are included in the model, equation (2) if EGARCH terms are included, and (3) if any power ARCH terms are included; again, see *Details of syntax*.

Priming values

The above model is recursive with potentially long memory. It is necessary to assume pre-estimation sample values for ϵ_t, ϵ_t^2, and σ_t^2 to begin the recursions, and the remaining computations are therefore conditioned on these priming values. These priming values can be controlled using the `arch0()` and `arma0()` options, and most of the computations used to compute these priming values are presented in *Options affecting conditioning (priming) values*.

The `arch0(xb0wt)` and `arch0(xbwt)` options compute a weighted sum of estimated disturbances with more weight on the early observations. With either of these options,

$$\sigma_{t_0-i}^2 = \epsilon_{t_0-i}^2 = (1 - .7) \sum_{t=0}^{T-1} .7^{T-t-1} \epsilon_{T-t}^2 \qquad \forall i$$

where t_0 is the first observation for which the likelihood is computed; see *Options affecting conditioning (priming) values*. The ϵ_t^2 are all computed from the conditional mean equation. If `arch0(xb0wt)` is specified, $\boldsymbol{\beta}$, ψ_i, ρ_j, and θ_k are taken from initial regression estimates and held constant during optimization. If `arch0(xbwt)` is specified, the current estimates of $\boldsymbol{\beta}$, ψ_i, ρ_j, and θ_k are used to compute ϵ_t^2 on every iteration. Note that if any ψ_i is in the mean equation (ARCH-in-mean is specified), the estimates of ϵ_t^2 from the initial regression estimates are not consistent.

Likelihood from prediction error decomposition

The likelihood function for ARCH has a particularly simple form. Given priming (or conditioning) values of ϵ_t, ϵ_t^2, and σ_t^2, the mean equation above can be solved recursively for every ϵ_t (prediction error decomposition). Likewise, the conditional variance can be computed recursively for each observation by using the variance equation. Using these predicted errors, their associated variances, and the assumption that $\epsilon_t \sim N(0, \sigma_t^2)$, the log likelihood for each observation t is

$$\ln L_t = -\frac{1}{2} w_t \left\{ \ln(2\pi\sigma_t^2) + \sum_{t=t_0}^{T} \frac{\epsilon_t^2}{\sigma_t^2} \right\}$$

where $w_t = 1$ if weights are not specified.

Missing data

ARCH will allow missing data or missing observations, but makes no attempt to condition on the surrounding data. If a dynamic component cannot be computed—ϵ_t, ϵ_t^2, and/or σ_t^2—its priming value is substituted. If a covariate, the dependent variable, or entire observation is missing, the observation does not enter the likelihood, and its dynamic components are set to their priming values for that observation. This is only acceptable asymptotically, and should not be used with large amounts of missing data.

References

Baum, C. F. 2000. sts15: Tests for stationarity of a time series. *Stata Technical Bulletin* 57: 36–39. Reprinted in *Stata Technical Bulletin Reprints*, vol. 10, pp. 356–360.

Baum, C. F. and R. Sperling. 2000. sts15.1: Tests for stationarity of a time series: update. *Stata Technical Bulletin* 58: 35–36. Reprinted in *Stata Technical Bulletin Reprints*, vol. 10, pp. 360–362.

Baum, C. F. and V. L. Wiggins. 2000. sts16: Tests for long memory in a time series. *Stata Technical Bulletin* 57: 39–44. Reprinted in *Stata Technical Bulletin Reprints*, vol. 10, pp. 362–368.

Berndt, E. K., B. H. Hall, R. E. Hall, and J. A. Hausman. 1974. Estimation and inference in nonlinear structural models. *Annals of Economic and Social Measurement* 3/4: 653–665.

Black, F. 1976. Studies of stock price volatility changes. *Proceedings from the American Statistical Association, Business and Economics Statistics*, 653–665.

Bollerslev, T. 1986. Generalized autoregressive conditional heteroskedasticity. *Journal of Econometrics* 31: 307–327.

Bollerslev, T., R. Y. Chou, and K. F. Kroner. 1992. ARCH modeling in finance. *Journal of Econometrics* 52: 5–59.

Bollerslev, T., R. F. Engle, and D. B. Nelson. 1994. ARCH models. In *Handbook of Econometrics, Volume IV*, ed. R. F. Engle and D. L. McFadden. New York: Elsevier.

Bollerslev, T. and J. M. Wooldridge. 1992. Quasi maximum likelihood estimation and inference in dynamic models with time varying covariances. *Econometric Reviews* 11: 143–172.

Davidson, R. and J. G. MacKinnon. 1993. *Estimation and Inference in Econometrics*. Oxford: Oxford University Press.

Ding, Z., C. W. J. Granger, and R. F. Engle. 1993. A long memory property of stock market returns and a new model. *Journal of Empirical Finance* 1: 83–106.

Enders, W. 1995. *Applied Econometric Time Series*. New York: John Wiley & Sons.

Engle, R. F. 1982. Autoregressive conditional heteroskedasticity with estimates of the variance of U.K. inflation. *Econometrica* 50: 987–1008.

——. 1990. Discussion: Stock market volatility and the crash of 87. *Review of financial studies* 3: 103–106.

Engle, R. F., D. M. Lilien, and R. P. Robins. 1987. Estimating time varying risk premia in the term structure: the ARCH-M model. *Econometrica* 55: 391–407.

Glosten, L. R., R. Jagannathan, and D. Runkle. 1993. On the relation between the expected value and the volatility of the nominal excess return on stocks. *Journal of Finance* 48: 1779–1801.

Greene, W. H. 2003. *Econometric Analysis*. 5th ed. Upper Saddle River, NJ: Prentice–Hall.

Hamilton, J. D. 1994. *Time Series Analysis*. Princeton: Princeton University Press.

Harvey, A. C. 1989. *Forecasting, structural time series models and the Kalman filter*. Cambridge: Cambridge University Press.

——. 1990. *The Econometric Analysis of Time Series*. 2d ed. Cambridge, MA: MIT Press.

Higgins, M. L. and A. K. Bera. 1992. A class of nonlinear ARCH models. *International Economic Review* 33: 137–158.

Judge, G. G., W. E. Griffiths, R. C. Hill, H. Lütkepohl, and T.-C. Lee. 1985. *The Theory and Practice of Econometrics*. 2d ed. New York: John Wiley & Sons.

Johnston, J. and J. DiNardo. 1997. *Econometric Methods*. 3d ed. New York: McGraw–Hill.

Kmenta, J. 1997. *Elements of Econometrics*. 2d ed. Ann Arbor: University of Michigan Press.

Nelson, D. B. 1991. Conditional heteroskedasticity in asset returns: a new approach. *Econometrica* 59: 347–370.

Press, W. H., S. A. Teukolsky, W. T. Vetterling, and B. P. Flannery. 1992. *Numerical Recipes in C: The Art of Scientific Computing*. 2d ed. Cambridge: Cambridge University Press.

Wooldridge, J. M. 2002. *Introductory Econometrics: A Modern Approach*. 2d ed. Cincinnati, OH: South-Western College Publishing.

Zakoian, J. M. 1990. Threshold heteroskedastic models. Unpublished manuscript. *CREST, INSEE*.

Also See

Complementary:	[TS] **tsset**,
	[R] **adjust**, [R] **lincom**, [R] **lrtest**, [R] **mfx**, [R] **nlcom** [R] **predict**,
	[R] **predictnl**, [R] **suest**, [R] **test**, [R] **testnl**, [R] **vce**, [R] **xi**
Related:	[TS] **arima**, [TS] **prais**,
	[R] **regress**, [R] **regression diagnostics**
Background:	[U] **14.4.3 Time-series varlists**,
	[U] **16.5 Accessing coefficients and standard errors**,
	[U] **23 Estimation and post-estimation commands**,
	[U] **23.14 Obtaining robust variance estimates**,
	[U] **23.15 Obtaining scores**,
	[U] **27.3 Time-series dates**

Title

arima — Autoregressive integrated moving average models

Syntax

Basic syntax for a regression/structural model with ARMA *disturbances*

> arima *depvar* $[varlist]$, ar(*numlist*) ma(*numlist*)

Basic syntax for an ARIMA(p,d,q) *model*

> arima *depvar* , arima($\#_p,\#_d,\#_q$)

Full syntax

> arima *depvar* $[varlist]$ $[weight]$ $[if\ exp]$ $[in\ range]$ $[$, ar(*numlist*) ma(*numlist*)
>
> arima($\#_p,\#_d,\#_q$) noconstant constraints(*numlist*) hessian opg robust
>
> score(*newvarlist* | *stub*) diffuse p0($\#$ | *matname*) state0($\#$ | *matname*)
>
> condition savespace detail level($\#$)
>
> *maximize_options* from(*initial_values*) gtolerance($\#$)
>
> bhhh dfp bfgs nr bhhhbfgs($\#,\#$) bhhhdfp($\#,\#$) $]$

by ... : may be used with arima; see [R] **by**.
You must tsset your data before using arima; see [TS] **tsset**.
depvar and *varlist* may contain time-series operators; see [U] **14.4.3 Time-series varlists**.
iweights are allowed; see [U] **14.1.6 weight**.
arima shares the features of all estimation commands; see [U] **23 Estimation and post-estimation commands**.

Syntax for predict

> predict $[type]$ *newvarname* $[if\ exp]$ $[in\ range]$ $[$, *statistic*
>
> structural dynamic(*time_constant*) t0(*time_constant*) $]$

where *statistic* is

xb	predicted values for the model—the differenced series; the default
y	fitted values in *y*—the undifferenced series
mse	mean square error of the prediction xb
residuals	residuals or predicted innovations
yresiduals	residuals or predicted innovations in *y*—the undifferenced series

and *time_constant* is a $\#$ or a time literal such as d(1jan1995) or q(1995q1), etc.

These statistics are available both in and out of sample; type predict ... if e(sample) ... if wanted only for the estimation sample.

Description

arima fits a model of *depvar* on *varlist* where the disturbances are allowed to follow a linear autoregressive moving-average (ARMA) specification. The dependent and independent variables may be differenced or seasonally differenced to any degree. When independent variables are not specified, these models reduce to autoregressive integrated moving average (ARIMA) models in the dependent variable. Missing data are allowed and are handled using the Kalman filter and methods suggested by Harvey (1989 and 1993); see *Methods and Formulas*.

Referring to the full syntax, *depvar* is the variable being modeled and the structural or regression part of the model is specified in *varlist*. The options ar() and ma() are used to specify the lags of autoregressive and moving average terms, respectively.

arima allows time-series operators in the dependent variable and independent variable lists, and it is often convenient to make extensive use of these operators; see [U] **27.3 Time-series dates** for an extended discussion of time-series operators.

arima typed without arguments redisplays the previous estimates.

Options

Material in the *Remarks* section may be helpful for understanding some of the options.

ar(*numlist*) specifies the autoregressive terms to be included in the model. These are the autoregressive terms of the structural model disturbance. For example, ar(1/3) specifies that lags of 1, 2, and 3 of the structural disturbance are included in the model, and ar(1 4) specifies that lags 1 and 4 are included, possibly to account for quarterly effects.

If the model does not contain any regressors, these terms can also be considered autoregressive terms for the dependent variable.

ma(*numlist*) specifies the moving average terms to be included in the model. These are the terms for the lagged innovations—white-noise disturbances.

arima($\#_p$,$\#_d$,$\#_q$) is an alternate, shorthand notation for specifying models that are autoregressive in the dependent variable. The dependent variable and any independent variables are differenced $\#_d$ times, 1 through $\#_p$ lags of autocorrelations are included, and 1 through $\#_q$ lags of moving averages are included. For example, the specification

        . arima D.y, ar(1/3) ma(1/3)

is equivalent to

        . arima y, arima(2,1,3)

The latter is easier to write for "classic" ARIMA models, but is not nearly as expressive as the former. If gaps in the AR or MA lags are to be modeled, or if different operators are to be applied to independent variables, the first syntax will generally be required.

noconstant suppresses the constant term (intercept) in the structural model.

constraints(*numlist*) specifies the constraint numbers of the linear constraints to be applied during estimation. The default is to perform unconstrained estimation. Constraints are specified using the constraint command; see [R] **constraint** (also see [R] **reg3** for use of constraint in multiple-equation contexts).

If constraints are placed between structural model parameters and ARMA terms, it is not uncommon for the first few iterations to attempt steps into nonstationary areas. This can be ignored if the final solution is well within the bounds of stationary solutions.

hessian and opg specifies how standard errors are to be calculated. The default is opg unless one of the options bfgs, dfp, or nr is specified, in which case, the default is hessian.

hessian specifies that the standard errors and coefficient covariance matrix be estimated from the full Hessian—the matrix of negative second derivatives of the log-likelihood function. These are the estimates produced by most of Stata's maximum likelihood estimators.

opg specifies that the standard errors and coefficient covariance matrix be estimated using the outer product of the coefficient gradients with respect to the observation likelihoods.

hessian and opg provide asymptotically equivalent estimates of the standard errors and covariance matrix and there is no theoretical justification for preferring either estimate.

robust specifies that the Huber/White/sandwich estimator of variance is to be used in place of the traditional calculation; see [U] **23.14 Obtaining robust variance estimates**.

For state-space models in general and ARIMA in particular, the robust or quasi-maximum likelihood estimates (QMLE) of variance are robust to symmetric nonnormality in the disturbances—including, as a special case, heteroskedasticity. The robust variance estimates are not generally robust to functional misspecification of the structural or ARMA components of the model; see Hamilton (1994, 389) for a brief discussion.

score(*newvarlist* | *stub**) creates a new variable for each parameter in the model. Each new variable contains the derivative of the model log-likelihood with respect to the parameter for each observation in the estimation sample: $\partial L_t / \partial \beta_k$, where L_t is the log likelihood for observation t and β_k is the kth parameter in the model.

If score(*newvarlist*) is specified, the *newvarlist* must contain a new variable for each parameter in the model. If score(*stub**) is specified, variables named *stub#* are created for each parameter in the model. The *newvarlist* is filled, or the #'s in *stub#* are created, in the order in which the estimated parameters are reported in the estimation results table.

Unlike scores for most other models, the scores from arima are individual gradients of the log likelihood with respect to the variables, not with respect to $\mathbf{x_t}\beta$. Since the general ARIMA model is inherently nonlinear, especially when it includes MA terms, the scores with respect to $\mathbf{x_t}\beta$ could not be used to reconstruct the gradients for the individual parameters.

diffuse specifies that a diffuse prior (see Harvey 1989 or 1993) be used as a starting point for the Kalman filter recursions. Using diffuse, nonstationary models may be fitted with arima (also see option p0() below; diffuse is equivalent to specifying p0(1e9)).

By default, arima uses the unconditional expected value of the state vector ξ_t (see *Methods and Formulas*) and the mean square error (MSE) of the state vector to initialize the filter. When the process is stationary, this corresponds to the expected value and expected variance of a random draw from the state vector, and produces unconditional maximum likelihood estimates of the parameters. This default is not appropriate, however, and the unconditional MSE cannot be computed when the process is not stationary. For a nonstationary process, an alternate starting point must be used for the recursions.

In the absence of nonsample or pre-sample information, diffuse may be specified to start the recursions from a state vector of zero and a state MSE matrix corresponding to an effectively infinite variance on this initial state. This amounts to an uninformative and improper prior that is updated to a proper MSE as data from the sample become available; see Harvey (1989).

Note that nonstationary models may also correspond to models with infinite variance given a particular specification. This and other problems with nonstationary series make convergence difficult and sometimes impossible.

diffuse can also be useful if a model contains one or more long AR or MA lags. Computation of the unconditional MSE of the state vector (see *Methods and Formulas*) requires construction and inversion of a square matrix that is of dimension $\max(p, q+1)$, where p and q are the maximum AR and MA lags, respectively. For a maximum lag of 28, this would require a 784 by 784 matrix. Estimation with diffuse does not require this matrix.

For large samples, there is little difference between using the default starting point and the diffuse starting point. Unless the series has a very long memory, the initial conditions affect the likelihood of only the first few observations.

p0(# | *matname*) is a rarely specified option that can be used for nonstationary series or when an alternate prior for starting the Kalman recursions is desired (see diffuse above for a discussion of the default starting point and *Methods and Formulas* for background).

If *matname* is given, it specifies a matrix to be used as the MSE of the state vector for starting the Kalman filter recursions—$\mathbf{P}_{1|0}$. Alternately, a single number, #, may be supplied, and the MSE of the initial state vector $\mathbf{P}_{1|0}$ will have this number on its diagonal and all off-diagonal values set to zero.

This option may be used with nonstationary series to specify a larger or smaller diagonal for $\mathbf{P}_{1|0}$ than that supplied by diffuse. It may also be used in conjunction with state0() when the user believes they have a better prior for the initial state vector and its MSE.

state0(# | *matname*) is a rarely used option that specifies an alternate initial state vector, $\boldsymbol{\xi}_{1|0}$ (see *Methods and Formulas*), for starting the Kalman filter recursions. If # is specified, all elements of the vector are taken to be #. The default initial state vector is state0(0).

condition specifies that conditional rather than full maximum likelihood estimates be produced. The pre-sample values for ϵ_t and μ_t are taken to be their expected value of zero, and the estimate of the variance of ϵ_t is taken to be constant over the entire sample; see Hamilton (1994, 132). This estimation method is not appropriate for nonstationary series, but may be preferable for long series or for models that have one or more long AR or MA lags. diffuse, p0(), and state0() have no meaning for models fitted from the conditional likelihood, and may not be specified with condition.

If the series is long and stationary and if the underlying data generating process does not have a long memory, then estimates will be similar whether estimated by unconditional maximum likelihood (the default), conditional maximum likelihood (condition), or maximum likelihood from a diffuse prior (diffuse).

In small samples, however, results of conditional versus unconditional maximum likelihood may differ substantially; see Ansley and Newbold (1980). Whereas the default unconditional maximum likelihood estimates make the most use of sample information when all the assumptions of the model are met, Harvey (1989) and Ansley and Kohn (1985) argue for diffuse priors in many cases, particularly in ARIMA models corresponding to an underlying structural model.

The condition or diffuse options may also be preferred when the model contains one or more long AR or MA lags; this avoids inverting potentially large matrices (see diffuse above).

When condition is specified, estimation is performed by the arch command (see [TS] **arch**), and more control of the estimation process can be obtained by using arch directly.

savespace specifies that memory use be conserved by retaining only those variables required for estimation. The original dataset is restored after estimation. This option is rarely used, and should be used only if there is insufficient space to fit a model without the option. Note, however, that arima requires considerably more temporary storage during estimation than most estimation commands in Stata.

detail specifies that a detailed list of any gaps in the series be reported. These include gaps due to missing observations or missing data for the dependent variable or independent variables.

level(#) specifies the confidence level in percent for confidence intervals of the coefficients.

maximize_options control the maximization process; see [R] **maximize**. These options are sometimes more important for ARIMA models than most maximum likelihood models because of potential convergence problems with ARIMA models, particularly if the specified model combined with the sample data imply a nonstationary model.

Several alternate optimization methods such as Berndt–Hall–Hall–Hausman (BHHH) and Broyden–Fletcher–Goldfarb–Shanno (BFGS) are provided for arima models. Whereas arima models are not as difficult to optimize as ARCH models, their likelihoods are generally not quadratic and often pose optimization difficulties; this is particularly true if a model is nonstationary or nearly nonstationary. Since each method attacks the optimization differently, some problems can be successfully optimized by an alternate method when one method fails.

The default optimization method for arima is a hybrid method combining BHHH and BFGS iterations. This hybrid method has been found operationally to provide good convergence properties on difficult likelihoods.

The following options are all related to maximization and are either particularly important in fitting ARIMA models or not available for most other estimators.

from(*initial_values*) allows the starting values of the model coefficients to be set by the user; see [R] **maximize** for a general discussion and syntax options.

The standard syntax for from() accepts a matrix, a list of values, or coefficient name value pairs; see [R] **maximize**. In addition, arima accepts from(armab0), and this specifies that the starting value for all ARMA parameters in the model be set to zero prior to optimization.

ARIMA models may be sensitive to initial conditions and may have coefficient values that correspond to local maxima. The default starting values for arima are generally very good, particularly in large samples for stationary series.

gtolerance(#) is a rarely used maximization option that specifies a threshold for the relative size of the gradient; see [R] **maximize**. The default gradient tolerance for arima is .05.

gtolerance(999) may be specified to effectively disable the gradient criterion when convergence is difficult to achieve. If the optimizer becomes stuck with repeated "(backed up)" messages, it is likely that the gradient still contains substantial values, but an uphill direction cannot be found for the likelihood. Using gtolerance(999), results will often be obtained, but it is unclear whether the global maximum likelihood has been found. It is usually better to set the maximum number of iterations (see [R] **maximize**) to the point where the optimizer appears to be stuck and then inspect the estimation results.

bhhh, dfp, bfgs, nr, bhhhbfgs(), and bhhhdfp() specify how the likelihood function is to be maximized. bhhhbfgs(5,10) is the default.

bhhh specifies that the Berndt–Hall–Hall–Hausman (BHHH, Berndt et al. 1974) method be used. While it is difficult to make general statements about convergence properties of nonlinear optimization methods, BHHH tends to do well in areas far from the maximum, but does not have quadratic convergence in areas near the maximum.

dfp specifies that the Davidon–Fletcher–Powell (DFP) method be used; see Press et al. (1992). As currently implemented, dfp requires substantially less temporary storage space than the other methods (with the exception of bfgs), and this may be an advantage for models with many parameters.

bfgs specifies that the Broyden–Fletcher–Goldfarb–Shanno (BFGS) method be used; see Press et al. (1992). BFGS optimization is similar to DFP with second-order terms included when updating the Hessian. bfgs, just as dfp, requires little memory.

nr specifies that Stata's modified Newton–Raphson method be used. Since all derivatives for arima are taken numerically, this method may be very slow for models with many parameters.

bhhhbfgs($\#_1$,$\#_2$) specifies BHHH and BFGS be combined. $\#_1$ designates the number of BHHH steps and $\#_2$ designates the number of BFGS steps. Optimization alternates between these sets of BHHH and BFGS steps until convergence is achieved. The default optimization method is bhhhbfgs(5, 10).

bhhhdfp($\#_1$,$\#_2$) specifies that BHHH and DFP be combined. $\#_1$ designates the number of BHHH steps and $\#_2$ designates the number of DFP steps. The optimization alternates between these sets of BHHH and DFP steps until convergence is achieved.

Options for predict

Five statistics can be computed by using predict after arima: the predictions from the model (the default also given by option xb), the undifferenced predictions (option y), the MSE of xb (option mse), the predictions of residuals or innovations (option residual), and the predicted residuals or innovations in terms of y (option yresiduals). Given the dynamic nature of the ARMA component and that the dependent variable might be differenced, there are alternate ways of computing each. We can use all the data on the dependent variable available right up to the time of each prediction (the default, which is often called a one-step prediction), or we can use the data up to a particular time, after which the predicted value of the dependent variable is used recursively to make subsequent predictions (option dynamic()). Either way, we can consider or ignore the ARMA disturbance component (the component is considered by default and is ignored if you specify option structural).

All calculations can be made in or out of sample.

xb (the default) calculates the predictions from the model. If D.*depvar* is the dependent variable, these predictions are of D.*depvar* and not of *depvar* itself.

y specifies that predictions of *depvar* are to be made even if the model was specified in terms of, say, D.*depvar*.

mse calculates the MSE of xb.

residuals calculates the residuals. If no other options are specified, these are the predicted innovations ϵ_t; i.e., they include the ARMA component. If option structural is specified, these are the residuals μ_t from the structural equation; see structural below.

yresiduals calculates the residuals in terms of *depvar*, even if the model was specified in terms of, say, D.*depvar*. As with residuals, the yresiduals are computed from the model including any ARMA component. If option structural is specified, any ARMA component is ignored and yresiduals are the residuals from the structural equation; see structural below.

structural specifies that the calculation is to be made considering the structural component only, ignoring the ARMA terms, thus producing the steady-state equilibrium predictions.

dynamic(*time_constant*) specifies how lags of y_t in the model are to be handled. If dynamic() is not specified, actual values are used everywhere lagged values of y_t appear in the model to produce one-step ahead forecasts.

dynamic(*time_constant*) produces dynamic (also known as recursive) forecasts. *time_constant* specifies when the forecast is to switch from one-step ahead to dynamic. In dynamic forecasts, references to y evaluate to the prediction of y for all periods at or after *time_constant*; they evaluate to the actual value of y for all prior periods.

dynamic(10) would calculate predictions where any reference to y_t with $t < 10$ evaluates to the actual value of y_t and any reference to y_t with $t \geq 10$ evaluates to the prediction of y_t. This means that one-step ahead predictions are calculated for $t < 10$ and dynamic predictions thereafter. Depending on the lag structure of the model, the dynamic predictions might still reference some actual values of y_t.

In addition, you may specify dynamic(.) to have predict automatically switch from one-step to dynamic predictions at $p + q$, where p is the maximum AR lag and q is the maximum MA lag.

t0(*time_constant*) specifies the starting point for the recursions to compute the predicted statistics; disturbances are assumed to be 0 for $t <$ t0(). The default is to set t0() to the minimum t observed in the estimation sample, meaning that observations prior to that are assumed to have disturbances of 0.

t0() is irrelevant if structural is specified because in that case, all observations are assumed to have disturbances of 0.

t0(5) would begin recursions at $t = 5$. If you were quarterly, you might instead type t0(q(1961q2)) to obtain the same result.

Note that the ARMA component of arima models is recursive and depends on the starting point of the predictions. This includes one-step ahead predictions.

Remarks

arima fits both standard ARIMA models that are autoregressive in the dependent variable and structural models with ARMA disturbances. Good introductions to the former models can be found in Box, Jenkins, and Reinsel (1994), Hamilton (1994), Harvey (1993), Newton (1988), Diggle (1990), and many others. The latter models are developed fully in Hamilton (1994) and Harvey (1989), both of which provide extensive treatment of the Kalman filter (Kalman 1960) and the state-space form used by arima to fit the models.

Considering a first-order autoregressive AR(1) and a moving average MA(1) process, arima estimates all of the parameters in the model

$$y_t = \mathbf{x_t}\boldsymbol{\beta} + \mu_t \qquad\qquad \text{\textit{structural equation}}$$
$$\mu_t = \rho\mu_{t-1} + \theta\epsilon_{t-1} + \epsilon_t \qquad\qquad \text{\textit{disturbance, ARMA}}(1,1)$$

where

ρ is the first-order autocorrelation parameter
θ is the first-order moving average parameter
ϵ_t $\sim$ *i.i.d.* $N(0, \sigma^2)$; which is to say, ϵ_t is taken to be a white-noise disturbance

We can combine the two equations and write a general ARMA(p, q) in the disturbances process as

$$y_t = \mathbf{x_t}\boldsymbol{\beta} + \rho_1(y_{t-1} - \mathbf{x}_{t-1}\boldsymbol{\beta}) + \rho_2(y_{t-2} - \mathbf{x}_{t-2}\boldsymbol{\beta}) + \cdots + \rho_p(y_{t-p} - \mathbf{x}_{t-p}\boldsymbol{\beta})$$
$$+ \theta_1\epsilon_{t-1} + \theta_2\epsilon_{t-2} + \cdots + \theta_q\epsilon_{t-q} + \epsilon_t$$

It is also common to write the general form of the ARMA model succinctly using lag operator notation

$$\boldsymbol{\rho}(L^p)(y_t - \mathbf{x}_t\boldsymbol{\beta}) = \boldsymbol{\theta}(L^q)\epsilon_t \qquad\qquad \text{ARMA}(p, q)$$

where

$$\boldsymbol{\rho}(L^p) = 1 - \rho_1 L - \rho_2 L^2 - \cdots - \rho_2 L^p$$
$$\boldsymbol{\theta}(L^q) = 1 + \theta_1 L + \theta_2 L^2 + \cdots + \theta_2 L^q$$

For stationary series, full or unconditional maximum likelihood estimates are obtained via the Kalman filter. For nonstationary series, if some prior information is available, initial values for the filter can be specified using state0() and p0() as suggested by Hamilton (1994), or an uninformative prior can be assumed using the option diffuse as suggested by Harvey (1989).

Time-series models without a structural component do not have the $\mathbf{x_t}\beta$ terms and are often written as autoregressions in the dependent variable, rather than autoregressions in the disturbances from a structural equation. Other than a scale factor for the constant, these models are exactly equivalent to the ARMA in the disturbances formulation estimated by arima, but the latter are more flexible and allow a wider class of models.

❏ Technical Note

Proof: Without loss of generality consider a model that is ARMA(1, 1) in the dependent variable

$$y_t = \alpha + \rho y_{t-1} + \theta \epsilon_{t-1} + \epsilon_t \tag{1a}$$

We can combine the structural and disturbance equations of the ARMA(1,1) in the disturbances formulation and replace the structural $\mathbf{x_t}\beta$ with the constant β_0 by writing

$$y_t = \beta_0 + \rho \mu_{t-1} + \theta \epsilon_{t-1} + \epsilon_t \tag{1b}$$

From the simplified structural equation we have $\mu_t = y_t - \beta_0$, so 1b can be rewritten as

$$y_t = \beta_0 + \rho(y_{t-1} - \beta_0) + \theta \epsilon_{t-1} + \epsilon_t$$

or

$$y_t = (1 - \rho)\beta_0 + \rho y_{t-1} + \theta \epsilon_{t-1} + \epsilon_t \tag{1c}$$

Equations $(1a)$ and $(1b)$ are equivalent with the constant in $(1b)$ scaled by $(1 - \rho)$. arima fits models as autoregressive in the disturbances, and we have just seen that these subsume models that are autoregressive in the dependent variable.

❏

▷ Example

Enders (1995, 106–110) considers an ARIMA model of the US Wholesale Price Index (WPI) using quarterly data over the period 1960q1 through 1990q4. The simplest ARIMA model that includes differencing, autoregressive, and moving average components is the ARIMA(1,1,1) specification. We can fit this model using arima by typing

```
. use http://www.stata-press.com/data/r8/wpi1

. arima wpi, arima(1,1,1)

(setting optimization to BHHH)
Iteration 0:    log likelihood = -139.80133
Iteration 1:    log likelihood =  -135.6278
Iteration 2:    log likelihood = -135.41838
Iteration 3:    log likelihood = -135.36691
Iteration 4:    log likelihood = -135.35892
(switching optimization to BFGS)
Iteration 5:    log likelihood = -135.35471
Iteration 6:    log likelihood = -135.35135
Iteration 7:    log likelihood = -135.35132
Iteration 8:    log likelihood = -135.35131
```

```
ARIMA regression
Sample:  1960q2 to 1990q4                    Number of obs    =        123
                                             Wald chi2(2)     =     310.64
Log likelihood = -135.3513                   Prob > chi2      =     0.0000
```

D.wpi	Coef.	OPG Std. Err.	z	P>|z|	[95% Conf. Interval]
wpi					
_cons	.7498197	.3340968	2.24	0.025	.0950019 1.404637
ARMA					
ar					
L1	.8742288	.0545435	16.03	0.000	.7673266 .981132
ma					
L1	-.4120458	.1000284	-4.12	0.000	-.6080979 -.2159938
/sigma	.7250436	.0368065	19.70	0.000	.6529042 .7971829

Examining the estimation results, we see that the AR(1) coefficient is .87 and the MA(1) coefficient is −.41 and that both are highly significant. The estimated standard deviation of the white-noise disturbance ϵ is .725.

This model could also have been fitted by typing

```
. arima D.wpi, ar(1) ma(1)
```

The D. placed in front of the dependent variable wpi is the Stata time-series operator for differencing. Thus, we would be modeling the first difference in WPI from 2nd quarter 1960 through 4th quarter 1990. The advantage of this second syntax is that it allows a richer choice of models. The arima($\#_p$, $\#_d$, $\#_q$) option does not provide for seasonal differencing or seasonal AR and MA terms.

◁

▷ Example

After examining first differences of WPI, Enders chose a model of differences in the natural logarithms to stabilize the variance in the differenced series. The raw data and first difference of the logarithms are graphed below.

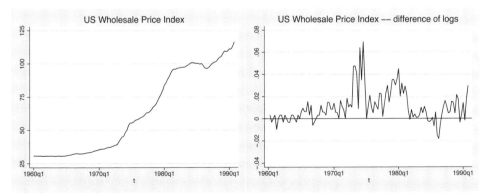

On the basis of the autocorrelations, partial autocorrelations (see graphs below), and the results of preliminary estimations, Enders identified an ARMA model in the log-differenced series.

```
. ac D.ln_wpi, ylabels(-.4(.2).6)
. pac D.ln_wpi, ylabels(-.4(.2).6)
```

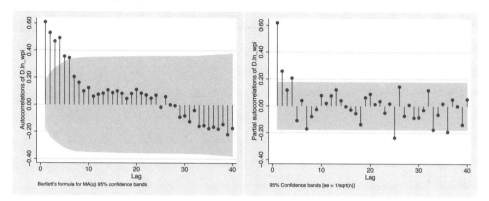

In addition to an autoregressive term and an MA(1) term, a seasonal MA(4) term at lag 4 is included to account for a remaining quarterly effect. Thus, the model to be fitted is

$$\Delta \ln(wpi_t) = \beta_0 + \rho\{\Delta \ln(wpi_{t-1}) - \beta_0\} + \theta_1 \epsilon_{t-1} + \theta_4 \epsilon_{t-4} + \epsilon_t$$

where $\Delta x \equiv x_t - x_{t-1}$ or, using lag operators $\Delta x \equiv (1 - L)x$.

We can fit this model using `arima` and Stata's standard difference operator:

```
. arima D.ln_wpi, ar(1) ma(1 4)
(setting optimization to BHHH)
Iteration 0:   log likelihood = 382.67447
Iteration 1:   log likelihood = 384.80754
Iteration 2:   log likelihood = 384.84749
Iteration 3:   log likelihood = 385.39213
Iteration 4:   log likelihood = 385.40983
(switching optimization to BFGS)
Iteration 5:   log likelihood =  385.9021
Iteration 6:   log likelihood = 385.95646
Iteration 7:   log likelihood = 386.02979
Iteration 8:   log likelihood = 386.03326
Iteration 9:   log likelihood = 386.03354
Iteration 10:  log likelihood = 386.03357
```

(*Continued on next page*)

```
ARIMA regression

Sample:  1960q2 to 1990q4                    Number of obs      =        123
                                             Wald chi2(3)       =     333.60
Log likelihood =  386.0336                   Prob > chi2        =     0.0000
```

		Coef.	OPG Std. Err.	z	P>|z|	[95% Conf.	Interval]
D.ln_wpi							
ln_wpi							
_cons		.0110493	.0048349	2.29	0.022	.0015731	.0205255
ARMA							
ar							
	L1	.7806991	.0944946	8.26	0.000	.5954931	.965905
ma							
	L1	-.3990039	.1258753	-3.17	0.002	-.6457149	-.1522928
	L4	.3090813	.1200945	2.57	0.010	.0737003	.5444622
/sigma		.0104394	.0004702	22.20	0.000	.0095178	.0113609

In this final specification, the log-differenced series is still highly autocorrelated at a level of .78, while innovations have a negative impact in the ensuing quarter ($-.40$) and a positive seasonal impact of .31 in the following year.

◁

❑ Technical Note

We also note one item where the results differ from most of Stata's estimation commands—the standard error of the coefficients is reported as OPG Std. Err. As noted in the *Options* section, the default standard errors and covariance matrix for arima estimates are derived from the outer product of gradients (OPG). This is one of three asymptotically equivalent methods of estimating the covariance matrix of the coefficients (only two of which are usually tractable to derive). Discussions and derivations of all three estimates can be found in Davidson and MacKinnon (1993), Greene (2003), and Hamilton (1994). Bollerslev, Engle, and Nelson (1994) suggest that the OPG estimates may be more numerically stable in time series regressions when the likelihood and its derivatives depend on recursive computations, certainly the case for the Kalman filter. To date, we have not found any numerical instabilities in either estimate of the covariance matrix—subject to the stability and convergence of the overall model.

Most of Stata's estimation commands provide covariance estimates derived from the Hessian of the likelihood function. These alternate estimates can also be obtained from arima by specifying the hessian option.

❑

▷ Example

As a simple example of a model including covariates, we can estimate an update of Friedman and Meiselman's (1963) equation representing the quantity theory of money. They postulated a straight-forward relationship between personal consumption expenditures (consump) and the money supply as measured by M2 (m2).

$$\text{consump}_t = \beta_0 + \beta_1 m2_t + \mu_t$$

Friedman and Meiselman fitted the model over a period ending in 1956; we will refit the model over the period 1959q1 through 1981q4. We restrict our attention to the period prior to 1982 because the Federal Reserve manipulated the money supply extensively in the latter 1980s to control inflation, and the relationship between consumption and the money supply becomes much more complex during the latter part of the decade.

Since our purpose is to demonstrate `arima`, we will include both an autoregressive term and a moving average term for the disturbances in the model; the original estimates included neither. Thus, we model the disturbance of the structural equation as

$$\mu_t = \rho\mu_{t-1} + \theta\epsilon_{t-1} + \epsilon_t$$

Following the original authors, the relationship is estimated on seasonally adjusted data, so there is no need to explicitly include seasonal effects. It might be preferable to obtain seasonally unadjusted data and simultaneously model the structural and seasonal effects.

The estimation will be restricted to the desired sample by using the `tin()` function in an `if` expression; see [R] **functions** and [U] **27.3 Time-series dates**. By leaving the first argument of `tin()` blank, we are including all available data up to and including the second date (1981q4). We fit the model by typing

```
. use http://www.stata-press.com/data/r8/friedman2
. arima consump m2, ar(1) ma(1), if tin( , 1981q4)
 (output omitted )
Iteration 10:  log likelihood = -340.50774

ARIMA regression

Sample:  1959q1 to 1981q4                      Number of obs    =        92
                                               Wald chi2(3)     =   4394.80
Log likelihood = -340.5077                     Prob > chi2      =    0.0000
```

consump	Coef.	OPG Std. Err.	z	P>|z|	[95% Conf. Interval]	
consump						
m2	1.122029	.0363563	30.86	0.000	1.050772	1.193286
_cons	-36.09872	56.56703	-0.64	0.523	-146.9681	74.77062
ARMA						
ar						
L1	.9348486	.0411323	22.73	0.000	.8542308	1.015467
ma						
L1	.3090592	.0885883	3.49	0.000	.1354293	.4826891
/sigma	9.655308	.5635157	17.13	0.000	8.550837	10.75978

We find a relatively small money velocity with respect to consumption (1.122029) over this period, although consumption is only one facet of the income velocity. We also note a very large first-order autocorrelation in the disturbances as well as a statistically significant first-order moving average.

We might be concerned that our specification has led to disturbances that are heteroskedastic or non-Gaussian. We refit the model using the `robust` option.

```
. arima consump m2, ar(1) ma(1) robust, if tin( , 1981q4)
 (output omitted )
Iteration 10:  log pseudo-likelihood = -340.50774
```

```
ARIMA regression
Sample:  1959q1 to 1981q4                      Number of obs    =        92
                                               Wald chi2(3)     =   1176.26
Log pseudo-likelihood = -340.5077              Prob > chi2      =    0.0000
```

consump	Coef.	Semi-robust Std. Err.	z	P>|z|	[95% Conf. Interval]	
consump						
m2	1.122029	.0433302	25.89	0.000	1.037103	1.206954
_cons	-36.09872	28.10478	-1.28	0.199	-91.18308	18.98564
ARMA						
ar						
L1	.9348486	.0493428	18.95	0.000	.8381385	1.031559
ma						
L1	.3090592	.1605359	1.93	0.054	-.0055854	.6237038
/sigma	9.655308	1.082639	8.92	0.000	7.533375	11.77724

We do note a substantial increase in the estimated standard errors, and our once clearly significant moving average term is now only marginally significant.

◁

Saved Results

arima saves in e():

Scalars

e(N)	number of observations	e(ic)	number of iterations
e(k)	number of variables	e(rank)	rank of e(V)
e(k_eq)	number of equations	e(sigma)	standard error of the disturbance
e(k_dv)	number of dependent variables	e(tmin)	minimum time
e(df_m)	model degrees of freedom	e(tmax)	maximum time
e(ll)	log likelihood	e(N_gaps)	number of gaps
o(rc)	return code	e(ar_max)	maximum AR lag
e(chi2)	χ^2	e(ma_max)	maximum MA lag
e(p)	significance		

Macros

e(cmd)	arima	e(chi2type)	Wald; type of model χ^2 test
e(depvar)	name of dependent variable	e(tech)	maximization technique
e(title)	title in estimation output	e(ma)	lags for moving average terms
e(eqnames)	names of equations	e(ar)	lags for autoregressive terms
e(wtype)	weight type	e(unsta)	unstationary or blank
e(wexp)	weight expression	e(tmins)	formatted minimum time
e(vcetype)	covariance estimation method	e(tmaxs)	formatted maximum time
e(user)	name of likelihood-evaluator program	e(scorevars)	variables containing scores
e(opt)	type of optimization	e(predict)	program used to implement predict

Matrices

e(b)	coefficient vector	e(V)	variance–covariance matrix of the estimators
e(ilog)	iteration log (up to 20 iterations)		

Functions

e(sample)	marks estimation sample

Methods and Formulas

arima is implemented as an ado-file.

Estimation is by maximum likelihood using the Kalman filter via the prediction error decomposition; see Hamilton (1994), Gourieroux and Monfort (1997) or, in particular, Harvey (1989). Any of these sources will serve as excellent background for the estimation of these models using the state-space form; each also provides considerable detail on the method outlined below.

ARIMA model

The model to be fitted is

$$y_t = \boldsymbol{\xi}_t \boldsymbol{\beta} + \mu_t$$

$$\mu_t = \sum_{i=1}^{p} \rho_i \mu_{t-i} + \sum_{j=1}^{q} \theta_j \epsilon_{t-j} + \epsilon_t$$

which can be written as the single equation:

$$y_t = \mathbf{x_t}\boldsymbol{\beta} + \sum_{i=1}^{p} \rho_i (y_{t-i} - x_{t-i}\boldsymbol{\beta}) + \sum_{j=1}^{q} \theta_j \epsilon_{t-j} + \epsilon_t$$

Kalman filter equations

We will roughly follow Hamilton's (1994) notation and write the Kalman filter

$$\boldsymbol{\xi}_t = \mathbf{F}\boldsymbol{\xi}_{t-1} + \mathbf{v}_t \qquad (state\ equation)$$

$$\mathbf{y}_t = \mathbf{A}'\mathbf{x}_t + \mathbf{H}'\boldsymbol{\xi}_t + \mathbf{w}_t \qquad (observation\ equation)$$

and

$$\begin{pmatrix} \mathbf{v}_t \\ \mathbf{w}_t \end{pmatrix} \sim N \left\{ \mathbf{0}, \begin{pmatrix} \mathbf{Q} & \mathbf{0} \\ \mathbf{0} & \mathbf{R} \end{pmatrix} \right\}$$

We maintain the standard Kalman filter matrix and vector notation, although for univariate models, $\mathbf{y}_t$, $\mathbf{w}_t$, and $\mathbf{R}$ are scalars.

Kalman filter or state-space representation of the ARIMA model

A univariate ARIMA model can be cast in state-space form by defining the Kalman filter matrices as follows (see Hamilton 1994, or Gourieroux and Monfort 1997, for details):

$$F = \begin{bmatrix} \rho_1 & \rho_2 & \cdots & \rho_{p-1} & \rho_p \\ 1 & 0 & \cdots & 0 & 0 \\ 0 & 1 & \cdots & 0 & 0 \\ 0 & 0 & \cdots & 1 & 0 \end{bmatrix}$$

$$\mathbf{v}_t = \begin{bmatrix} \epsilon_{t-1} \\ 0 \\ \cdots \\ \cdots \\ \cdots \\ 0 \end{bmatrix}$$

$$\mathbf{A}' = \beta$$

$$\mathbf{H}' = \begin{bmatrix} 1 & \theta_1 & \theta_2 & \cdots & \theta_q \end{bmatrix}$$

$$\mathbf{w}_t = 0$$

Note that the Kalman filter representation does not require that the moving average terms be invertible.

Kalman filter recursions

In order to see how missing data are handled, the updating recursions for the Kalman filter will be left in two steps. It is also common to write the updating equations as a single step using the gain matrix $\mathbf{K}$. We will provide the updating equations with little justification; see the sources listed above for details.

As a linear combination of a vector of random variables, the state $\boldsymbol{\xi}_t$ can be updated to its expected value based on the prior state as

$$\boldsymbol{\xi}_{t|t-1} = \mathbf{F}\boldsymbol{\xi}_{t-1} + \mathbf{v}_{t-1} \tag{1}$$

and this state is a quadratic form that has the covariance matrix

$$\mathbf{P}_{t|t-1} = \mathbf{F}\mathbf{P}_{t-1}\mathbf{F}' + \mathbf{Q} \tag{2}$$

and the estimator of $\mathbf{y}_t$ is

$$\widehat{\mathbf{y}}_{t|t-1} = \mathbf{x}_t\beta + \mathbf{H}'\boldsymbol{\xi}_{t|t-1}$$

which implies an innovation or prediction error

$$\widehat{\iota}_t = \mathbf{y}_t - \widehat{\mathbf{y}}_{t|t-1}$$

and this value or vector has mean square error (MSE)

$$\mathbf{M}_t = \mathbf{H}'\mathbf{P}_{t|t-1}\mathbf{H} + \mathbf{R}$$

Now, the expected value of $\boldsymbol{\xi}_t$ conditional on a realization of $\mathbf{y}_t$ is

$$\boldsymbol{\xi}_t = \boldsymbol{\xi}_{t|t-1} + \mathbf{P}_{t|t-1}\mathbf{H}\mathbf{M}_t^{-1}\widehat{\iota}_t \tag{3}$$

with MSE

$$\mathbf{P}_t = \mathbf{P}_{t|t-1} - \mathbf{P}_{t|t-1}\mathbf{H}\mathbf{M}_t^{-1}\mathbf{H}'\mathbf{P}_{t|t-1} \tag{4}$$

This gives the full set of Kalman filter recursions.

Kalman filter initial conditions

When the series, conditional on $\mathbf{x}_t\boldsymbol{\beta}$, is stationary, the initial conditions for the filter can be considered a random draw from the stationary distribution of the state equation. The initial values of the state and the state MSE will be the expected values from this stationary distribution. For an ARIMA model, these can be written as

$$\xi_{1|0} = \mathbf{0}$$

and

$$\text{vec}(\mathbf{P}_{1|0}) = (\mathbf{I}_{r^2} - \mathbf{F} \otimes \mathbf{F})^{-1}\text{vec}(\mathbf{Q})$$

where vec() is an operator representing the column matrix resulting from stacking each successive column of the target matrix.

If the series is not stationary, the above does not constitute a random draw from a stationary distribution, and some other values must be chosen for initial state conditions. Hamilton (1994) suggests that they be specified based on prior expectations, while Harvey suggests a diffuse and improper prior having a state vector of $\mathbf{0}$ and with an infinite variance. This corresponds to $\mathbf{P}_{1|0}$ with diagonal elements of ∞. Stata allows either approach to be taken for nonstationary series—initial priors may be specified with state0() and p0(), and a diffuse prior may be specified with diffuse.

Likelihood from prediction error decomposition

Given the outputs from the Kalman filter recursions and the assumption that the state and observation vectors are Gaussian, the likelihood for the state space model follows directly from the resulting multivariate normal in the predicted innovations. The log likelihood for observation t is

$$\ln L_t = -\frac{1}{2}\Big\{ \ln(2\pi) + \ln(|\mathbf{M}_t|) - \boldsymbol{\iota}_t'\mathbf{M}_t^{-1}\boldsymbol{\iota}_t \Big\}$$

Missing data

Missing data, whether a missing dependent variable y_t, one or more missing covariates $\mathbf{x}_t$, or completely missing observations, are handled by continuing the state updating equations without any contribution from the data; see Harvey (1989 and 1993). That is to say, (1) and (2) are iterated for every missing observation, while (3) and (4) are ignored. Thus, for observations with missing data, $\boldsymbol{\xi}_t = \boldsymbol{\xi}_{t|t-1}$ and $\mathbf{P}_t = \mathbf{P}_{t|t-1}$. In the absence of any information from the sample, this effectively assumes the prediction error for the missing observations is 0. Alternate methods of handling missing data based on the EM algorithm have been suggested; e.g., Shumway (1984, 1988).

References

Ansley, C. F. and R. Kohn. 1985. Estimation, filtering and smoothing in state space models with incompletely specified initial conditions. *Annals of Statistics* 13: 1286–1316.

Ansley, C. F. and P. Newbold. 1980. Finite sample properties of estimators for auto-regressive moving average processes. *Journal of Econometrics* 13: 159–184.

Baum, C. F. 2000. sts15: Tests for stationarity of a time series. *Stata Technical Bulletin* 57: 36–39. Reprinted in *Stata Technical Bulletin Reprints*, vol. 10, pp. 356–360.

———. 2001. sts18: A test for long-range dependence in a time series. *Stata Technical Bulletin* 60: 37–39. Reprinted in *Stata Technical Bulletin Reprints*, vol. 10, pp. 370–373.

Baum, C. F. and R. Sperling. 2001. sts15.1: Tests for stationarity of a time series: update. *Stata Technical Bulletin* 58: 35–36. Reprinted in *Stata Technical Bulletin Reprints*, vol. 10, pp. 360–362.

Baum, C. F. and V. L. Wiggins. 2000. sts16: Tests for long memory in a time series. *Stata Technical Bulletin* 57: 39–44. Reprinted in *Stata Technical Bulletin Reprints*, vol. 10, pp. 362–368.

Berndt, E. K., B. H. Hall, R. E. Hall, and J. A. Hausman. 1974. Estimation and inference in nonlinear structural models. *Annals of Economic and Social Measurement* 3/4: 653–665.

Bollerslev, T., R. F. Engle, D. B. Nelson. 1994. ARCH Models. In *Handbook of Econometrics, Volume IV*, ed. R. F. Engle and D. L. McFadden. New York: Elsevier.

Box, G. E. P., G. M. Jenkins, G. C. Reinsel. 1994. *Time Series Analysis: Forecasting and Control*. 3d ed. Englewood Cliffs, NJ: Prentice–Hall.

David, J. S. 1999. sts14: Bivariate Granger causality test. *Stata Technical Bulletin* 51: 40–41. Reprinted in *Stata Technical Bulletin Reprints*, vol. 9, pp. 350–351.

Davidson, R. and J. G. MacKinnon. 1993. *Estimation and Inference in Econometrics*. Oxford: Oxford University Press.

Diggle, P. J. 1990. *Time Series. A Biostatistical Introduction*. Oxford: Oxford University Press.

Enders, W. 1995. *Applied Econometric Time Series*. New York: John Wiley & Sons.

Friedman, M. and D. Meiselman. 1963. The relative stability of monetary velocity and the investment multiplier in the United States, 1987–1958. In *Stabilization Policies*, Commission on Money and Credit. Englewood Cliffs, NJ: Prentice–Hall.

McDowell, A. W. 2002. From the help desk: Transfer functions. *The Stata Journal* 2: 71–85.

Gourieroux, C. and A. Monfort. 1997. *Time Series and Dynamic Models*. Cambridge: Cambridge University Press.

Greene, W. H. 2003. *Econometric Analysis*. 5th ed. Upper Saddle River, NJ: Prentice–Hall.

Hamilton, J. D. 1994. *Time Series Analysis*. Princeton: Princeton University Press.

Harvey, A. C. 1989. *Forecasting, structural time series models and the Kalman filter*. Cambridge: Cambridge University Press.

——. 1993. *Time Series Models*. Cambridge, MA: MIT Press.

Hipel, K. W. and A. I. McLeod. 1994. *Time Series Modelling of Water Resources and Environmental Systems*. Amsterdam: Elsevier.

Kalman, R. E. 1960. A new approach to linear filtering and prediction problems. *Journal of Basic Engineering, Transactions of the ASME* Series D, 82: 35–45.

Newton, H. J. 1988. *TIMESLAB: A Time Series Analysis Laboratory*. Belmont, CA: Wadsworth & Brooks/Cole.

Press, W. H., S. A. Teukolsky, W. T. Vetterling, B. P. Flannery. 1992. *Numerical Recipes in C: The Art of Scientific Computing*. 2d ed. Cambridge: Cambridge University Press.

Shumway, R. H. 1984. Some applications of the EM algorithm to analyzing incomplete time series data. In *Time Series Analysis of Irregularly Observed Data*, ed. E. Parzen, 290–324. New York: Springer.

——. 1988. *Applied Statistical Time Series Analysis*. Upper Saddle River, NJ: Prentice–Hall.

Also See

Title

---
corrgram — Correlogram

Syntax

> corrgram *varname* [if *exp*] [in *range*] [, noplot lags(#)]
>
> ac *varname* [if *exp*] [in *range*] [, lags(#) fft nograph generate(*newvarname*)
>
>   level(#) ciopts(*rarea_options*) plot(*plot*) dropline_options twoway_options]
>
> pac *varname* [if *exp*] [in *range*] [, lags(#) nograph generate(*newvarname*)
>
>   level(#) srv srvopts(*scatter_options*) ciopts(*rarea_options*) plot(*plot*)
>
>   dropline_options twoway_options]

These commands are for use with time-series data; see [TS] **tsset**. You must tsset your data before using corrgram, ac, or pac. In addition, the time series must be dense (nonmissing and no gaps in the time variable) in the sample if you specify the fft option.

varname may contain time-series operators; see [U] **14.4.3 Time-series varlists**.

Description

corrgram lists a table of the autocorrelations, partial autocorrelations, and Q statistics. It will also list a character-based plot of the autocorrelations and partial autocorrelations.

The ac command produces a correlogram (the autocorrelations) with pointwise confidence intervals obtained from the Q statistic; see [TS] **wntestq**.

The pac command produces a graph of the partial correlogram (the partial autocorrelations) with confidence intervals calculated using a standard error of $1/\sqrt{n}$. The residual variances for each lag may optionally be included on the graph.

Options

noplot prevents the character-based plots from being in the listed table of autocorrelations and partial autocorrelations.

lags(#) specifies the number of autocorrelations to calculate. The default is to use $\min([n/2]-2, 40)$ where $[n/2]$ is the greatest integer less than or equal to $n/2$.

fft (ac only) specifies that the autocorrelations should be calculated using two Fourier transforms. This technique can be faster than simply iterating over the requested number of lags.

nograph prevents ac and pac from constructing a graph. This option is required by the generate() option.

generate(*newvarname*) specifies a new variable to contain the autocorrelation (ac command) or partial autocorrelation (pac command) values. This option requires the nograph option.

level(#) specifies the confidence level, in percent, for the confidence bands in the ac or pac graph. The default is level(95) or as set by set level; see [R] **level**.

srv (pac only) specifies that the standardized residual variances also be plotted with the partial autocorrelations.

srvopts(*scatter_options*) affect the rendition of the plotted standardized residual variances; see [G] **graph twoway scatter**. This option implies the srv option.

ciopts(*rarea_options*) affect the rendition of the confidence bands; see [G] **graph twoway rarea**.

plot(*plot*) provides a way to add other plots to the generated graph; see [G] *plot_option*.

dropline_options affect the rendition of the plotted (partial) autocorrelations; see [G] **graph twoway dropline**.

twoway_options are any of the options documented in [G] *twoway_options* excluding by(). These include options for titling the graph (see [G] *title_options*) and options for saving the graph to disk (see [G] *saving_option*).

Remarks

The Q statistics provided in the output are the same statistics that you would get by running the wntestq command for each of the lags in the table; see [TS] **wntestq**.

corrgram provides an easy means to obtain lists of autocorrelations and partial autocorrelations. By default, character-based plots of these values are provided, but if you are going to cut and paste these values to a report, you may want to use the noplot option to suppress these character-based plots.

▷ Example

Here we use the international airline passengers dataset (Box, Jenkins, and Reinsel 1994, Series G). This dataset has 144 observations on the monthly number of international airline passengers from 1949 through 1960. We can list the autocorrelations and partial autocorrelations using

(Continued on next page)

```
. use http://www.stata-press.com/data/r8/air2
(TIMESLAB: Airline passengers)

. corrgram air, lags(20)
```

					-1 0 1 -1 0 1
LAG	AC	PAC	Q	Prob>Q	[Autocorrelation] [Partial Autocor]
1	0.9480	0.9589	132.14	0.0000	|—————— |——————
2	0.8756	-0.3298	245.65	0.0000	|—————— ——|
3	0.8067	0.2018	342.67	0.0000	|————— |—
4	0.7526	0.1450	427.74	0.0000	|————— |—
5	0.7138	0.2585	504.8	0.0000	|————— |—
6	0.6817	-0.0269	575.6	0.0000	|—————	
7	0.6629	0.2043	643.04	0.0000	|————— |—
8	0.6556	0.1561	709.48	0.0000	|————— |—
9	0.6709	0.5686	779.59	0.0000	|————— |———
10	0.7027	0.2926	857.07	0.0000	|————— |—
11	0.7432	0.8402	944.39	0.0000	|————— |—————
12	0.7604	0.6127	1036.5	0.0000	|————— |————
13	0.7127	-0.6660	1118	0.0000	|————— ————|
14	0.6463	-0.3846	1185.6	0.0000	|———— ——|
15	0.5859	0.0787	1241.5	0.0000	|————	
16	0.5380	-0.0266	1289	0.0000	|————	
17	0.4997	-0.0581	1330.4	0.0000	|————	
18	0.4687	-0.0435	1367	0.0000	|———	
19	0.4499	0.2773	1401.1	0.0000	|——— |—
20	0.4416	-0.0405	1434.1	0.0000	|———	

If we wished to produce a high-quality graph instead of the character-based plot, we could type

```
. ac air, lags(20)
```

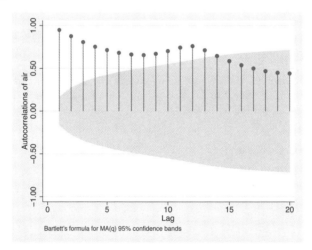

In the preceding example, we have not removed the trend or annual cycle from the data. We can do that by taking first and twelfth differences. Below, we plot the partial autocorrelations of the transformed data:

(*Continued on next page*)

. pac DS12.air, lags(20)

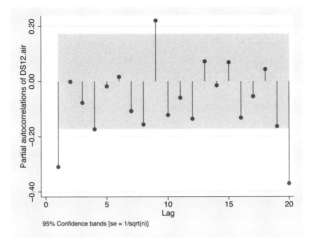

We could also use the `srv` option to request that the standardized residual variances be added to the plot of partial autocorrelations.

. pac DS12.air, lags(20) srv

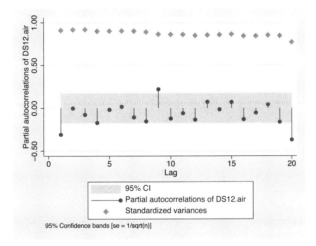

◁

Saved Results

`corrgram` saves in `r()`:

Scalars

    `r(lags)`   number of lags

Matrices

    `r(AC)`   vector of autocorrelations
    `r(PAC)`   vector of partial autocorrelations
    `r(Q)`   vector of Q statistics

Methods and Formulas

corrgram, ac, and pac are implemented as ado-files.

Box, Jenkins, and Reinsel (1994), Newton (1988), Chatfield (1996), and Hamilton (1994) provide excellent descriptions of correlograms. Newton (1988) provides additional discussion on the calculation of the various quantities.

The autocovariance function for a time series $x_1, x_2, \ldots, x_n$ is defined for $|v| < n$ as

$$\widehat{R}(v) = \frac{1}{n} \sum_{i=1}^{n-|v|} (x_i - \overline{x})(x_{i+v} - \overline{x})$$

where $\overline{x}$ is the sample mean, and the autocorrelation function is then defined as

$$\widehat{\rho}(v) = \frac{\widehat{R}(v)}{\widehat{R}(0)}$$

The partial autocorrelation of lag v is the autocorrelation between $x(t)$ and $x(t+v)$ after having removed the common linear effect of the data in between; the lag-1 partial autocorrelation is then asymptotically the same as the lag-1 autocorrelation.

The residual variances (which optionally appear on the graph produced with the pac command) are the sample multiple correlation coefficients of the decomposition of the overall variability due to the autocovariance at the various lags.

In other words, for a given lag v, we regress x on lags 1 through v of x. The partial autocorrelation coefficient $\widehat{\theta}(v)$ is the coefficient on lag v of x in the regression, and the residual variance is the estimated variance of the regression—these residual variances are then standardized by dividing them by the sample variance, $\widehat{R}(0)$, of the time series.

Acknowledgment

The ac and pac commands are based on the ac and pac commands written by Sean Becketti (1992), a past editor of the *Stata Technical Bulletin*.

References

Becketti, S. 1992. sts1: Autocorrelation and partial autocorrelation graphs. *Stata Technical Bulletin* 5: 27–28. Reprinted in *Stata Technical Bulletin Reprints*, vol. 1, pp. 221–223.

Box, G. E. P., G. M. Jenkins, and G. C. Reinsel. 1994. *Time Series Analysis: Forecasting and Control*. 3d ed. Englewood Cliffs, NJ: Prentice–Hall.

Chatfield, C. 1996. *The Analysis of Time Series: An Introduction*. 5th ed. London: Chapman & Hall.

Hamilton, J. D. 1994. *Time Series Analysis*. Princeton: Princeton University Press.

Newton, H. J. 1988. *TIMESLAB: A Time Series Laboratory*. Pacific Grove, CA: Wadsworth & Brooks/Cole.

Also See

Complementary:	[TS] **tsset**, [TS] **wntestq**
Related:	[TS] **pergram**
Background:	*Stata Graphics Reference Manual*

Title

cumsp — Cumulative spectral distribution

Syntax

> cumsp *varname* $\left[\text{if } exp\right]$ $\left[\text{in } range\right]$ $\left[\text{, } \underline{\text{gen}}\text{erate}(varname) \text{ plot}(plot)\right.$
>
>   *connected_options twoway_options* $\left.\right]$

cumsp is for use with time-series data; see [TS] **tsset**. You must tsset your data before using cumsp. In addition, the time series must be dense (nonmissing and no gaps in the time variable) in the sample specified.

varname may contain time-series operators; see [U] **14.4.3 Time-series varlists**.

Description

cumsp plots the cumulative sample spectral distribution function evaluated at the natural frequencies for a (dense) time series.

Options

generate(*varname*) specifies a new variable to contain the estimated cumulative spectral distribution values.

plot(*plot*) provides a way to add other plots to the generated graph; see [G] ***plot_option***.

connected_options affect the rendition of the plotted points connected by lines; see [G] **graph twoway connected**.

twoway_options are any of the options documented in [G] ***twoway_options*** excluding by(). These include options for titling the graph (see [G] ***title_options***) and options for saving the graph to disk (see [G] ***saving_option***).

Remarks

▷ Example

Here we use the international airline passengers dataset (Box, Jenkins, and Reinsel 1994, Series G). This dataset has 144 observations on the monthly number of international airline passengers from 1949 through 1960. In the cumulative sample spectral distribution function for these data, we also request a vertical line at frequency $1/12$. Since the data are monthly, there will be a pronounced jump in the cumulative sample spectral distribution plot at the $1/12$ value if there is an annual cycle in the data.

(Continued on next page)

```
. use http://www.stata-press.com/data/r8/air2
(TIMESLAB: Airline passengers)

. cumsp air, xline(.083333333)
```

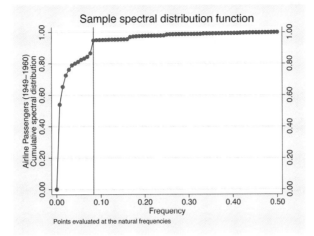

The cumulative sample spectral distribution function clearly illustrates the annual cycle.

◁

Methods and Formulas

cumsp is implemented as an ado-file.

A time series of interest is decomposed into a unique set of sinusoids of various frequencies and amplitudes.

A plot of the sinusoidal amplitudes versus the frequencies for the sinusoidal decomposition of a time series gives us the spectral density of the time series. If we calculate the sinusoidal amplitudes for a discrete set of "natural" frequencies $(1/n, 2/n, \ldots, q/n)$, we obtain the periodogram.

Let $x(1), \ldots, x(n)$ be a time series, and let $\omega_k = (k-1)/n$ denote the natural frequencies for $k = 1, \ldots, [n/2] + 1$ where $[\,]$ indicates the greatest integer function. Define

$$C_k^2 = \frac{1}{n^2} \left| \sum_{t=1}^{n} x(t) e^{2\pi i (t-1)\omega_k} \right|^2$$

A plot of nC_k^2 versus ω_k is then called the periodogram.

The sample spectral density may then be defined as $\widehat{f}(\omega_k) = nC_k^2$.

(Continued on next page)

If we let $\widehat{f}(\omega_1), \ldots, \widehat{f}(\omega_Q)$ be the sample spectral density function of the time series evaluated at the frequencies $\omega_j = (j-1)/Q$ for $j = 1, \ldots, Q$ and we let $q = \lfloor Q/2 \rfloor + 1$, then

$$\widehat{F}(\omega_k) = \frac{\sum_{i=1}^{k} \widehat{f}(\omega_j)}{\sum_{i=1}^{q} \widehat{f}(\omega_j)}$$

is the sample spectral distribution function of the time series.

References

Box, G. E. P., G. M. Jenkins, and G. C. Reinsel. 1994. *Time Series Analysis: Forecasting and Control.* 3d ed. Englewood Cliffs, NJ: Prentice–Hall.

Newton, H. J. 1988. *TIMESLAB: A Time Series Laboratory.* Pacific Grove, CA: Wadsworth & Brooks/Cole.

Also See

Complementary:	[TS] **tsset**
Related:	[TS] **corrgram**, [TS] **pergram**
Background:	*Stata Graphics Reference Manual*

Title

---
dfgls — Perform DF-GLS unit-root test

Syntax

dfgls *varname* [if *exp*] [in *range*] [, <u>m</u>axlag(*#*) <u>not</u>rend ers]

dfgls is for use with time-series data; see [TS] **tsset**. You must tsset your data before using dfgls.
varname may contain time-series operators; see [U] **14.4.3 Time-series varlists**.

Description

dfgls performs the modified Dickey–Fuller t test proposed by Elliot, Rothenberg, and Stock (1996). Essentially, the test is an augmented Dickey–Fuller test, similar to the test performed by Stata's dfuller command, only the time series is transformed via a generalized least squares (GLS) regression prior to performing the test. This test has become known as the DF-GLS test. Elliot, Rothenberg, and Stock (1996) and subsequent studies have shown that this test has significantly higher power than the previous versions of the augmented Dickey–Fuller test. See Stock and Watson (2003, 549–552) for a discussion of this test.

dfgls performs the DF-GLS test for the series of models that include 1 to k lags of the first-differenced, detrended variable, where k can be set by the user or by the method of Schwert (1989).

Options

maxlag(*#*) sets the value of k, the highest lag order for the first-differenced, detrended variable in the Dickey–Fuller regression. By default, dfgls sets k according to the method proposed by Schwert (1989); i.e., dfgls sets $k_{max} = \text{int}[12\{(T+1)/100\}^{0.25}]$.

notrend specifies that the alternative hypothesis is that the series is stationary around a mean instead of around a linear time trend. By default, a trend is included.

ers specifies that dfgls should present interpolated critical values from tables presented by Elliot, Rothenberg, and Stock (1996), which they obtained from simulations. See *Critical values* under *Methods and Formulas* for details.

Remarks

Elliot, Rothenberg, and Stock (1996), known as ERS, use local to unity analysis to show that an augmented Dickey–Fuller test on the GLS transform that they derive will be close to the asymptotic power envelope for this class of tests. While the critical values for the mean-only case remain identical to those of the augmented Dickey–Fuller Test, ERS used simulations to determine the critical values for the case with a linear time trend. Stock and Watson (2003) provides an excellent discussion of the methodology and the remarks below closely follows their treatment.

The Dickey–Fuller test, due to Dickey and Fuller (1979), and Said and Dickey (1984), tests for an autoregressive unit root in the series y_t by testing the null hypothesis that $\rho = 0$ in either the regression

$$\Delta y_t = \beta_0 + \rho y_{t-1} + \sum_{j=1}^{k} \beta_j \Delta y_{t-j} + e_t$$

or

$$\Delta y_t = \beta_0 + \gamma t + \rho y_{t-1} + \sum_{j=1}^{k} \beta_j \Delta y_{t-j} + e_t$$

The DG-GLS test is performed analogously, but on GLS detrended data. The null hypothesis of the test is that y_t has a random walk trend, possibly with drift. There are two possible alternative hypotheses. The first alternative is that y_t is stationary about a linear time trend. The second alternative is that y_t is stationary with a (possibly) non-zero mean, but with no linear time trend.

Under the first alternative hypothesis, the DG-GLS test is performed by first estimating the intercept and trend via GLS. The GLS estimation is performed by generating the new variables, $\widetilde{y}_t$, x_t, and z_t, where

$$
\begin{aligned}
\widetilde{y}_1 &= y_1 \\
\widetilde{y}_t &= y_t - \alpha^* y_{t-1}, \qquad t = 2, \ldots, T \\
x_1 &= 1 \\
x_t &= 1 - \alpha^*, \qquad t = 2, \ldots, T \\
z_1 &= 1 \\
z_t &= t - \alpha^*(t-1)
\end{aligned}
$$

where $\alpha^* = 1 - (13.5/T)$.

An OLS regression is then estimated for the equation

$$\widetilde{y}_t = \delta_0 x_t + \delta_1 z_t + \epsilon_t$$

The OLS estimators $\widehat{\delta}_0$ and $\widehat{\delta}_1$ are then used to remove the trend from y_t; i.e., we generate

$$y^* = y_t - (\widehat{\delta}_0 + \widehat{\delta}_1 t)$$

Finally, we perform an augmented Dickey–Fuller test on the transformed variable by fitting the OLS regression

$$\Delta y_t^* = \beta_0 + \rho y_{t-1}^* + \sum_{j=1}^{k} \beta_j \Delta y_{t-j}^* + e_t \qquad (1)$$

and performing a test of the null hypothesis that $\rho = 0$ using tabulated critical values.

To perform the DF-GLS test under the second alternative hypothesis, proceed as before, but define $\alpha^* = 1 - (7/T)$, eliminate z from the GLS regression, compute $y^* = y_t - \delta_0$, fit the augmented Dickey–Fuller regression using the newly transformed variable and perform a test of the null hypothesis that $\rho = 0$ using the tabulated critical values.

dfgls reports the DF-GLS statistic and its critical values obtained from the regression in (1) for $k \in \{1, 2, \ldots, k_{max}\}$. By default, dfgls sets $k_{max} = \text{int}[12\{(T+1)/100\}^{0.25}]$ as proposed by Schwert (1989), although users may override this choice with another value. The sample size available with k_{max} lags is used in all the regressions. Since there are k_{max} lags of the first-differenced series, $k_{max} + 1$ observations are lost leaving $T - k_{max}$ observations. Note that dfgls requires that the sample of $T + 1$ observations on $y_t = (y_0, y_1, \ldots, y_T)$ not have any gaps.

▷ Example

Consider the following example which uses the data from Lütkepohl (1993). We model the first-differenced series under the assumption that there is a unit root in the natural log series. Here we show that the log of income might not contain a unit root. We begin by running dfgls with its default values:

```
. use http://www.stata-press.com/data/r8/lutkepohl
(Quarterly SA West German macro data, Bil DM, from Lutkepohl 1993 Table E.1)

. dfgls linvestment

DF-GLS for linvestment          Number of obs =     80
Maxlag = 11 chosen by Schwert criterion
               DF-GLS tau     1% Critical    5% Critical    10% Critical
      [lags]   Test Statistic    Value          Value          Value

        11       -2.925         -3.610         -2.763         -2.489
        10       -2.671         -3.610         -2.798         -2.523
         9       -2.766         -3.610         -2.832         -2.555
         8       -3.259         -3.610         -2.865         -2.587
         7       -3.536         -3.610         -2.898         -2.617
         6       -3.115         -3.610         -2.929         -2.646
         5       -3.054         -3.610         -2.958         -2.674
         4       -3.016         -3.610         -2.986         -2.699
         3       -2.071         -3.610         -3.012         -2.723
         2       -1.675         -3.610         -3.035         -2.744
         1       -1.752         -3.610         -3.055         -2.762

Opt Lag (Ng-Perron seq t) =  7 with RMSE   .0388771
Min SC    = -6.169137 at lag  4 with RMSE   .0398949
Min MAIC = -6.136371 at lag  1 with RMSE   .0440319
```

The null of a unit root is rejected at the 5% level for lags 4–8 and 11 and at the 10% level for lags 9 and 10. For comparison, we run dfuller on this same series. We must explicitly include the trend in dfuller, whereas dfgls includes a trend by default. The conclusions from dfuller, although similar, are not as strong as those produced by dfgls. This result is intuitive since the DF-GLS test with a trend is known to be more powerful than the standard augmented Dickey–Fuller test.

```
. dfuller linvestment, lag(4) trend
Augmented Dickey-Fuller test for unit root        Number of obs    =      87

                             ————— Interpolated Dickey-Fuller ————
                 Test        1% Critical    5% Critical    10% Critical
               Statistic        Value          Value          Value

  Z(t)          -3.133         -4.069         -3.463         -3.158

* MacKinnon approximate p-value for Z(t) = 0.0987
```

```
. dfuller linvestment, lag(5) trend
```

| Augmented Dickey-Fuller test for unit root | | Number of obs | = | 86 |

	Test Statistic	1% Critical Value	Interpolated Dickey-Fuller 5% Critical Value	10% Critical Value
Z(t)	-3.296	-4.071	-3.464	-3.158

```
* MacKinnon approximate p-value for Z(t) = 0.0669
. dfuller linvestment, lag(6) trend
```

| Augmented Dickey-Fuller test for unit root | | Number of obs | = | 85 |

	Test Statistic	1% Critical Value	Interpolated Dickey-Fuller 5% Critical Value	10% Critical Value
Z(t)	-3.498	-4.073	-3.465	-3.159

```
* MacKinnon approximate p-value for Z(t) = 0.0396
. dfuller linvestment, lag(7) trend
```

| Augmented Dickey-Fuller test for unit root | | Number of obs | = | 84 |

	Test Statistic	1% Critical Value	Interpolated Dickey-Fuller 5% Critical Value	10% Critical Value
Z(t)	-3.994	-4.075	-3.466	-3.160

```
* MacKinnon approximate p-value for Z(t) = 0.0090
```

Since none of the lag length selection routines select $k > 7$, let's set the maximum lag length to 8:

```
. dfgls linvestment, maxlag(8)
```

DF-GLS for linvestment   Number of obs =   83

[lags]	DF-GLS tau Test Statistic	1% Critical Value	5% Critical Value	10% Critical Value
8	-3.329	-3.610	-2.867	-2.588
7	-3.635	-3.610	-2.898	-2.617
6	-3.213	-3.610	-2.928	-2.645
5	-3.119	-3.610	-2.956	-2.671
4	-3.076	-3.610	-2.982	-2.695
3	-2.136	-3.610	-3.006	-2.718
2	-1.719	-3.610	-3.029	-2.738
1	-1.801	-3.610	-3.048	-2.755

```
Opt Lag (Ng-Perron seq t) =  7 with RMSE   .0382228
Min SC   = -6.21015 at lag  4 with RMSE   .0392355
Min MAIC = -6.171507 at lag  1 with RMSE   .0432556
```

◁

(Continued on next page)

Saved Results

dfgls saves in r():

Scalars

r(maxlag)	highest lag order k
r(N)	number of observations
r(sclag)	lag chosen by Schwartz criteria
r(maiclag)	lag chosen by modified AIC method
r(optlag)	lag chosen by sequential-t method

Matrices

r(results)	k, MAIC, SIC, RMSE, and DF-GLS statistics

Methods and Formulas

dfgls is implemented as an ado-file.

dfgls reports the results of three different methods for choosing which value of k to use. These methods are (1) the Ng–Perron sequential t, (2) the minimum Schwartz Information Criteria (SIC), and (3) the Ng–Perron Modified Akaike Information Criteria (MAIC). While the SIC has a long history in time-series modeling, the Ng–Perrron sequential t was developed by Ng and Perron (1995) and the MAIC was developed by Ng and Perron (2000).

The SIC can be calculated using either the log likelihood or the sum-of-squared errors from a regression; DF-GLS uses the latter definition. Specifically, for each k

$$\text{SIC} = \ln(\widehat{\text{rmse}}^2) + (k+1)\frac{\ln(T - k_{\max})}{(T - k_{\max})}$$

where

$$\widehat{\text{rmse}} = \frac{1}{(T - k_{\max})}\sum_{t=k_{\max}+1}^{T}\widehat{e}_t^2$$

DF-GLS reports the value of the smallest SIC and the k that produced it.

Ng and Perron (1995) derived a sequential-t algorithm for choosing k. This algorithm is

i. Set $n = 0$ and run the regression in (2) with all $k_{\max} - n$ lags. If the coefficient on $\beta_{k_{\max}}$ is significantly different from zero at level α, choose k to $k_{\max}$. Otherwise, if $\beta_{k_{\max}}$ is not significantly different from zero, continue on to ii.

ii. If $n < k_{\max}$, set $n = n + 1$ and continue on to iii. Otherwise, set $k = 0$ and stop.

iii. Run the regression in (2) with $k_{\max} - n$ lags. If the coefficient on $\beta_{k_{\max}-n}$ is significantly different from zero at level α, choose k to $k_{\max} - n$. Otherwise, if $\beta_{k_{\max}}$ is not significantly different from zero, return to ii.

Following Ng and Perron (1995), dfgls uses $\alpha = 10\%$. dfgls reports the k selected by this sequential-t algorithm and the $\widehat{\text{rmse}}$ from the regression.

Method (3) is based on choosing k to minimize the MAIC. The MAIC is calculated as

$$\text{MAIC}(k) = \ln(\widehat{\text{rmse}}^2) + \frac{2\{\tau(k) + k\}}{T - k_{\max}}$$

where

$$\tau(k) = \frac{1}{\widehat{\text{rmse}}^2} \widehat{\beta}_0^2 \sum_{t=k_{\max}+1}^{T} \widetilde{y}_t^2$$

and $\widetilde{y}$ is defined in the text.

Critical values

By default, `dfgls` uses the 5% and 10% critical values computed from the response surface analysis of Cheung and Kim (1995). Since Cheung and Kim (1995) did not present results for 1% case, the 1% critical values are always interpolated from the critical values presented by ERS.

ERS presented critical values, obtained from simulations, for the DF-GLS test with a linear trend and showed that the critical values for the mean only DF-GLS test were the same as those for the ADF test. If `dfgls` is run with the `ers` option, then `dfgls` will present interpolated critical values from these tables. The method of interpolation is standard. For the trend case, below 50 observations and above 200, there is no interpolation; the values for 50 and ∞ are from the tables are reported. For a value N that lies between two values in the table, say, N_1 and N_2, with corresponding critical values CV_1 and CV_2, the critical value

$$\text{cv} = CV_1 + \frac{N - N_1}{N_1}(CV_2 - CV_1)$$

is presented. The same method is used for the mean-only case, except that interpolation is possible for values between 50 and 500.

Acknowledgments

We wish to thank Christopher Baum of Boston College and Richard Sperling of Boston College for a previous version of `dfgls`.

References

Cheung, Y. and K. S. Lai. 1995. Lag order and critical values of a modified Dickey–Fuller test. *Oxford Bulletin of Economics and Statistics* (57)3: 411–419.

Dickey, D. A. and W. A. Fuller. 1979. Distribution of the estimators for autoregressive time series with a unit root. *Journal of the American Statistical Association* 74: 427–431.

Elliot, G., T. Rothenberg, and J. H. Stock. 1996. Efficient tests for an autoregressive unit root. *Econometrica* 64: 813–836.

Lütkepohl, H. 1993. *Introduction to Multiple Time Series Analysis*. 2d ed. New York: Springer.

Ng, S. and P. Perron. 1995. Unit root tests in ARMA models with data-dependent methods for the selection of the truncation lag. *Journal of the American Statistical Association* 90: 268–281.

——. 2000. Lag length selection and the construction of unit root tests with good size and power. Working paper. Department of Economics, Boston College.

Said, S. E. and D. A. Dickey. 1984. Testing for unit roots in autoregressive-moving average models of unknown order. *Biometrika* 71: 599–607.

Schwert, G. W. 1989. Tests for unit roots: A Monte Carlo investigation. *Journal of Business and Economic Statistics* 2: 147–159.

Stock, J. H. and M. W. Watson. 2003. *Introduction to Econometrics.* Boston: Addison–Wesley.

Also See

Related: [TS] **dfuller**, [TS] **pperron**, [TS] **tsset**

Title

dfuller — Augmented Dickey–Fuller test for a unit root

Syntax

dfuller *varname* [if *exp*] [in *range*] [, <u>nocon</u>stant <u>lags</u>(#) <u>trend</u> <u>regr</u>ess]

dfuller is for use with time-series data; see [TS] **tsset**. You must tsset your data before using dfuller. *varname* may contain time-series operators; see [U] **14.4.3 Time-series varlists**.

Description

dfuller performs the augmented Dickey–Fuller test for unit roots on a variable. The user may optionally exclude the constant, include a trend term, and/or include lagged values of the difference of the variable in the regression.

Options

<u>nocon</u>stant suppresses the constant term (intercept) in the model.

<u>lags</u>(#) specifies the number of lagged difference terms to include in the covariate list.

<u>trend</u> specifies that a trend term should be included in the associated regression. This option may not be used with the noconstant option.

<u>regr</u>ess specifies that the associated regression table should appear in the output. By default, the regression table is not produced.

Remarks

Hamilton (1994) and Fuller (1976) give excellent overviews of this topic; see especially Chapter 17 of the former. Dickey and Fuller (1979) proposed a collection of tests for unit roots that relied on the derived asymptotic distributions of test statistics for AR(1) random walks (standard Brownian motion). See their paper for details.

▷ Example

In this example, we examine the international airline passengers dataset from Box, Jenkins, and Reinsel (1994, Series G). This dataset has 144 observations on the monthly number of international airline passengers from 1949 through 1960.

```
. use http://www.stata-press.com/data/r8/air2
(TIMESLAB: Airline passengers)
```

```
. dfuller air
Dickey-Fuller test for unit root                 Number of obs   =      143
```

	Test Statistic	1% Critical Value	5% Critical Value	10% Critical Value
Z(t)	-1.748	-3.496	-2.887	-2.577

```
* MacKinnon approximate p-value for Z(t) = 0.4065
```

If we wanted to see the associated regression, we could type

```
. dfuller air, regress
Dickey-Fuller test for unit root                 Number of obs   =      143
```

	Test Statistic	1% Critical Value	5% Critical Value	10% Critical Value
Z(t)	-1.748	-3.496	-2.887	-2.577

```
* MacKinnon approximate p-value for Z(t) = 0.4065
```

D.air		Coef.	Std. Err.	t	P>|t|	[95% Conf. Interval]
air						
	L1	-.041068	.023493	-1.75	0.083	-.0875122 .0053761
_cons		13.7055	7.133673	1.92	0.057	-.3972779 27.80829

Note that we fail to reject the hypothesis that there is a unit root in this time series by looking either at the MacKinnon approximate asymptotic p-value or the interpolated Dickey–Fuller critical values.

◁

▷ Example

In this example, we examine the Canadian lynx data from Newton (1988, 587). Here we include a time trend and two lags of the differenced time series in the calculation of the statistic.

```
. use http://www.stata-press.com/data/r8/lynx2
(TIMESLAB: Canadian lynx)
. dfuller lynx, lags(2) trend
Augmented Dickey-Fuller test for unit root         Number of obs   =      111
```

	Test Statistic	1% Critical Value	5% Critical Value	10% Critical Value
Z(t)	-6.388	-4.036	-3.449	-3.149

```
* MacKinnon approximate p-value for Z(t) = 0.0000
```

We reject the hypothesis that there is a unit root in this time series.

◁

Saved Results

dfuller saves in r():

Scalars

r(N)	number of observations	r(Zt)	Dickey–Fuller test statistic
r(lags)	Number of lagged differences	r(p)	MacKinnon approximate p-value (if there is a constant or trend in associated regression)

Methods and Formulas

In the OLS estimation of an AR(1) process with Gaussian errors,

$$y_t = \rho y_{t-1} + \epsilon_t$$

where ϵ_t are independent and identically distributed as $N(0, \sigma^2)$ and $y_0 = 0$, the OLS estimate (based on an n-observation time series) of the autocorrelation parameter ρ is given by

$$\widehat{\rho}_n = \frac{\sum_{t=1}^{n} y_{t-1} y_t}{\sum_{t=1}^{n} y_t^2}$$

We know that if $|\rho| < 1$, then

$$\sqrt{n}(\widehat{\rho}_n - \rho) \to N(0, 1 - \rho^2)$$

If this result is valid when $\rho = 1$, then the resulting distribution collapses to a point mass (the variance is zero).

It is this motivation that drives one to check for the possibility of a unit root in an autoregressive process.

In order to compute the test statistics, we fit either the Dickey–Fuller regression

$$y_t = \beta_0 + \rho y_{t-1} + \epsilon_t$$

or

$$y_t = \beta_0 + \gamma t + \rho y_{t-1} + \epsilon_t$$

Including the constant, β_0, in either regression is optional.

The augmented Dickey–Fuller regression instead uses the differenced time series $\Delta y_t = y_t - y_{t-1}$ and fits either the regression

$$\Delta y_t = \beta_0 + \rho y_{t-1} + \sum_{j=1}^{k} \beta_j \Delta y_{t-j} + e_t$$

or

$$\Delta y_t = \beta_0 + \gamma t + \rho y_{t-1} + \sum_{j=1}^{k} \beta_j \Delta y_{t-j} + e_t$$

where again we may optionally exclude the constant from either regression. We also, in this case, specify the number of lagged difference terms to include in the list of covariates. The lagged differences are included in order to eliminate any serial correlation in the ϵ_t values.

The critical values included in the output are linearly interpolated from the table of values that appears in Fuller (1976), and the MacKinnon approximate p-values use the regression surface published in MacKinnon (1994).

References

Box, G. E. P., G. M. Jenkins, and G. C. Reinsel. 1994. *Time Series Analysis: Forecasting and Control*. 3d ed. Englewood Cliffs, NJ: Prentice–Hall.

Dickey, D. A. and W. A. Fuller. 1979. Distribution of the estimators for autoregressive time series with a unit root. *Journal of the American Statistical Association* 74: 427–431.

Fuller, W. A. 1976. *Introduction to Statistical Time Series*. New York: John Wiley & Sons.

Hamilton, J. D. 1994. *Time Series Analysis*. Princeton: Princeton University Press.

MacKinnon, J. G. 1994. Approximate asymptotic distribution functions for unit-root and cointegration tests. *Journal of Business and Economic Statistics* 12: 167–176.

Newton, H. J. 1988. *TIMESLAB: A Time Series Laboratory*. Pacific Grove, CA: Wadsworth & Brooks/Cole.

Also See

Complementary: [TS] **tsset**

Related: [TS] **pperron**

Title

newey — Regression with Newey–West standard errors

Syntax

newey *depvar* [*varlist*] [*weight*] [if *exp*] [in *range*] , lag(*#*)

    [<u>nocon</u>stant <u>l</u>evel(*#*)]

newey is for use with time-series data. You must tsset your data before using newey; see [TS] **tsset**.

depvar and *varlist* may contain time-series operators; see [U] **14.4.3 Time-series varlists**.

aweights are allowed; see [U] **14.1.6 weight**.

newey shares the features of all estimation commands; see [U] **23 Estimation and post-estimation commands**.

Syntax for predict

predict [*type*] *newvarname* [if *exp*] [in *range*] [, [xb | stdp | <u>r</u>esiduals]]

These statistics are available both in and out of sample; type predict ... if e(sample) ... if wanted only for the estimation sample.

Description

newey produces Newey–West standard errors for coefficients estimated by OLS regression. The error structure is assumed to be heteroskedastic and is possibly autocorrelated up to some lag.

Note that if lag(0) is specified, the variance estimates produced by newey are simply the Huber/White/sandwich robust variances estimates calculated by regress, robust; see [R] **regress**.

Options

lag(*#*) is not optional; it specifies the maximum lag to be considered in the autocorrelation structure. If you specify lag(0), the output is exactly the same as regress, robust.

noconstant specifies that the fitted regression should not include an intercept term.

level(*#*) specifies the confidence level, in percent, for confidence intervals. The default is level(95) or as set by set level; see [U] **23.6 Specifying the width of confidence intervals**.

Options for predict

xb, the default, calculates the linear prediction.

stdp calculates the standard error of the linear prediction.

residuals calculates the residuals.

Remarks

The Huber/White/sandwich robust variance estimator (see, for example, White 1980) produces consistent standard errors for OLS regression coefficient estimates in the presence of heteroskedasticity. The Newey–West (1987) variance estimator is an extension that produces the consistent estimates when there is autocorrelation in addition to possible heteroskedasticity.

The Newey–West variance estimator handles autocorrelation up to and including a lag of m, where m is specified by stipulating the lag() option. Thus, it assumes that any autocorrelation at lags greater than m can be ignored.

▷ Example

nowoy, lag(0) is equivalent to regress, robust:

```
. use http://www.stata-press.com/data/r8/auto
(1978 Automobile Data)
. regress price weight displ, robust
```

Regression with robust standard errors

				Number of obs =	74
				F(2, 71) =	14.44
				Prob > F =	0.0000
				R-squared =	0.2909
				Root MSE =	2518.4

price	Coef.	Robust Std. Err.	t	P>|t|	[95% Conf. Interval]	
weight	1.823366	.7808755	2.34	0.022	.2663445	3.380387
displacement	2.087054	7.436967	0.28	0.780	-12.74184	16.91595
_cons	247.907	1129.602	0.22	0.827	-2004.455	2500.269

```
. generate t = _n
. tsset t
        time variable:  t, 1 to 74
. newey price weight displ, lag(0)
```

Regression with Newey-West standard errors
maximum lag: 0

				Number of obs =	74
				F(2, 71) =	14.44
				Prob > F =	0.0000

price	Coef.	Newey-West Std. Err.	t	P>|t|	[95% Conf. Interval]	
weight	1.823366	.7808755	2.34	0.022	.2663445	3.380387
displacement	2.087054	7.436967	0.28	0.780	-12.74184	16.91595
_cons	247.907	1129.602	0.22	0.827	-2004.455	2500.269

Since newey requires the dataset to be tsset, we generated a fake time variable t, which, in this example, played no role in the estimation.

◁

▷ Example

We have time-series measurements on variables usr and idle and now wish to fit an OLS model, but obtain Newey–West standard errors allowing for a lag of up to 3:

```
. use http://www.stata-press.com/data/r8/idle2
```

```
. tsset time
        time variable:  time, 1 to 30

. newey usr idle, lag(3)
```

Regression with Newey-West standard errors   Number of obs =   30
maximum lag: 3   F(1,   28) =   10.90
                                                     Prob > F   =   0.0026

usr	Coef.	Newey-West Std. Err.	t	P>|t|	[95% Conf. Interval]	
idle	-.2281501	.0690927	-3.30	0.003	-.3696801	-.08662
_cons	23.13483	6.327031	3.66	0.001	10.17449	36.09516

◁

Saved Results

newey saves in e():

Scalars

e(N)	number of observations		e(F)	F statistic
e(df_m)	model degrees of freedom		e(lag)	maximum lag
e(df_r)	residual degrees of freedom			

Macros

e(cmd)	newey		e(wexp)	weight expression
e(depvar)	name of dependent variable		e(vcetype)	covariance estimation method
e(wtype)	weight type		e(predict)	program used to implement predict

Matrices

e(b)	coefficient vector		e(V)	variance–covariance matrix of the estimators

Functions

e(sample)	marks estimation sample

Methods and Formulas

newey is implemented as an ado-file.

newey calculates the estimates

$$\widehat{\beta}_{\mathrm{OLS}} = (\mathbf{X}'\mathbf{X})^{-1}\mathbf{X}'\mathbf{y}$$
$$\widehat{\mathrm{Var}}(\widehat{\beta}_{\mathrm{OLS}}) = (\mathbf{X}'\mathbf{X})^{-1}\mathbf{X}'\widehat{\boldsymbol{\Omega}}\mathbf{X}(\mathbf{X}'\mathbf{X})^{-1}$$

That is, the coefficient estimates are simply those of OLS linear regression.

For the case of lag(0) (no autocorrelation), the variance estimates are calculated using the White formulation:

$$\mathbf{X}'\widehat{\boldsymbol{\Omega}}\mathbf{X} = \mathbf{X}'\widehat{\boldsymbol{\Omega}}_0\mathbf{X} = \frac{n}{n-k}\sum_i \widehat{e}_i^2 \mathbf{x}_i'\mathbf{x}_i$$

Here $\widehat{e}_i = y_i - \mathbf{x}_i \widehat{\boldsymbol{\beta}}_{\text{OLS}}$, where $\mathbf{x}_i$ is the ith row of the $\mathbf{X}$ matrix, n is the number of observations, and k is the number of predictors in the model, including the constant if there is one. Note that the above formula is exactly the same as that used by `regress, robust` with the regression-like formula (the default) for the multiplier q_c; see the *Methods and Formulas* section of [R] **regress**.

For the case of `lag(#)`, $\# > 0$, the variance estimates are calculated using the Newey–West (1987) formulation

$$\mathbf{X}'\widehat{\boldsymbol{\Omega}}\mathbf{X} = \mathbf{X}'\widehat{\boldsymbol{\Omega}}_0\mathbf{X} + \frac{n}{n-k}\sum_{l=1}^{m}\left(1 - \frac{l}{m+1}\right)\sum_{t=l+1}^{n}\widehat{e}_t\widehat{e}_{t-l}(\mathbf{x}'_t\mathbf{x}_{t-l} + \mathbf{x}'_{t-l}\mathbf{x}_t)$$

where m is the maximum lag and $\mathbf{x}_t$ is one row of the $\mathbf{X}$ matrix observed at time t.

References

Hardin, J. W. 1997. sg72: Newey–West standard errors for probit, logit, and poisson models. *Stata Technical Bulletin* 39: 32–35. Reprinted in *Stata Technical Bulletin Reprints*, vol. 7, pp. 182–186.

Newey, W. K. and K. D. West. 1987. A simple, positive semi-definite, heteroskedasticity and autocorrelation consistent covariance matrix. *Econometrica* 55: 703–708.

White, H. 1980. A heteroskedasticity-consistent covariance matrix estimator and a direct test for heteroskedasticity. *Econometrica* 48: 817–838.

Also See

Complementary:	[TS] **tsset**,
	[R] **adjust**, [R] **lincom**, [R] **linktest**, [R] **mfx**, [R] **nlcom**,
	[R] **test**, [R] **testnl**, [R] **vce**
Related:	[R] **regress**,
	[SVY] **svy estimators**,
	[XT] **xtgls**, [XT] **xtpcse**
Background:	[U] **16.5 Accessing coefficients and standard errors**,
	[U] **23 Estimation and post-estimation commands**

Title

> **pergram** — Periodogram

Syntax

> pergram *varname* [if *exp*] [in *range*] [, generate(*newvarname*) nograph
>
> plot(*plot*) *connected_options twoway_options*]

pergram is for use with time-series data; see [TS] **tsset**. You must tsset your data before using pergram. In addition, the time series must be dense (nonmissing and no gaps in the time variable) in the specified sample.

varname may contain time-series operators; see [U] **14.4.3 Time-series varlists**.

Description

pergram plots the log standardized periodogram for a (dense) time series.

Options

generate(*newvarname*) specifies a new variable to contain the raw periodogram values. Note that the generated graph log-transforms and scales the values by the sample variance, and then truncates them to the [−6, 6] interval prior to graphing them.

nograph prevents pergram from constructing a graph.

plot(*plot*) provides a way to add other plots to the generate graph; see [G] **plot_option**.

connected_options affect the rendition of the plotted points connected by lines; see [G] **graph twoway connected**.

twoway_options are any of the options documented in [G] **twoway_options** excluding by(). These include options for titling the graph (see [G] **title_options**) and options for saving the graph to disk (see [G] **saving_option**).

Remarks

A good discussion of the periodogram is provided in Chatfield (1996), Hamilton (1994), and Newton (1988). Chatfield is also a very good introductory reference for time-series analysis. Another classic reference is Box, Jenkins, and Reinsel (1994).

❑ Technical Note

pergram produces a scatterplot where the points of the scatterplot are connected. The points themselves represent the log-standardized periodogram, and the connections between points represent the (continuous) log-standardized sample spectral density. Although the periodogram is asymptotically unbiased for the spectral density, it is not consistent, and many analysts will obtain the raw ordinates from this command with the gen() option and smooth them prior to plotting.

❑

In the following examples, we present the periodograms together with an interpretation of the main features of the plots.

▷ Example

We have time-series data consisting of 144 observations on the monthly number (in thousands) of international airline passengers between 1949 and 1960 (Box, Jenkins, and Reinsel 1994, Series G). We can graph the raw series and the log-periodogram for these data by typing

```
. use http://www.stata-press.com/data/r8/air2
(TIMESLAB: Airline passengers)
. scatter air time, m(o) c(l)
```

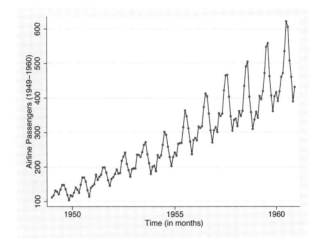

```
. pergram air
```

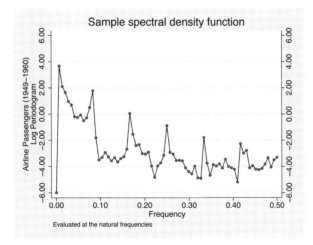

The periodogram clearly indicates the annual cycle together with the harmonics. The similarity in shape of each group of twelve observations reveals the annual cycle. The magnitude of the cycle is increasing, resulting in the peaks in the periodogram at the harmonics of the principal annual cycle.

◁

▷ Example

In this example, the data consist of 215 observations on the annual number of sunspots from 1749 to 1963 (Box and Jenkins 1976, Series E). The graph of the raw series and the log-periodogram for these data are given as

```
. use http://www.stata-press.com/data/r8/sunspot
(TIMESLAB: Wolfer sunspot data)
. scatter spot time, m(o) c(l)
```

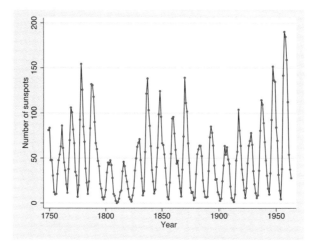

```
. pergram spot
```

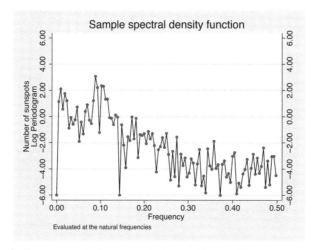

The periodogram indicates a peak frequency between 10 and 12 years.

◁

▷ Example

Here we examine the number of trapped Canadian lynx (Newton 1988, 587). The raw series and the log-periodogram are given as

```
. use http://www.stata-press.com/data/r8/lynx2
(TIMESLAB: Canadian lynx)

. scatter lynx time, m(o) c(l)
```

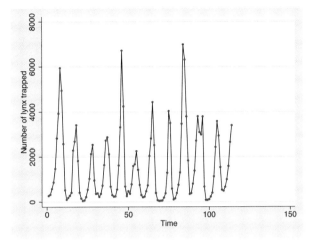

```
. pergram lynx
```

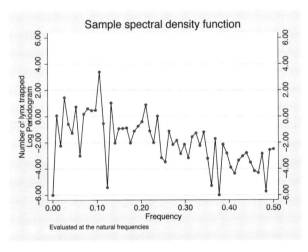

The periodogram indicates that there may be a periodicity at 15 years for these data, but is otherwise random in nature. In [TS] **corrgram**, we see evidence of the ARMA (autoregressive moving average) nature of this time series.

◁

▷ Example

In order to more clearly highlight what the periodogram depicts, we present the result of analyzing a time series of the sum of four sinusoids (of different periods). The periodogram should be able to decompose the time series into four different sinusoids whose periods may be determined from the plot.

```
. use http://www.stata-press.com/data/r8/cos4
(TIMESLAB: Sum of 4 Cosines)

. scatter sumfc time, m(o) c(l)
```

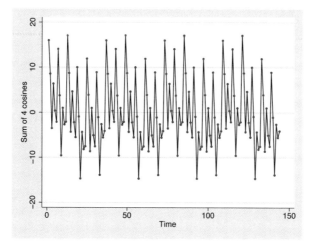

```
. pergram sumfc, gen(ordinate)
```

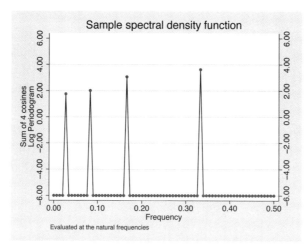

The periodogram clearly shows the four contributions to the original time series. From the plot, we can see that the periods of the summands were 3, 6, 12, and 36, although you can confirm this by using

```
. generate double omega = (_n-1)/144

. generate double period = 1/omega
(1 missing value generated)
```

```
. list period omega if ordinate> 1e-5 & omega <=.5
```

	period	omega
5.	36	.02777778
13.	12	.08333333
25.	6	.16666667
49.	3	.33333333

◁

Methods and Formulas

pergram is implemented as an ado-file.

We use the notation of Newton (1988) in the following:

A time series of interest is decomposed into a unique set of sinusoids of various frequencies and amplitudes.

A plot of the sinusoidal amplitudes (ordinates) versus the frequencies for the sinusoidal decomposition of a time series gives us the spectral density of the time series. If we calculate the sinusoidal amplitudes for a discrete set of "natural" frequencies $(1/n, 2/n, \ldots, q/n)$, then we obtain the periodogram.

Let $x(1), \ldots, x(n)$ be a time series, and let $\omega_k = (k-1)/n$ denote the natural frequencies for $k = 1, \ldots, (n/2) + 1$. Define

$$C_k^2 = \frac{1}{n^2} \left| \sum_{t=1}^{n} x(t) e^{2\pi i(t-1)\omega_k} \right|^2$$

A plot of nC_k^2 versus ω_k is then called the periodogram.

The sample spectral density is defined for a continuous frequency ω as

$$\widehat{f}(\omega) = \begin{cases} \dfrac{1}{n} \left| \displaystyle\sum_{t=1}^{n} x(t) e^{2\pi i(t-1)\omega} \right|^2 & \text{if } \omega \in [0, .5] \\ \widehat{f}(1-\omega) & \text{if } \omega \in [.5, 1] \end{cases}$$

Note that the periodogram (and sample spectral density) is symmetric about $\omega = .5$. Further standardize the periodogram such that

$$\frac{1}{n} \sum_{k=2}^{n} \frac{nC_k^2}{\widehat{\sigma}^2} = 1$$

(where $\widehat{\sigma}^2$ is the sample variance of the time series) so that the average value of the ordinate is one.

Once the amplitudes are standardized, we may then take the natural log of the values and produce the log-periodogram. In doing so, we truncate the graph at ± 6. Note that one frequently drops the prefix "log-" and simply refers to the "log-periodogram" as the "periodogram" in text.

References

Box, G. E. P. and G. M. Jenkins. 1976. *Time Series Analysis: Forecasting and Control*. Oakland, CA: Holden–Day.

Box, G. E. P., G. M. Jenkins, and G. C. Reinsel. 1994. *Time Series Analysis: Forecasting and Control*. 3d ed. Englewood Cliffs, NJ: Prentice–Hall.

Chatfield, C. 1996. *The Analysis of Time Series: An Introduction*. 5th ed. London: Chapman & Hall.

Hamilton, J. D. 1994. *Time Series Analysis*. Princeton: Princeton University Press.

Newton, H. J. 1988. *TIMESLAB: A Time Series Analysis Laboratory*. Pacific Grove, CA: Wadsworth & Brooks/Cole.

Also See

Complementary:	[TS] **tsset**
Related:	[TS] **corrgram**, [TS] **cumsp**, [TS] **wntestb**
Background:	*Stata Graphics Reference Manual*

Title

pperron — Phillips–Perron test for unit roots

Syntax

pperron *varname* [if *exp*] [in *range*] [, no̱constant la̱gs(*#*) tre̱nd regress]

pperron is for use with time-series data; see [TS] **tsset**. You must tsset your data before using pperron. *varname* may contain time-series operators; see [U] **14.4.3 Time-series varlists**.

Description

pperron performs the Phillips–Perron test for a unit root on a variable. The user may optionally exclude the constant, include a trend term, and/or include lagged values of the difference of the variable in the regression.

Options

noconstant suppresses the constant term (intercept) in the model.

lags(*#*) specifies the number of Newey–West lags to use in the calculation of the standard error.

trend specifies that a trend term should be included in the associated regression. This option may not be specified if noconstant is specified.

regress specifies that the associated regression table should appear in the output. By default, the regression table is not produced.

Remarks

Hamilton (1994) and Fuller (1976) give excellent overviews of this topic; see especially chapter 17 of the former. Phillips (1986) and Phillips and Perron (1988) present statistics for testing whether a time series had an autoregressive unit-root component.

▷ Example

Here we use the international airline passengers dataset (Box, Jenkins, and Reinsel 1994, Series G). This dataset has 144 observations on the monthly number of international airline passengers from 1949 through 1960.

```
. use http://www.stata-press.com/data/r8/air2
(TIMESLAB: Airline passengers)
```

(Continued on next page)

```
. pperron air
Phillips-Perron test for unit root               Number of obs    =        143
                                                 Newey-West lags =          4
                          ──────────── Interpolated Dickey-Fuller ────────────
                Test        1% Critical      5% Critical       10% Critical
              Statistic        Value            Value             Value
         ─────────────────────────────────────────────────────────────────────
Z(rho)        -6.564         -19.943          -13.786           -11.057
Z(t)          -1.844          -3.496           -2.887            -2.577
```

* MacKinnon approximate p-value for Z(t) = 0.3588

Note that we fail to reject the hypothesis that there is a unit root in this time series by looking either at the MacKinnon approximate asymptotic p-value or at the interpolated Dickey–Fuller critical values.

◁

▷ Example

In this example, we examine the Canadian lynx data from Newton (1988, 587). Here we include a time trend in the calculation of the statistic.

```
. use http://www.stata-press.com/data/r8/lynx2
(TIMESLAB: Canadian lynx)

. pperron lynx, trend
Phillips-Perron test for unit root               Number of obs    =        113
                                                 Newey-West lags =          4
                          ──────────── Interpolated Dickey-Fuller ────────────
                Test        1% Critical      5% Critical       10% Critical
              Statistic        Value            Value             Value
         ─────────────────────────────────────────────────────────────────────
Z(rho)       -38.365         -27.487          -20.752           -17.543
Z(t)          -4.585          -4.036           -3.448            -3.148
```

* MacKinnon approximate p-value for Z(t) = 0.0011

We reject the hypothesis that there is a unit root in this time series.

◁

Saved Results

pperron saves in r():

Scalars

r(N)	number of observations	r(Zt)	Phillips–Perron τ test statistic
r(lags)	Number of lagged differences used	r(Zrho)	Phillips–Perron ρ test statistic
r(pval)	MacKinnon approximate p-value		
	(not included if noconstant specified)		

Methods and Formulas

pperron is implemented as an ado-file.

In the OLS estimation of an AR(1) process with Gaussian errors,

$$y_i = \rho y_{i-1} + \epsilon_i$$

where ϵ_i are independent and identically distributed as $N(0, \sigma^2)$ and $y_0 = 0$, the OLS estimate (based on an n-observation time series) of the autocorrelation parameter ρ is given by

$$\widehat{\rho}_n = \frac{\displaystyle\sum_{i=1}^{n} y_{i-1} y_i}{\displaystyle\sum_{i=1}^{n} y_i^2}$$

We know that if $|\rho| < 1$, then $\sqrt{n}(\widehat{\rho}_n - \rho) \to N(0, 1 - \rho^2)$. If this result were valid for the case that $\rho = 1$, then the resulting distribution collapses to a point mass (the variance is zero).

It is this motivation that drives one to check for the possibility of a unit root in an autoregressive process. In order to compute the test statistics, we compute the Phillips–Perron regression

$$y_i = \alpha + \rho y_{i-1} + \epsilon_i$$

where we may exclude the constant or include a trend term (i). There are two statistics, Z_ρ and Z_τ, calculated as

$$Z_\rho = n(\widehat{\rho}_n - 1) - \frac{1}{2} \frac{n^2 \widehat{\sigma}^2}{s_n^2} \left(\widehat{\lambda}_n^2 - \widehat{\gamma}_{0,n} \right)$$

$$Z_\tau = \sqrt{\frac{\widehat{\gamma}_{0,n}}{\widehat{\lambda}_n^2}} \frac{\widehat{\rho}_n - 1}{\widehat{\sigma}} - \frac{1}{2} \left(\widehat{\lambda}_n^2 - \widehat{\gamma}_{0,n} \right) \frac{1}{\widehat{\lambda}_n} \frac{n\widehat{\sigma}}{s_n}$$

$$\widehat{\gamma}_{j,n} = \frac{1}{n} \sum_{i=j+1}^{n} \widehat{u}_i \widehat{u}_{i-j}$$

$$\widehat{\lambda}_n^2 = \widehat{\gamma}_{0,n} + 2 \sum_{j=1}^{q} \left(1 - \frac{j}{q+1} \right) \widehat{\gamma}_{j,n}$$

$$s_n^2 = \frac{1}{n-k} \sum_{i=1}^{n} \widehat{u}_i^2$$

where u_i is the OLS residual, k is the number of covariates in the regression, q is the number of Newey–West lags to use in the calculation of $\widehat{\lambda}_n^2$, and $\widehat{\sigma}$ is the OLS standard error of $\widehat{\rho}$.

The critical values (which have the same distribution as the Dickey–Fuller statistic; see Dickey and Fuller (1979)) included in the output are linearly interpolated from the table of values that appear in Fuller (1976), and the MacKinnon approximate p-values use the regression surface published in MacKinnon (1994).

References

Box, G. E. P., G. M. Jenkins, and G. C. Reinsel. 1994. *Time Series Analysis: Forecasting and Control.* 3d ed. Englewood Cliffs, NJ: Prentice–Hall.

Dickey, D. A. and W. A. Fuller. 1979. Distribution of the estimators for autoregressive time series with a unit root. *Journal of the American Statistical Association* 74: 427–431.

Fuller, W. A. 1976. *Introduction to Statistical Time Series*. New York: John Wiley & Sons.

Hakkio, C. S. 1994. sts6: Approximate *p*-values for unit root and cointegration tests. *Stata Technical Bulletin* 17: 25–28. Reprinted in *Stata Technical Bulletin Reprints*, vol. 3, pp. 219–224.

Hamilton, J. D. 1994. *Time Series Analysis*. Princeton: Princeton University Press.

MacKinnon, J. G. 1994. Approximate asymptotic distribution functions for unit root and cointegration tests. *Journal of Business and Economic Statistics* 12: 167–176.

Newton, H. J. 1988. *TIMESLAB: A Time Series Laboratory*. Pacific Grove, CA: Wadsworth & Brooks/Cole.

Phillips, P. C. B. 1986. Time series regression with a unit root. *Econometrica* 56: 1021–1043.

Phillips, P. C. B. and P. Perron. 1988. Testing for a unit root in time series regression. *Biometrika* 75: 335–346.

Also See

Complementary:	[TS] **tsset**
Related:	[TS] **dfuller**

Title

> **prais** — Prais–Winsten regression and Cochrane–Orcutt regression

Syntax

>   prais *depvar* [*varlist*] [if *exp*] [in *range*] [, corc <u>sse</u>search
>
>   <u>rho</u>type(*rhomethod*) <u>two</u>step <u>robust</u> <u>c</u>luster(*varname*) hc2 hc3 <u>nocon</u>stant
>
>   <u>h</u>ascons <u>save</u>space nodw <u>level</u>(#) <u>nolog</u> *maximize_options*]

prais is for use with time-series data; see [TS] **tsset**. You must tsset your data before using prais.

depvar and *varlist* may contain time-series operators; see [U] **14.4.3 Time-series varlists**.

prais shares the features of all estimation commands; see [U] **23 Estimation and post-estimation commands**.

Syntax for predict

>   predict [*type*] *newvarname* [if *exp*] [in *range*] [, [xb | <u>r</u>esiduals | stdp]]

These statistics are available both in and out of sample; type predict ... if e(sample) ... if wanted only for the estimation sample.

Description

prais fits a linear regression of *depvar* on *varlist* that is corrected for first-order serially-correlated residuals using the Prais–Winsten (1954) transformed regression estimator, the Cochrane–Orcutt (1949) transformed regression estimator, or a version of the search method suggested by Hildreth and Lu (1960).

Options

corc specifies that the Cochrane–Orcutt transformation be used to estimate the equation. With this option, the Prais–Winsten transformation of the first observation is not performed, and the first observation is dropped when estimating the transformed equation; see *Methods and Formulas* below.

ssesearch specifies that a search be performed for the value of ρ that minimizes the sum of squared errors of the transformed equation (Cochrane–Orcutt or Prais–Winsten transformation). The search method employed is a combination of quadratic and modified bisection search using golden sections.

rhotype(*rhomethod*) selects a specific computation for the autocorrelation parameter ρ, where *rhomethod* can be

<u>reg</u>ress	$\rho_{\text{reg}} = \beta$ from the residual regression $\epsilon_t = \beta\epsilon_{t-1}$
freg	$\rho_{\text{freg}} = \beta$ from the residual regression $\epsilon_t = \beta\epsilon_{t+1}$
<u>ts</u>corr	$\rho_{\text{tscorr}} = \epsilon'\epsilon_{t-1}/\epsilon'\epsilon$, where ϵ is the vector of residuals
dw	$\rho_{\text{dw}} = 1 - \text{dw}/2$, where dw is the Durbin–Watson d statistic
<u>t</u>heil	$\rho_{\text{theil}} = \rho_{\text{tscorr}}(N - k)/N$
<u>na</u>gar	$\rho_{\text{nagar}} = (\rho_{\text{dw}} * N^2 + k^2)/(N^2 - k^2)$

The prais estimator can use any consistent estimate of ρ to transform the equation, and each of these estimates meets that requirement. The default is regress, and it produces the minimum sum of squares solution (ssesearch option) for the Cochrane–Orcutt transformation—no computation will produce the minimum sum of squares solution for the full Prais–Winsten transformation. See Judge, Griffiths, Hill, Lütkepohl, and Lee (1985) for a discussion of each of the estimates of ρ.

twostep specifies that prais will stop on the first iteration after the equation is transformed by ρ—the two-step efficient estimator. Although it is customary to iterate these estimators to convergence, they are efficient at each step.

robust specifies that the Huber/White/sandwich estimator of variance is to be used in place of the traditional calculation. robust combined with cluster() further allows observations that are not independent within cluster (although they must be independent between clusters). See [U] **23.14 Obtaining robust variance estimates**.

Note that all estimates from prais are conditional on the estimated value of ρ. This means that robust variance estimates in this case are only robust to heteroskedasticity and are not generally robust to misspecification of the functional form or omitted variables. The estimation of the functional form is intertwined with the estimate of ρ, and all estimates are conditional on ρ. Thus, we cannot be robust to misspecification of functional form. For these reasons, it is probably best to interpret robust in the spirit of White's (1980) original paper on estimation of heteroskedastic consistent covariance matrices.

cluster(*varname*) specifies that the observations are independent across groups (clusters) but not necessarily within groups. *varname* specifies to which group each observation belongs. cluster() affects the estimated standard errors and variance–covariance matrix of the estimators (VCE), but not the estimated coefficients. Specifying cluster() implies robust.

hc2 and hc3 specify an alternative bias correction for the robust variance calculation; for more information, see [R] **regress**. hc2 and hc3 may not be specified with cluster(). Specifying hc2 or hc3 implies robust.

noconstant suppresses estimation of the constant term (intercept).

hascons indicates that a user-defined constant, or a set of variables that in linear combination forms a constant, has been included in the regression. For some computational concerns, see the discussion in [R] **regress**.

savespace specifies that prais attempt to save as much space as possible by retaining only those variables required for estimation. The original data are restored after estimation. This option is rarely used, and should generally be used only if there is insufficient space to fit a model without the option.

nodw suppresses reporting of the Durbin–Watson statistic.

level(#) specifies the confidence level, in percent, for confidence intervals. The default is level(95) or as set by set level; see [U] **23.6 Specifying the width of confidence intervals**.

nolog suppresses the iteration log.

maximize_options control the maximization process; see [R] **maximize**. You should never have to specify them.

Options for predict

xb, the default, calculates the fitted values—the prediction of $x_j b$ for the specified equation. This is the linear predictor from the fitted regression model; it does not apply the estimate of ρ to prior residuals.

residuals calculates the residuals from the linear prediction.

stdp calculates the standard error of the prediction for the specified equation. It can be thought of as the standard error of the predicted expected value or mean for the observation's covariate pattern. This is also referred to as the standard error of the fitted value.

As computed for prais, this is strictly the standard error from the variance in the estimates of the parameters of the linear model and assumes that ρ is estimated without error.

Remarks

The most common autocorrelated error process is the first-order autoregressive process. Under this assumption, the linear regression model may be written as

$$y_t = \mathbf{x}_t \boldsymbol{\beta} + u_t$$

where the errors satisfy

$$u_t = \rho \, u_{t-1} + e_t$$

and the e_t are independent and identically distributed as $N(0, \sigma^2)$. The covariance matrix $\boldsymbol{\Psi}$ of the error term e may then be written as

$$\boldsymbol{\Psi} = \frac{1}{1 - \rho^2} \begin{bmatrix} 1 & \rho & \rho^2 & \cdots & \rho^{T-1} \\ \rho & 1 & \rho & \cdots & \rho^{T-2} \\ \rho^2 & \rho & 1 & \cdots & \rho^{T-3} \\ \vdots & \vdots & \vdots & \ddots & \vdots \\ \rho^{T-1} & \rho^{T-2} & \rho^{T-3} & \cdots & 1 \end{bmatrix}$$

The Prais–Winsten estimator is a generalized least squares (GLS) estimator. The Prais–Winsten method (as described in Judge et al. 1985) is derived from the AR(1) model for the error term described above. Whereas the Cochrane–Orcutt method uses a lag definition and loses the first observation in the iterative method, the Prais–Winsten method preserves that first observation. In small samples, this can be a significant advantage.

❑ Technical Note

To fit a model with autocorrelated errors, you must specify your data as time series and have (or create) a variable denoting the time at which an observation was collected. The data for the regression should be equally spaced in time.

❑

▷ Example

You wish to fit a time-series model of usr on idle, but are concerned that the residuals may be serially correlated. We will declare the variable t to represent time by typing

```
. use http://www.stata-press.com/data/r8/idle
. tsset t
        time variable: t, 1 to 30
```

We can obtain Cochrane–Orcutt estimates by specifying the corc option:

```
. prais usr idle, corc
Iteration 0:  rho = 0.0000
Iteration 1:  rho = 0.3518
(output omitted)
Iteration 13:  rho = 0.5708
```

Cochrane-Orcutt AR(1) regression -- iterated estimates

Source	SS	df	MS
Model	40.1309584	1	40.1309584
Residual	166.898474	27	6.18142498
Total	207.029433	28	7.39390831

Number of obs =   29
F(1,   27) =   6.49
Prob > F   =   0.0168
R-squared   =   0.1938
Adj R-squared =   0.1640
Root MSE   =   2.4862

usr	Coef.	Std. Err.	t	P>|t|	[95% Conf. Interval]	
idle	-.1254511	.0492356	-2.55	0.017	-.2264742	-.024428
_cons	14.54641	4.272299	3.40	0.002	5.78038	23.31245
rho	.5707918					

```
Durbin-Watson statistic (original)     1.295766
Durbin-Watson statistic (transformed) 1.466222
```

The fitted model is

$$\text{usr}_t = -.1254\,\text{idle}_t + 14.55 + u_t \quad \text{and} \quad u_t = .5708\,u_{t-1} + e_t$$

We can also fit the model with the Prais–Winsten method,

```
. prais usr idle
Iteration 0:  rho = 0.0000
Iteration 1:  rho = 0.3518
(output omitted)
Iteration 14:  rho = 0.5535
```

Prais-Winsten AR(1) regression -- iterated estimates

Source	SS	df	MS
Model	43.0076941	1	43.0076941
Residual	169.165739	28	6.04163354
Total	212.173433	29	7.31632528

Number of obs =   30
F(1,   28) =   7.12
Prob > F   =   0.0125
R-squared   =   0.2027
Adj R-squared =   0.1742
Root MSE   =   2.458

usr	Coef.	Std. Err.	t	P>|t|	[95% Conf. Interval]	
idle	-.1356522	.0472195	-2.87	0.008	-.2323769	-.0389275
_cons	15.20415	4.160391	3.65	0.001	6.681978	23.72633
rho	.5535476					

```
Durbin-Watson statistic (original)     1.295766
Durbin-Watson statistic (transformed) 1.476004
```

where the Prais–Winsten fitted model is

$$\text{usr}_t = -.1357\,\text{idle}_t + 15.20 + u_t \quad \text{and} \quad u_t = .5535\,u_{t-1} + e_t$$

As the results indicate, for these data there is little to choose between the Cochrane–Orcutt and Prais–Winsten estimators, whereas the OLS estimate of the slope parameter is substantially different.

◁

▷ Example

We have data on quarterly sales, in millions of dollars, for five years, and we would like to use this information to model sales for company X. First, we fit a linear model by OLS and obtain the Durbin–Watson statistic using `dwstat`; see [TS] **regression diagnostics**.

```
. use http://www.stata-press.com/data/r8/qsales

. regress csales isales
```

Source	SS	df	MS		Number of obs =	20
					F(1, 18) =	14888.15
Model	110.256901	1	110.256901		Prob > F =	0.0000
Residual	.133302302	18	.007405683		R-squared =	0.9988
					Adj R-squared =	0.9987
Total	110.390204	19	5.81001072		Root MSE =	.08606

csales	Coef.	Std. Err.	t	P>|t|	[95% Conf. Interval]	
isales	.1762828	.0014447	122.02	0.000	.1732475	.1793181
_cons	-1.454753	.2141461	-6.79	0.000	-1.904657	-1.004849

```
. dwstat
Durbin-Watson d-statistic(  2,    20) =  .7347276
```

Noting that the Durbin–Watson statistic is far from 2 (the expected value under the null hypothesis of no serial correlation) and well below the 5% lower limit of 1.2, we conclude that the disturbances are serially correlated. (Upper and lower bounds for the d statistic can be found in most econometrics texts; e.g., Harvey, 1993. The bounds have been derived for only a limited combination of regressors and observations.) To reinforce this conclusion, we use another two tests to test for serial correlation in the error distribution.

```
. bgodfrey, lags(1)
Breusch-Godfrey LM test for autocorrelation
```

lags(p)	chi2	df	Prob > chi2
1	7.998	1	0.0047

H0: no serial correlation

```
. durbina
Durbin's alternative test for autocorrelation
```

lags(p)	chi2	df	Prob > chi2
1	11.329	1	0.0008

H0: no serial correlation

bgodfrey reports the Breusch–Godfrey Lagrange multiplier test statistic, and durbina reports the Durbin's alternative test statistic. Both tests give a small *p*-value, and thus reject the null hypothesis of no serial correlation. These two tests are asymptotically equivalent when testing for AR(1) process. See [TS] **regression diagnostics** if you are not familiar with these two tests.

We correct for the autocorrelation using the ssesearch option of prais to search for the value of ρ that minimizes the sum of squared residuals of the Cochrane–Orcutt transformed equation. Normally, the default Prais–Winsten transformations would be used with such a small dataset, but the less efficient Cochrane–Orcutt transformation will allow us to demonstrate an aspect of the estimator's convergence.

```
. prais csales isales, corc ssesearch
Iteration 1:  rho = 0.8944 , criterion =  -.07298558
Iteration 2:  rho = 0.8944 , criterion =  -.07298558
  (output omitted )
Iteration 15:  rho = 0.9588 , criterion =  -.07167037

Cochrane-Orcutt AR(1) regression -- SSE search estimates
```

Source	SS	df	MS			
Model	2.33199178	1	2.33199178			
Residual	.071670369	17	.004215904			
Total	2.40366215	18	.133536786			

				Number of obs =	19
				F(1, 17) =	553.14
				Prob > F =	0.0000
				R-squared =	0.9702
				Adj R-squared =	0.9684
				Root MSE =	.06493

csales	Coef.	Std. Err.	t	P>|t|	[95% Conf. Interval]	
isales	.1605233	.0068253	23.52	0.000	.1461233	.1749234
_cons	1.738946	1.432674	1.21	0.241	-1.283732	4.761624
rho	.9588209					

```
Durbin-Watson statistic (original)      0.734728
Durbin-Watson statistic (transformed) 1.724419
```

It was noted in the *Options* section that with the default computation of ρ the Cochrane–Orcutt method produces an estimate of ρ that minimizes the sum of squared residuals—the same criterion as the ssesearch option. Given that the two methods produce the same results, why would the search method ever be preferred? It turns out that the back-and-forth iterations employed by Cochrane–Orcutt can often have difficulty converging if the value of ρ is large. Using the same data, the Cochrane–Orcutt iterative procedure requires over 350 iterations to converge, and a higher tolerance must be specified to prevent premature convergence:

(Continued on next page)

```
. prais csales isales, corc tol(1e-9) iterate(500)
Iteration 0:   rho = 0.0000
Iteration 1:   rho = 0.6312
Iteration 2:   rho = 0.6866
 (output omitted )
Iteration 377:  rho = 0.9588
Iteration 378:  rho = 0.9588
Iteration 379:  rho = 0.9588
Cochrane-Orcutt AR(1) regression -- iterated estimates
```

Source	SS	df	MS
Model	2.33199171	1	2.33199171
Residual	.071670369	17	.004215904
Total	2.40366208	18	.133536782

```
Number of obs =       19
F(  1,    17) =  553.14
Prob > F      =  0.0000
R-squared     =  0.9702
Adj R-squared =  0.9684
Root MSE      =   .06493
```

csales	Coef.	Std. Err.	t	P>|t|	[95% Conf. Interval]	
isales	.1605233	.0068253	23.52	0.000	.1461233	.1749234
_cons	1.738946	1.432674	1.21	0.241	-1.283732	4.761625
rho	.9588209					

```
Durbin-Watson statistic (original)    0.734728
Durbin-Watson statistic (transformed) 1.724419
```

Once convergence is achieved, the two methods produce identical results.

◁

Saved Results

prais saves in e():

Scalars

e(N)	number of observations	e(N_clust)	number of clusters
e(mss)	model sum of squares	e(rho)	autocorrelation parameter ρ
e(df_m)	model degrees of freedom	e(dw)	Durbin–Watson d statistic for
e(rss)	residual sum of squares		untransformed regression
e(df_r)	residual degrees of freedom	e(dw_0)	Durbin–Watson d statistic of
e(r2)	R-squared		transformed regression
e(r2_a)	adjusted R-squared	e(tol)	target tolerance
e(F)	F statistic	e(max_ic)	maximum number of iterations
e(rmse)	root mean square error	e(ic)	number of iterations
e(ll)	log likelihood	e(N_gaps)	number of gaps

Macros

e(cmd)	prais	e(vcetype)	covariance estimation method
e(depvar)	name of dependent variable	e(tranmeth)	corc or prais
e(clustvar)	name of cluster variable	e(cons)	noconstant or not reported
e(rhotype)	method specified in rhotype option	e(predict)	program used to implement
e(method)	twostep, iterated, or SSE search		predict

Matrices

e(b)	coefficient vector	e(V)	variance–covariance matrix
			of the estimators

Functions

e(sample)	marks estimation sample

Methods and Formulas

prais is implemented as an ado-file.

Consider the command 'prais y x z'. The 0th iteration is obtained by estimating a, b, and c from the standard linear regression:

$$y_t = ax_t + bz_t + c + u_t$$

An estimate of the correlation in the residuals is then obtained. By default, prais uses the auxiliary regression:

$$u_t = \rho u_{t-1} + e_t$$

This can be changed to any of the computations noted in the rhotype() option.

Next, we apply a Cochrane–Orcutt transformation (1) for observations $t = 2, \ldots, n$

$$y_t - \rho y_{t-1} = a(x_t - \rho x_{t-1}) + b(z_t - \rho z_{t-1}) + c(1 - \rho) + v_t \tag{1}$$

and the transformation $(1')$ for $t = 1$

$$\sqrt{1 - \rho^2}\, y_1 = a(\sqrt{1 - \rho^2}\, x_1) + b(\sqrt{1 - \rho^2}\, z_1) + c\sqrt{1 - \rho^2} + \sqrt{1 - \rho^2}\, v_1 \tag{1'}$$

Thus, the differences between the Cochrane–Orcutt and the Prais–Winsten methods are that the latter uses $(1')$ in addition to (1), whereas the former uses only (1) and necessarily decreases the sample size by one.

Equations (1) and $(1')$ are used to transform the data and obtain new estimates of a, b, and c.

When the twostep option is specified, the estimation process is halted at this point, and these are the estimates reported. Under the default behavior of iterating to convergence, this process is repeated until the change in the estimate of ρ is within a specified tolerance.

The new estimates are used to produce fitted values

$$\widehat{y}_t = \widehat{a}x_t + \widehat{b}z_t + \widehat{c}$$

and then ρ is re-estimated, by default using the regression defined by

$$y_t - \widehat{y}_t = \rho(y_{t-1} - \widehat{y}_{t-1}) + u_t \tag{2}$$

We then re-estimate (1) using the new estimate of ρ, and continue to iterate between (1) and (2) until the estimate of ρ converges.

Convergence is declared after iterate() iterations or when the absolute difference in the estimated correlation between two iterations is less than tol(); see [R] **maximize**. Sargan (1964) has shown that this process will always converge.

Under the ssesearch option, a combined quadratic and bisection search using golden sections is used to search for the value of ρ that minimizes the sum of squared residuals from the transformed equation. The transformation may be either the Cochrane–Orcutt (1 only) or the Prais–Winsten (1 and $1'$).

All reported statistics are based on the ρ-transformed variables, and there is an assumption that ρ is estimated without error. See Judge et al. (1985) for details.

The Durbin–Watson d statistic reported by `prais` and `dwstat` is

$$d = \frac{\sum\limits_{j=1}^{n-1} (u_{j+1} - u_j)^2}{\sum\limits_{j=1}^{n} u_j^2}$$

where u_j represents the residual of the jth observation.

Acknowledgment

We thank Richard Dickens of the Centre for Economic Performance at the London School of Economics and Political Science for testing and assistance with an early version of this command.

References

Chatterjee, S., A. S. Hadi, and B. Price. 2000. *Regression Analysis by Example*. 3d ed. New York: John Wiley & Sons.

Cochrane, D. and G. H. Orcutt. 1949. Application of least-squares regression to relationships containing autocorrelated error terms. *Journal of the American Statistical Association* 44: 32–61.

Durbin, J. and G. S. Watson. 1950 and 1951. Testing for serial correlation in least-squares regression. *Biometrika* 37: 409–428 and 38: 159–178.

Hardin, J. W. 1995. sts10: Prais–Winsten regression. *Stata Technical Bulletin* 25: 26–29. Reprinted in *Stata Technical Bulletin Reprints*, vol. 5, pp. 234–237.

Harvey, A. C. 1993. *The Econometric Analysis of Time Series*. Cambridge, MA: MIT Press.

Hildreth, C. and J. Y. Lu. 1960. Demand relations with autocorrelated disturbances. *Agricultural Experiment Station Technical Bulletin* 276. East Lansing, MI: Michigan State University.

Johnston, J. and J. DiNardo. 1997. *Econometric Methods*. 4th ed. New York: McGraw–Hill.

Judge, G. G., W. E. Griffiths, R. C. Hill, H. Lütkepohl, and T.-C. Lee. 1985. *The Theory and Practice of Econometrics*. 2d ed. New York: John Wiley & Sons.

Kmenta, J. 1997. *Elements of Econometrics*. 2d ed. Ann Arbor: University of Michigan Press.

Prais, S. J. and C. B. Winsten. 1954. Trend Estimators and Serial Correlation. *Cowles Commission Discussion Paper No. 383*, Chicago.

Sargan, J. D. 1964. Wages and prices in the United Kingdom: a study in econometric methodology. In *Econometric Analysis for National Economic Planning*, ed. P. E. Hart, G. Mills, J. K. Whitaker, 25–64. London: Butterworths.

Theil, H. 1971. *Principles of Econometrics*. New York: John Wiley & Sons.

White, H. 1980. A heteroskedasticity-consistent covariance matrix estimator and a direct test for heteroskedasticity. *Econometrica* 48: 817–838.

Also See

Complementary:	[TS] **regression diagnostics**, [TS] **tsset**,
	[R] **adjust**, [R] **lincom**, [R] **mfx**, [R] **nlcom**, [R] **predict**, [R] **predictnl**,
	[R] **test**, [R] **testnl**, [R] **vce**, [R] **xi**
Related:	[R] **regress**
Background:	[U] **16.5 Accessing coefficients and standard errors**,
	[U] **23 Estimation and post-estimation commands**,
	[U] **23.14 Obtaining robust variance estimates**

Title

> **regression diagnostics** — Regression diagnostics for time series

Syntax

dwstat

durbina [, lags(*numlist*) nomiss0 [robust | small] force]

bgodfrey [, lags(*numlist*) nomiss0 small]

archlm [, lags(*numlist*) force]

These statistics are for use after regress; see [R] **regress**.
These commands are for use with time-series data; see [TS] **tsset**. You must tsset your data before using them.

Description

dwstat computes the Durbin–Watson d statistic (Durbin and Watson 1950) to test for first-order serial correlation in the disturbance when all the regressors are strictly exogenous.

durbina performs Durbin's alternative test for serial correlation in the disturbance. This test does not require that all the regressors be strictly exogenous.

bgodfrey performs the Breusch–Godfrey test for higher order of serial correlation in the disturbance. This test does not require that all the regressors be strictly exogenous.

archlm performs Engle's LM test for the presence of autoregressive conditional heteroskedasticity.

Options

lags(*numlist*) specifies a list of numbers, which are the lag orders to be tested. The test will be performed separately for each order. The default is order one.

nomiss0 specifies not to use Davidson and MacKinnon's approach, which replaces the missing values in the initial observations on the lagged residuals in the auxiliary regression with zeros.

robust requests that the Huber/White/sandwich robust estimator of the variance–covariance matrix be used in Durbin's alternative test.

small specifies that the p-values of the test statistics are to be obtained using the F or t distribution instead of the default chi-squared or normal distribution. This option may not be specified with robust, which always uses an F or t distribution.

force allows the durbina and archlm tests to be run after regress, robust. In addition, durbina, force will execute after [TS] **newey**, although archlm will not. Note that none of these commands will work if the cluster option was specified on the [R] **regress** command.

Tests in time-series linear regression models

The Durbin–Watson test computes the d statistic from OLS residuals to test whether the error distribution follows an AR(1) scheme. For a linear model,

$$y_t = X_t \beta + u_t$$

the AR(1) process can be written as

$$u_t = \rho u_{t-1} + \epsilon_t$$

In general, the definition of an AR(1) process only requires that ϵ_t be independently and identically distributed (i.i.d.). However, in order for the Durbin–Watson statistic to have its exact distribution, it is required that $\epsilon_t \sim N(0, \sigma^2)$. It should also be noted that the Durbin–Watson test can only be applied when the covariates in X_t are strictly exogenous. As discussed in Wooldridge (2002), a covariate x_t is strictly exogenous if $\text{Corr}(x_s, u_t) = 0 \quad \forall s, t$. In particular, note that this requires that the error at time t is not correlated with future values of the covariate. This requirement excludes the case of lagged dependent variables.

The null and alternative hypotheses of the test are

$$H_0: \rho = 0 \qquad \text{versus} \qquad H_1: \rho \neq 0$$

The Durbin–Watson d statistic takes on values in the range $[0, 4]$. The center of this range ($d = 2$) corresponds to the null hypothesis of no first-order serial correlation. Values of the d statistic much greater than 2 favor a positive autocorrelation ($\rho < 0$), and values much less than 2 favor negative autocorrelation ($\rho < 0$). The calculation of the exact distribution of the d statistic is computationally expensive, but the empirical upper and lower bounds are established based on the sample size and the number of regressors. Extended tables for the d statistic have been published by Savin and White (1977). As an example, with 30 observations and 3 regressors (including the constant term), the upper bound of d statistic is 1.284 and the lower bound is 1.567 for a test at 0.05 significance level. A value falling within (1.284, 1.567) leads to no conclusion of whether or not to reject the null hypothesis.

As mentioned above, the Durbin–Watson test is based on the assumption that all covariates are strictly exogenous. When the regressors include lagged dependent variables,

$$y_t = \beta_1 y_{t-1} + \cdots + \beta_r y_{t-r} + \beta_{r+1} x_{1t} + \cdots + \beta_{r+s} x_{st} + u_t$$

the past values of the error term are correlated with the lagged dependent variables at time t, implying that the lagged dependent variables are not strictly exogenous. The inclusion of covariates that are not strictly exogenous causes the d statistic to be biased toward the acceptance of the null hypothesis. Durbin (1970) suggested an alternative test for the more general case, and extended it for higher order serial correlation AR(p), where the AR(p) process can be expressed as

$$u_t = \rho_1 u_{t-1} + \cdots + \rho_p u_{t-p} + \epsilon_t,$$

where ϵ_t is independent and identically distributed (i.i.d.) with variance σ^2 (ϵ_t it not assumed to have a normal distribution), and where the null hypothesis of the test is

$$H_0 : \rho_1 = 0, \ldots, \rho_p = 0$$

One of the interesting and useful aspects of this test is that although the null hypothesis was originally derived for an AR(p) process, this test has power against MA(p) processes as well. Hence, the actual null of this test is that there is no serial correlation up to order p. The reason for this serendipitous result is that the MA(p) and the AR(p) are locally equivalent alternatives under the null. See Godfrey (1988, 113–115) for a discussion of this result.

Durbin's alternative test is a Lagrange multiplier (LM) test, but it is most easily computed via a Wald test on the coefficients of the lagged residuals in an auxiliary-OLS regression of the residuals on their lags, and all the covariates in the original regression. Consider the linear regression model

$$y_t = \beta_1 x_{1t} + \cdots + \beta_k x_{kt} + u_t \tag{1}$$

in which the covariates x_1 through x_k are not assumed to strictly exogenous, and u_t is assumed to be i.i.d and to have finite variance. The process is also assumed to be stationary. (See Wooldridge (2002) for a discussion of stationarity.) After estimating the parameters in (1) by OLS, we obtain the residuals $\widehat{u}_t$. Next, we estimate another OLS regression of $\widehat{u}_t$ on $\widehat{u}_{t-1}, \ldots, \widehat{u}_{t-p}$ and the other regressors,

$$\widehat{u}_t = \gamma_1 \widehat{u}_{t-1} + \cdots + \gamma_p \widehat{u}_{t-p} + \beta_1 x_{1t} + \cdots + \beta_k x_{kt} + \epsilon_t \tag{2}$$

where ϵ_t stands for the random error term in this auxiliary OLS regression. Durbin's alternative test is then obtained by performing a Wald test that $\gamma_1, \ldots, \gamma_p$ are jointly zero. The test can be made robust to an unknown form of heteroskedasticity by using a robust VCE estimator when estimating the regression in (2). When there are only strictly exogenous regressors and $p = 1$, this test is asymptotically equivalent to the Durbin–Watson test.

The Breusch–Godfrey test is also an (LM) test, but it is computed as the $N * R^2$, where N is the number of observations and R^2 is the R^2 from (2). This test and Durbin's alternative test are asymptotically equivalent. The test statistic $N * R^2$ has an asymptotic χ^2 distribution with p degrees of freedom. It is valid with or without the strict exogeneity assumption, but is not robust to conditional heteroskedasticity.

The values of the lagged residuals will be missing in the initial time periods. As noted by Davidson and MacKinnon (1993), the residuals will not be orthogonal to the other covariates in the model in this restricted sample. This implies that the R^2 from the auxiliary regression will not be zero when the lagged residuals are left out. Hence, the $N * R^2$ version of the test may over-reject in small samples. To correct this problem, Davidson and MacKinnon (1993) recommended setting the missing values of the lagged residuals to zero and running the auxiliary regression in (2) over the full sample used in (1). This small-sample correction has become conventional for both the Breusch–Godfrey and Durbin's alternative test, and it is the default for both commands. Specifying the nomiss0 option overrides this default behavior and causes the initial missing values generated by regressing on the lagged residuals to be treated as missing. Hence, miss0 causes these initial observations to be dropped from the sample of the auxiliary regression.

It should also be mentioned that Durbin's alternative test and the Breusch–Godfrey test were originally derived for the case covered by [R] **regress**, without either the robust or the cluster() options. However, since durbina, robust makes sense after regress, robust and [TS] **newey**, if the force option is specified, durbina will run after these two commands. Since bgodfrey does not have a robust option, it does not have a force option and it will not run after any command except regress. If the force option is specified, archlm will run after regress, robust, but not after newey. Also note that none of the three commands will run if the cluster() option was specified on the regression.

▷ Example

Using the Klein (1950) data, we can investigate the disturbance of an OLS model that regresses consumption on the government wage bill. Let's begin by estimating the regression in the output below.

```
. use http://www.stata-press.com/data/r8/klein

. tsset yr
        time variable:  yr, 1920 to 1941

. regress consump wagegovt
```

Source	SS	df	MS		Number of obs =	22
					F(1, 20) =	17.72
Model	532.567711	1	532.567711		Prob > F =	0.0004
Residual	601.207167	20	30.0603584		R-squared =	0.4697
					Adj R-squared =	0.4432
Total	1133.77488	21	53.9892799		Root MSE =	5.4827

consump	Coef.	Std. Err.	t	P>|t|	[95% Conf. Interval]	
wagegovt	2.50744	.5957173	4.21	0.000	1.264796	3.750085
_cons	40.84699	3.192183	12.80	0.000	34.18821	47.50577

If we assume that `wagegov` is a strictly exogenous variable, we can use the Durbin–Watson test to check for first-order serial correlation in the errors. In the output below, we perform the Durbin–Watson on our previously fitted model.

```
. dwstat
Durbin-Watson d-statistic(  2,    22) =  .3217998
```

The Durbin–Watson d statistic 0.32 is far away from the center of its distribution ($d = 2.0$). Given 22 observations and 2 regressors (including the constant term) in the model, the lower 5% bound is about 0.997, much greater than the computed d statistic. Assuming that `wagegov` is strictly exogenous, we could reject the null of no first-order serial correlation. Note that rejecting the null hypothesis does not necessarily mean an AR process. Other forms of misspecification may also lead to a significant test statistic. If we were to assume that the errors follow an AR(1) process and that the regressors are strictly exogenous, then the autocorrelation parameter ρ could be estimated, and the other coefficients could be more efficiently estimated by `arima` or `prais`; see [TS] **arima** or [TS] **prais**.

If we are not willing to assume that `wagegov` is strictly exogenous, we could use Durbin's alternative test or the Breusch–Godfrey to test for first order serial-correlation. As can be seen from the output below, both of these tests provide strong evidence against the null of no first-order serial correlation for this model.

```
. durbina, small
Durbin's alternative test for autocorrelation
```

lags(p)	F	df	Prob > F
1	35.035	(1, 19)	0.0000

H0: no serial correlation

```
. bgodfrey, small
Breusch-Godfrey LM test for autocorrelation
```

lags(p)	F	df	Prob > F
1	14.264	(1, 19)	0.0013

H0: no serial correlation

Given the degree to which the null of no first-order serial-correlation was rejected, we might consider adding one or more lags of consump to the original model. In the output below, we refit the model, adding two lags of consump to the model, and then run durbina and bgodfrey. Since including lags of the dependent variable violates the strict exogeneity assumption, we cannot use the Durbin–Watson test in this case. Still, both Durbin's alternative and the Breusch–Godfrey remain valid. Although wagegov and the constant term are no longer statistically different from zero at the 5% level, the output from durbina and bgodfrey indicates that including the two lags of consump has removed any serial correlation from the errors.

```
. regress consump wagegovt L.consump L2.consump
```

Source	SS	df	MS		
Model	702.660311	3	234.220104	Number of obs =	20
Residual	85.1596011	16	5.32247507	F(3, 16) =	44.01
				Prob > F =	0.0000
				R-squared =	0.8919
				Adj R-squared =	0.8716
Total	787.819912	19	41.4642059	Root MSE =	2.307

consump	Coef.	Std. Err.	t	P>|t|	[95% Conf. Interval]	
wagegovt	.6904282	.3295485	2.10	0.052	-.0081835	1.38904
consump						
L1	1.420536	.197024	7.21	0.000	1.002864	1.838208
L2	-.650888	.1933351	-3.37	0.004	-1.06074	-.241036
_cons	9.209073	5.006701	1.84	0.084	-1.404659	19.82281

```
. durbina, small lags(1/2)
Durbin's alternative test for autocorrelation
```

lags(p)	F	df	Prob > F
1	0.080	(1, 15)	0.7805
2	0.260	(2, 14)	0.7750

H0: no serial correlation

```
. bgodfrey, small lags(1/2)
Breusch-Godfrey LM test for autocorrelation
```

lags(p)	F	df	Prob > F
1	0.107	(1, 15)	0.7484
2	0.358	(2, 14)	0.7056

H0: no serial correlation

◁

Engle (1982) suggested an LM test for checking for autoregressive conditional heteroskedasticity (ARCH). The pth order of ARCH model can be written as

$$\sigma_t^2 = E(u_t^2 | u_{t-1}, \ldots, u_{t-p})$$
$$= \gamma_0 + \gamma_1 u_{t-1}^2 + \cdots + \gamma_p u_{t-p}^2$$

To test the null hypothesis of no autoregressive conditional heteroskedasticity (i.e., $\gamma_1 = \cdots = \gamma_p = 0$), we first fit the OLS model (1), obtain the residuals $\widehat{u}_t$, and run another OLS regression on the lagged residuals:

$$\widehat{u}_t^2 = \gamma_0 + \gamma_1 \widehat{u}_{t-1}^2 + \cdots + \gamma_p \widehat{u}_{t-p}^2 + \epsilon \tag{3}$$

The test statistic is NR^2, where R^2 is the R^2 from the regression in (3) and where N is the number of observations in the sample. Under the null hypothesis, the test statistic follows a χ_p^2 distribution.

▷ Example

In the output below, we refit our original model that does not include the two lags of consump. We then use archlm to see if there is any evidence that the errors might be autoregressive conditional heteroskedastic. The output below presents evidence that the errors from the original model are autoregressively conditionally heteroskedastic. archlm shows the results for tests of ARCH(1), ARCH(2), and ARCH(3) effects, respectively. At the 5% significance level, all the three tests reject the null hypothesis that the errors are not autoregressively conditionally heteroskedastic. Again, we should be cautious to accept the alternative hypothesis that there are ARCH effects in the disturbance. Other possible alternative models need to be considered in addition to this one. See [TS] **arch** for estimators of the ARCH model.

```
. regress consump wagegovt
```

Source	SS	df	MS		
Model	532.567711	1	532.567711	Number of obs =	22
Residual	601.207167	20	30.0603584	F(1, 20) =	17.72
				Prob > F =	0.0004
				R-squared =	0.4697
Total	1133.77488	21	53.9892799	Adj R-squared =	0.4432
				Root MSE =	5.4827

consump	Coef.	Std. Err.	t	P>|t|	[95% Conf. Interval]
wagegovt	2.50744	.5957173	4.21	0.000	1.264796 3.750085
_cons	40.84699	3.192183	12.80	0.000	34.18821 47.50577

```
. archlm, lags(1 2 3)
```

LM test for autoregressive conditional heteroskedasticity (ARCH)

lags(p)	chi2	df	Prob > chi2
1	5.543	1	0.0186
2	9.431	2	0.0090
3	9.039	3	0.0288

H0: no ARCH effects   *vs.* H1: ARCH(p) disturbance

◁

Saved Results

dwstat saves in r():

Scalars

r(N)	number of observations	r(N_gaps)	number of gaps
r(k)	number of regressors	r(dw)	Durbin–Watson statistic

durbina saves in r():

Scalars

r(N)	number of observations	r(N_gaps)	number of gaps
r(k)	number of regressors		

Macros

r(lags)	orders of lags

Matrices

r(chi2)	χ^2 statistic for each order of lags	r(p)	two-sided p-values
r(F)	F statistic for each order of lags, (small only)	r(df)	degrees of freedom
r(df_r)	residual degrees of freedom, (small only)		

bgodfrey saves in r():

Scalars

r(N)	number of observations	r(N_gaps)	number of gaps
r(k)	number of regressors		

Macros

r(lags)	orders of lag

Matrices

r(chi2)	χ^2 statistic for each order of lags	r(p)	two sided p-values
r(F)	F statistic for each order of lags, (small only)	r(df)	degrees of freedom
r(df_r)	residual degrees of freedom, (small only)		

archlm saves in r():

Scalars

r(N)	number of observations	r(N_gaps)	number of gaps
r(k)	number of regressors		

Macros

r(lags)	orders of lags

Matrices

r(arch)	test statistic for each order of lags	r(p)	two sided p-values
r(df)	degrees of freedom		

Methods and Formulas

dwstat, durbina, bgodfrey, and archlm are implemented as ado-files.

Consider the regression

$$y_t = \beta_1 x_{1t} + \cdots + \beta_k x_{kt} + u_t \tag{4}$$

in which some of the covariates are not strictly exogenous. In particular, some the x_{it} may be lags of the dependent variable. We are interested in whether the u_t are serially correlated.

As discussed in Wooldridge (2002) and Davidson and MacKinnon (1993), the Durbin–Watson test is not valid in the presence of covariates that are not strictly exogenous. However, the Durbin's alternative test and the Breusch–Godfrey test are both applicable to the case in which some of the covariates are not strictly exogenous. Both Durbin's alternative test and the Breusch–Godfrey test are LM tests are asymptotically equivalent, and are both computed via an auxiliary regression.

To compute Durbin's alternative test and the Breusch–Godfrey test against the null hypothesis of no pth order of serial correlation, first fit the regression in (4), compute the residuals, and then fit the following auxiliary regression of the residuals $\widehat{u}_t$ on p lags of $\widehat{u}_t$ and on all of the covariates in the original regression in (4):

$$\widehat{u}_t = \gamma_1 \widehat{u}_{t-1} + \cdots + \gamma_p \widehat{u}_{t-p} + \beta_1 x_{1t} + \cdots + \beta_k x_{kt} + \epsilon \tag{5}$$

Durbin's alternative test is computed by performing a Wald test for whether the coefficients of $\widehat{u}_{t-1}, \ldots, \widehat{u}_{t-p}$ are jointly different from zero. By default, the statistic is assumed to be distributed $\chi^2(p)$. When the option small is specified, the statistic is assumed to follow an $F(p, N - p - k)$ distribution. The reported p-value is two-sided p-value. When the robust option is specified, the Wald test is performed using the Huber/White/sandwich estimator of the variance–covariance matrix, and the test is robust to an unspecified form of heteroskedasticity. (See [P] _robust for more on this variance estimator.)

The Breusch–Godfrey test is computed as $N * R^2$, where N is the number of observations in the auxiliary regression (5) and R^2 is the R^2 from the same regression (5).

Like Durbin's alternative test, the Breusch–Godfrey test is asymptotically distributed $\chi^2(p)$, but specifying small will cause the p-value to be computed using an F($p, N - p - k$).

As discussed in the text, by default, the initial missing values of the lagged residuals are replaced with zeros, and the auxiliary regression is run over the full sample used in the original regression of (4). Specifying the nomiss0 option causes these missing values to be treated as missing values, and the observations are dropped from the sample.

Engle's LM test for ARCH(p) effects fits an OLS regression of $\widehat{u}_t^2$ on $\widehat{u}_{t-1}^2, \ldots, \widehat{u}_{t-p}^2$.

$$\widehat{u}_t^2 = \gamma_0 + \gamma_1 \widehat{u}_{t-1}^2 + \cdots + \gamma_p \widehat{u}_{t-p}^2 + \epsilon$$

The test statistic is nR^2 with an asymptotic $\chi^2(p)$ distribution.

The Durbin–Watson d statistic reported by dwstat is

$$d = \frac{\sum_{j=1}^{n-1} (\widehat{e}_{i+1} - \widehat{e}_i)^2}{\sum_{j=1}^{n} \widehat{e}_i^2}$$

where $\widehat{e}_i$ represents the residual of the ith observation.

Acknowledgment

The original versions of `bgodfrey`, `durbina`, and `archlm` were written by Christopher F. Baum, Boston College.

References

Baum, C. F. and V. L. Wiggins. 2000a. sg135: Test for autoregressive conditional heteroskedasticity in regression error distribution. *Stata Technical Bulletin* 55: 13–14. Reprinted in *Stata Technical Bulletin Reprints*, vol. 10, 143–144.

——. 2000b. sg136: Tests for serial correlation in regression error distribution. *Stata Technical Bulletin* 55: 14–15. Reprinted in *Stata Technical Bulletin Reprints*, vol. 10, 145–147.

Davidson, R. and J. G. MacKinnon. 1993. *Estimation and Inference in Econometrics*. New York: Oxford University Press.

Durbin, J. 1970. Testing for serial correlation in least squares regressions when some of the regressors are lagged dependent variables. *Econometrica* 38: 410–421.

Engle, R. F. 1982. Autoregressive conditional heteroskedasticity with estimates of the variance of United Kingdom inflation. *Econometrica* 50: 987–1007.

Godfrey, L. G. 1988. *Misspecification Tests in Econometrics*. Econometric Society Monographs, No. 16, Cambridge: Cambridge University Press.

Greene, W. H. 2003. *Econometric Analysis*. 5th ed. Upper Saddle River, NJ: Prentice–Hall.

Johnston, J. and J. DiNardo. 1997. *Econometric Methods*. 4th ed. New York: McGraw–Hill.

Klein, L. 1950. *Economic Fluctuations in the United States 1921-1941*. New York: John Wiley & Sons.

Savin, E. and K. J. White. 1977. The Durbin–Watson test for serial correlation with extreme sample sizes or many regressors. *Econometrica* 45(8): 1989–1996.

Wooldridge, J. M. 2002. *Introductory Econometrics: A Modern Approach*. 2d ed. Cincinnati, OH: South-Western College Publishing.

Also See

Complementary:	[TS] **tsset**
Related:	[R] **regression diagnostics**

Title

> **tsappend** — Add observations to time-series dataset

Syntax

tsappend , add(*integer*) [panel(*integer*)]

tsappend , last(*date*) tsfmt(*string*) [panel(*integer*)]

You must tsset your data before using tsappend; see [TS] **tsset**.

Description

tsappend appends observations to a time-series dataset. tsappend uses the information set by tsset, automatically fills in the time variable, and fills in the panel variable if the panel variable was set.

Options

add(*integer*) specifies the number of observations to add.

last(*date*) specifies the date of the last observation to add.

tsfmt(*string*) specifies the Stata time-series function to use in converting the date specified in last() to an integer.

panel(*integer*) specifies that observations should only be added to the panel with the id specified in panel().

Remarks

tsappend is used to add observations to a time-series or panel dataset. Since tsappend uses the information set by tsset, you must tsset your data before using tsappend. The time variable will automatically be filled in, as will the panel variable if it is present. Any gaps in the time variable will be filled in.

There are two ways to use tsappend: you can use the add() option to request that observations be added, or you can use the last() option to request that observations be appended until the date specified in last() is the last observation. The user must specify tsfmt() if last() is specified. tsfmt() gives the Stata time-series date function that will be used to specify how to convert the date held in last() to an integer.

tsappend works with panel data. By default, it will add the requested observations to all the panels. The panel() option can be used to add the requested observations only to the one panel specified in panel().

tsappend can be useful in appending observations prior to dynamically predicting a time series. Consider an example in which tsappend is used to add on the extra observations before dynamically predicting from an AR(1) regression:

```
. use http://www.stata-press.com/data/r8/tsappend1
. regress y l.y
```

Source	SS	df	MS
Model	115.349555	1	115.349555
Residual	461.241577	477	.966063473
Total	576.591132	478	1.2062576

```
Number of obs =      479
F(  1,   477) =   119.29
Prob > F      =   0.0000
R-squared     =   0.2001
Adj R-squared =   0.1984
Root MSE      =   .98334
```

y	Coef.	Std. Err.	t	P>|t|	[95% Conf. Interval]
y					
L1	.4493507	.0411417	10.92	0.000	.3685093 .5301921
_cons	11.11877	.8314581	13.37	0.000	9.484993 12.75254

```
. matrix b = e(b)
. matrix colnames b = L.xb one
. tsset
        time variable:  t2, 1960m2 to 2000m1
. tsappend, add(12)
. tsset
        time variable:  t2, 1960m2 to 2001m1
. predict xb if t2<=m(2000m2)
(option xb assumed; fitted values)
(12 missing values generated)
. generate one = 1
. matrix score xb=b if t2 >= m(2000m2), replace
```

Note the call to `tsset` before and after the call to `tsappend`. The output from these two calls to `tsset` illustrates that `tsappend` added on another year of observations. We then use `predict` and `matrix score` to obtain the dynamic predictions that lead to the following graph:

```
. line y xb t2 if t2>=m(1995m1), ytitle("") xtitle("time")
```

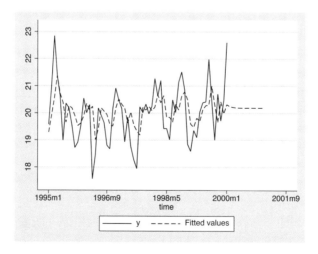

In the call to `tsappend`, instead of saying that we wanted to add 12 observations, we could have specified that we wanted to fill in observations through the first month of 2001:

```
. use tsappend1, clear

. tsset
        time variable:  t2, 1960m2 to 2000m1

. tsappend, last(2001m1) tsfmt(m)

. tsset
        time variable:  t2, 1960m2 to 2001m1
```

Note that we specified the m() function in the tsfmt() option in the above example. In [R] **functions**, there is a list of time-series functions for translating date literals to integers. Since we have monthly data, and since [R] **functions** tells us that we want to use the m() function, we specified the option tsfmt(m). Here is a table that gives the most common types of time-series data, their formats, the appropriate translation functions, and the corresponding options for tsappend:

Description	Format	Function	Option
daily	%td	d()	tsfmt(d)
weekly	%tw	w()	tsfmt(w)
monthly	%tm	m()	tsfmt(m)
quarterly	%tq	q()	tsfmt(q)
yearly	%ty	y()	tsfmt(y)
half-yearly	%th	h()	tsfmt(h)

If the end dates vary over panels, last() and add() will produce different results. add() will always add on the specified number of observations to each panel. If the data end at different time periods before calling tsappend, add(), the data still end at different periods after the call to tsappend, add(). In contrast, tsappend, last() tsfmt() will cause all the panels to end at the specified last date. Note that if the beginning dates differ across panels, using tsappend, last() tsfmt() to provide a uniform ending date will not create balanced panels because the number of observations per panel will still differ.

Consider the panel data summarized in the output below:

```
. use http://www.stata-press.com/data/r8/tsappend3

. xtdes
        id:  1, 2, ..., 3                               n =          3
        t2:  456, 457, ..., 480                         T =         25
             Delta(t2) = 1; (480-456)+1 = 25
             (id*t2 uniquely identifies each observation)
    Distribution of T_i:     min      5%     25%      50%     75%     95%     max
                              13      13      13       20      24      24      24

      Freq.  Percent    Cum. |  Pattern
 ----------------------------+-------------------------
          1    33.33   33.33 |  ............1111111111111
          1    33.33   66.67 |  1111.11111111111111111111
          1    33.33  100.00 |  11111111111111111111.....
 ----------------------------+-------------------------
          3   100.00         |  XXXXXXXXXXXXXXXXXXXXXXXXX
```

(Continued on next page)

```
. by id: sum t2
```

```
-> id = 1
    Variable |      Obs        Mean    Std. Dev.       Min        Max
-------------+--------------------------------------------------------
          t2 |       13         474     3.89444        468        480
```

```
-> id = 2
    Variable |      Obs        Mean    Std. Dev.       Min        Max
-------------+--------------------------------------------------------
          t2 |       20       465.5     5.91608        456        475
```

```
-> id = 3
    Variable |      Obs        Mean    Std. Dev.       Min        Max
-------------+--------------------------------------------------------
          t2 |       24    468.3333    7.322786        456        480
```

The output from xtdes and summarize on these data tell us that the one panel starts later than the other, that another panel ends prior to the other two, and that the remaining panel has a gap in the time variable but otherwise spans the whole time frame.

Now consider the data after a call to tsappend, add(6):

```
. tsappend, add(6)
. xtdes
      id:  1, 2, ..., 3                              n =          3
      t2:  456, 457, ..., 486                        T =         31
           Delta(t2) = 1; (486-456)+1 = 31
           (id*t2 uniquely identifies each observation)
Distribution of T_i:   min      5%     25%      50%     75%     95%     max
                        19      19      19       26      31      31      31

     Freq.  Percent    Cum. |  Pattern
 ---------------------------+-------------------------------------
         1    33.33   33.33 |  ............1111111111111111111111
         1    33.33   66.67 |  1111111111111111111111111111.....
         1    33.33  100.00 |  1111111111111111111111111111111111
 ---------------------------+-------------------------------------
         3   100.00         |  XXXXXXXXXXXXXXXXXXXXXXXXXXXXXXXXXX
. by id: sum t2
```

```
-> id = 1
    Variable |      Obs        Mean    Std. Dev.       Min        Max
-------------+--------------------------------------------------------
          t2 |       19         477    5.627314        468        486
```

```
-> id = 2
    Variable |      Obs        Mean    Std. Dev.       Min        Max
-------------+--------------------------------------------------------
          t2 |       26       468.5    7.648529        456        481
```

```
-> id = 3
    Variable |      Obs        Mean    Std. Dev.       Min        Max
-------------+--------------------------------------------------------
          t2 |       31         471    9.092121        456        486
```

The above output from xtdes and summarize after the call to tsappend shows that the call to tsappend, add(6) causes six observations to be added to each panel and fills in the gap in the time variable in the second panel. As noted above, tsappend, add() did not cause a uniform end date over the panels.

The following output illustrates the contrast between tsappend, add() and tsappend, last() tsfmt() with panel data that ends at different dates. The output from xtdes and summarize shows that the call to tsappend, last() tsfmt() filled in the gap in t2 and caused all the panels to end at the specified end date. The output also shows that the panels remain unbalanced because one panel has a later entry date than the other two.

```
. use http://www.stata-press.com/data/r8/tsappend2
. tsappend, last(2000m7) tsfmt(m)
. xtdes
      id:  1, 2, ..., 3                              n =          3
      t2:  456, 457, ..., 486                        T =         31
           Delta(t2) = 1; (486-456)+1 = 31
           (id*t2 uniquely identifies each observation)
Distribution of T_i:    min      5%     25%     50%     75%     95%     max
                         19      19      19      31      31      31      31

     Freq.  Percent    Cum.  |  Pattern
  ----------------------------+---------------------------------
        2    66.67   66.67   |  1111111111111111111111111111111
        1    33.33  100.00   |  ............1111111111111111111
  ----------------------------+---------------------------------
        3   100.00           |  XXXXXXXXXXXXXXXXXXXXXXXXXXXXXXX
. by id: sum t2
```

```
-> id = 1
     Variable |       Obs        Mean    Std. Dev.       Min        Max
  ------------+--------------------------------------------------------
           t2 |        19         477    5.627314        468        486
```

```
-> id = 2
     Variable |       Obs        Mean    Std. Dev.       Min        Max
  ------------+--------------------------------------------------------
           t2 |        31         471    9.092121        456        486
```

```
-> id = 3
     Variable |       Obs        Mean    Std. Dev.       Min        Max
  ------------+--------------------------------------------------------
           t2 |        31         471    9.092121        456        486
```

Saved Results

tsappend saves in r():

Scalars
  e(add)   number of observations added

Methods and Formulas

tsappend is implemented as an ado-file.

Also See

Complementary:	[TS] **tsset**
Related:	[R] **egen**, [R] **generate**
Background:	[U] **14.4.3 Time-series varlists**

Title

tsreport — Report time-series aspects of dataset or estimation sample

Syntax

tsreport [if *exp*] [in *range*] [, report report0 list panel]

tsreport typed without options produces no output, but does provide its standard saved results.

Description

tsreport reports on time gaps in a sample of observations. A one-line statement displaying a count of the gaps is provided by the report option, and a full list of records that follow gaps is provided by the list option. A return value r(N_gaps) is set to the number of gaps in the sample.

Options

report specifies that a count of the number of gaps in the time series be reported, if any gaps exist.

report0 specifies that the count of gaps be reported even if there are no gaps.

list specifies that a tabular list of gaps be displayed.

panel specifies that panel changes are not to be counted as gaps. Whether panel changes are counted as gaps usually depends on how the calling command handles panels.

Remarks

Time-series commands sometimes require that observations be on a fixed time interval with no gaps, or that the behavior of the command might be different if the time series contains gaps. tsreport provides a tool for reporting the gaps in a sample.

▷ Example

The following monthly panel data have two panels and a missing month (March) in the second panel.

```
. list edlevel month income in 1/6, sep(0)
```

	edlevel	month	income
1.	1	1998m1	687
2.	1	1998m2	783
3.	1	1998m3	790
4.	2	1998m1	1435
5.	2	1998m2	1522
6.	2	1998m4	1532

Invoking `tsreport` without the `panel` option gives us the following report:

```
. tsreport, report
Number of gaps in sample:  2   (gap count includes panel changes)
```

We could get a list of gaps and better see what has been counted as a gap by using the `list` option:

```
. tsreport, report list
Number of gaps in sample:  2   (gap count includes panel changes)
Observations with preceding time gaps
(gaps include panel changes)
```

Record	edlevel	month
4	2	1998m1
6	2	1998m4

We now see why `tsreport` is reporting two gaps. It is counting the known gap in March of the second panel and also counting the change from the first to the second panel. (If we are programmers writing a procedure that does not account for panels, a change from one panel to the next represents a break in the time series just as a gap in the data does.)

We may prefer that the changes in panels not be counted as gaps. We obtain a count without the panel change by using the `panel` option:

```
. tsreport, report panel
Number of gaps in sample:  1
```

To obtain a fuller report, we type

```
. tsreport, report list panel
Number of gaps in sample:  1
Observations with preceding time gaps
```

Record	edlevel	month
6	2	1998m4

◁

Saved Results

`tsreport` saves in `r()`:

Scalars

`r(N_gaps)`   number of gaps in sample

Also See

Complementary:	[TS] **tsset**
Background:	[U] **14.4.3 Time-series varlists**,
	[U] **15.5.4 Time-series formats**,
	[U] **27.3 Time-series dates**,
	[U] **29.13 Models with time-series data**

Title

tsrevar — Time-series operator programming command

Syntax

tsrevar [*varlist*] [if *exp*] [in *range*] [, substitute list]

tsrevar is for use with time-series data. You must tsset your data before using tsrevar; see [TS] **tsset**.

Description

tsrevar, substitute takes a variable list that might contain *op.varname* combinations and substitutes real variables that are equivalent for the combinations. For instance, the original *varlist* might be "gnp L.gnp r", and tsrevar, substitute would create *newvar* = L.gnp and create the equivalent varlist "gnp *newvar* r". This new varlist could then be used with commands that do not otherwise support time-series operators, or it could be used in a program to make execution faster at the expense of using more memory.

tsrevar, substitute might create no new variables, one new variable, or many new variables, depending on the number of *op.varname* combinations appearing in *varlist*. Any new variables created are temporary variables. The new, equivalent varlist is returned in r(varlist). Note that the new varlist has a one-to-one correspondence with the original *varlist*.

tsrevar, list does something different. It returns in r(varlist) the list of base variable names of *varlist* with the time-series operators removed. tsrevar, list creates no new variables. For instance, if the original *varlist* were "gnp l.gnp l2.gnp r l.cd", then r(varlist) would contain "gnp r cd". This is useful for programmers who might want to create programs to keep just the variables corresponding to *varlist*.

Options

substitute specifies that tsrevar is to resolve *op.varname* combinations by creating temporary variables as described above. substitute is the default action taken by tsrevar; you do not need to specify the option.

list specifies that tsrevar is to return a list of base variable names.

▷ Example

```
. tsrevar l.gnp d.gnp r
```

creates two new temporary variables containing the values for l.gnp and d.gnp. The variable r appears in the new variable list, but does not require a temporary variable.

The resulting variable list is

```
. display "'r(varlist)'"
__00014P __00014Q r
```

We can see the results by listing the new variables alongside the original value of gnp.

```
. list gnp 'r(varlist)' in 1/5
```

	gnp	__00014P	__00014Q	r
1.	128	.	.	3.2
2.	135	128	7	3.8
3.	132	135	-3	2.6
4.	138	132	6	3.9
5.	145	138	7	4.2

Remember that temporary variables automatically vanish when the program concludes.

If we had needed only the base variable names, we could have specified

```
. tsrevar l.gnp d.gnp r, list
. display "'r(varlist)'"
gnp r
```

The ordering of the list will probably be different from the original list; base variables are listed only once and will be listed in the order in which they appear in the dataset.

◁

❏ Technical Note

tsrevar, substitute is smart and avoids creating duplicate variables. Consider

```
. tsrevar gnp l.gnp r cd l.cd l.gnp
```

Note that l.gnp appears twice in the varlist. tsrevar will create only one new variable for l.gnp and the use that new variable twice in the resulting r(varlist). Moreover, tsrevar will even do this across multiple calls:

```
. tsrevar gnp l.gnp cd l.cd
. tsrevar cpi l.gnp
```

Note that l.gnp appears in two separate calls. At the first call, tsrevar will create a temporary variable corresponding to l.gnp. At the second call, tsrevar will remember what it has done and use that same previously created temporary variable for l.gnp again.

❏

Saved Results

tsrevar saves in r():

Macros

    r(varlist)   the modified variable list or list of base variable names

Also See

Related: [P] **syntax**, [P] **unab**

Background: [U] **14 Language syntax**,
                       [U] **14.4.3 Time-series varlists**,
                       [U] **21 Programming Stata**

Title

> **tsset** — Declare dataset to be time-series data

Syntax

tsset [*panelvar*] *timevar* [, format(%*fmt*)

  [daily | weekly | monthly | quarterly | halfyearly | yearly | generic]]

tsset

tsset, clear

tsfill [, full]

where *panelvar* is a variable that identifies the panels and *timevar* is a variable that identifies the time periods.

Description

tsset declares the data a time series and designates that *timevar* represents time. *timevar* must take on integer values. If *panelvar* is also specified, the dataset is declared to be a cross section of time series (e.g., time series of different countries). tsset must be used before time-series operators may be used in expressions and varlists. After tsset, the data will be sorted on *timevar* or on *panelvar timevar*.

If you have annual data with the variable year representing time, using tsset is as simple as typing

    . tsset year, yearly

although you could omit the yearly option because it affects only how results are displayed.

Note that if you are using one of the xt commands that require the data be tsset, specifying tsset will override any settings previously specified by iis and tis.

tsset without arguments displays how the dataset is currently tsset and re-sorts the data on *timevar* or on *panelvar timevar* if it is sorted differently from that.

tsset, clear is a rarely used programmer's command to declare that the data are no longer a time series.

tsfill is used after tsset to fill in missing times with missing observations. For instance, perhaps observations for *timevar* = 1, 3, 5, 6, ..., 22 exist. tsfill would create observations for *timevar* = 2 and *timevar* = 4 containing all missing values. There is seldom reason to do this because Stata's time-series operators work on the basis of *timevar* and not on observation number. Referring to L.gnp to obtain lagged gnp values would correctly produce a missing value for *timevar* = 3 even if the data were not filled in. Referring to L2.gnp would correctly return the value of gnp in the first observation for *timevar* = 3 even if the data were not filled in.

Options

format(%*fmt*), daily, weekly, monthly, quarterly, halfyearly, yearly, and generic deal with how *timevar* will be subsequently displayed—which %t format, if any, will be placed on *timevar*. Whether *timevar* is formatted is optional; all tsset requires is that *timevar* take on integer values.

The optional format states how *timevar* is measured (for instance, *timevar* = 5 might mean the fifth day of January 1960, or the fifth week of January 1960, or the fifth something else) and how you want it displayed (*timevar* = 5 might be displayed 05jan1960, 1960–5, or in a host of other ways). In addition, the format makes the `tin()` and `twithin()` selection functions work so that later, after `tsset`ing the data, you can type things like `regress ... if tin(1jan1998,1apr1998)` to run the regression on the subsample 1jan1998 ≤ *timevar* ≤ 1apr1998.

Formatting is accomplished by placing a `%t` format on *timevar*. You can do it yourself or you can ask `tsset` to do it.

The time scales `%t` understands are daily (`%td`, 0 = 1jan1960), weekly (`%tw`, 0 = 1960w1), monthly (`%tm`, 0 = 1960m1), quarterly (`%tq`, 0 = 1960q1), halfyearly (`%th`, 0 = 1960h1), yearly (`%ty`, 1960 = 1960), and generic (`%tg`, 0 = ?).

Say *timevar* is recorded in Stata-quarterly units, meaning 0 = 1960q1, 1 = 1960q2, etc. (Perhaps your data start in 1990q1; then the first observation has *timevar* = 120.) You could format *timevar* and then `tsset` your data,

    . format *timevar* %tq
    . tsset *timevar*

or you could `tsset` your data and then format *timevar*,

    . tsset *timevar*
    . format *timevar* %tq

or you could `tsset` your data specifying `tsset`'s `format()` option,

    . tsset *timevar*, format(%tq)

or you could `tsset` your data specifying `tsset`'s `quarterly` option,

    . tsset *timevar*, quarterly

These alternatives yield the same result; use whichever appeals to you. See [U] **27.3 Time-series dates** and [U] **15.5.4 Time-series formats** for more information on the `%t` format and the advantages of setting it.

`clear` is for use with `tsset`; `tsset, clear` is a rarely used programmer's command to declare that the data are no longer a time series.

`full` is for use with `tsfill` and affects the outcome only if a *panelvar* has been previously `tsset`.

By default, with panel data, `tsfill` fills in observations for each panel according to the minimum and maximum values of *timevar* for the panel. Thus, if the first panel spanned the times 5–20 and the second panel the times 1–15, after `tsfill`, they would still span the same time periods; observations would be created to fill in any missing times from 5 through 20 in the first panel and from 1 through 15 in the second.

If `full` is specified, observations will be created so that both panels span the period 1 through 20, the overall minimum and maximum of *timevar* across panels.

Remarks

`tsset` sets *timevar* so that Stata's time-series operators are understood in varlists and expressions. The time-series operators are

operator	meaning
L.	lag x_{t-1}
L2.	2-period lag x_{t-2}
...	
F.	lead x_{t+1}
F2.	2-period lead x_{t+2}
...	
D.	difference $x_t - x_{t-1}$
D2.	difference of difference $x_t - x_{t-1} - (x_{t-1} - x_{t-2}) = x_t - 2x_{t-1} + x_{t-2}$
...	
S.	"seasonal" difference $x_t - x_{t-1}$
S2.	lag-2 (seasonal) difference $x_t - x_{t-2}$
...	

Time-series operators may be repeated and combined. L3.gnp refers to the third lag of variable gnp, and so do LLL.gnp, LL2.gnp, and L2L.gnp. LF.gnp is the same as gnp. DS12.gnp refers to the one-period difference of the 12-period difference. LDS12.gnp refers to the same concept, lagged once.

Note that D1. = S1. but D2. ≠ S2., D3. ≠ S3., and so on. D2. refers to the difference of the difference. S2. refers to the two-period difference. If you wanted the difference of the difference of the 12-period difference of gnp, you would write D2S12.gnp.

Operators may be typed in uppercase or lowercase. Most users would type d2s12.gnp instead of D2S12.gnp.

You may type operators however you wish; Stata internally converts operators to their canonical form. If you typed 1d21s12d.gnp, Stata would present the operated variable as L2D3S12.gnp.

In addition to *operator#*, Stata understands *operator(numlist)* to mean a set of operated variables. For instance, typing L(1/3).gnp in a varlist is the same as typing 'L.gnp L2.gnp L3.gnp'. The operators can also be applied to a list of variables by enclosing the variables in parentheses; e.g.,

```
. list year L(1/3).(gnp cpi)
```

	year	L.gnp	L2.gnp	L3.gnp	L.cpi	L2.cpi	L3.cpi
1.	1989	.	.	.	.	.	.
2.	1990	5452.8	.	.	100	.	.
3.	1991	5764.9	5452.8	.	105	100	.
4.	1992	5932.4	5764.9	5452.8	108	105	100
			(output omitted)				
8.	1996	7330.1	6892.2	6519.1	122	119	112

In *operator#*, making # zero returns the variable itself. L0.gnp is gnp. Thus, the listing above could have been produced by typing list year 1(0/3).gnp.

The parentheses notation may be used with any operator. Typing D(1/3).gnp would return the first through third differences.

The parentheses notation may be used in operator lists with multiple operators, such as L(0/3)D2S12.gnp.

Operator lists may include up to one set of parentheses, and the parentheses may enclose a *numlist*; see [U] **14.1.8 numlist**.

Before you can use these time-series operators, there are two requirements the dataset must satisfy:

1. the dataset must be `tsset`, and

2. the dataset must be sorted by *timevar* or, if it is a cross-sectional time-series dataset, by *panelvar timevar*.

`tsset` handles both requirements. As you use Stata, however, you may later use a command that re-sorts that data, and, if you do, the time-series operators will refuse to work:

```
. tsset time
(output omitted)
. regress y x l.x
(output omitted)
. (you continue to use Stata and, sometime later:)
. regress y x l.x
not sorted
r(5);
```

In that case, typing `tsset` without arguments will re-establish the sort order:

```
. tsset
(output omitted)
. regress y x l.x
(output omitted)
```

In this case, typing `tsset` is the same as typing `sort time`. Had we previously `tsset country time`, however, typing `tsset` would be the same as typing `sort country time`. You can type the `sort` command or type `tsset` without arguments; it makes no difference.

▷ Example

You have monthly data on personal income. Variable `month` records the time of an observation:

```
. list t income
```

	t	income
1.	1	1153
2.	2	1181
	(output omitted)	
9.	9	1282

```
. tsset t
        time variable:  t, 1 to 9
. regress income l.income
(output omitted)
```

◁

▷ Example

In the example above, it is not important that t start at 1. The t variable could just as well be recorded 21, 22, ..., 29, or 426, 427, ..., 434, or any other way we liked. What is important is that the difference in t between observations when there are no gaps is 1.

Although how time is measured makes no difference, Stata has formats to display time prettily if it is recorded in certain ways. In particular, Stata likes time variables with 1jan1960 recorded as 0. In our example above, if our first observation is July 1995, such that t = 1 corresponds to July 1995, then we could make a time variable that fits Stata's preference by typing

```
. generate newt = m(1995m7) + t - 1
```

m() is the function that returns month equivalent; m(1995m6) evaluates to the constant 425, meaning 425 months after January 1960. We now have variable newt containing

```
. list t newt income
```

	t	newt	income
1.	1	426	1153
2.	2	427	1181
3.	3	428	1208
	(output omitted)		
9.	9	434	1282

If we put a %tm format on newt, it will list more prettily:

```
. format newt %tm
```

	t	newt	income
1.	1	1995m7	1153
2.	2	1995m8	1181
3.	3	1995m9	1208
	(output omitted)		
9.	9	1996m3	1282

We could now tsset newt rather than t:

```
. tsset newt
        time variable:  newt, 1995m7 to 1996m3
```

◁

▷ Example

Perhaps we have the same time-series data, but no time variable:

```
. list income
```

	income
1.	1153
2.	1181
3.	1208
4.	1272
5.	1236
6.	1297
7.	1265
8.	1230
9.	1282

Pretend we know that the first observation corresponds to July 1995 and continues without gaps. We can create a monthly time variable and format it by typing

```
. generate t = m(1995m7) + _n - 1
. format t %tm
```

We can now `tsset` our dataset and `list` it:

```
. tsset t
        time variable:  t, 1995m7 to 1996m3

. list t income
```

	t	income
1.	1995m7	1153
2.	1995m8	1181
3.	1995m9	1208
	(output omitted)	
9.	1996m3	1282

◁

❏ Technical Note

Your data do not have to be monthly. Stata understands daily, weekly, monthly, quarterly, halfyearly, and yearly data. Correspondingly, there are the `d()`, `w()`, `m()`, `q()`, `h()`, and `y()` functions, and there are the `%td`, `%tw`, `%tm`, `%tq`, `%th`, and `%ty` formats. Here is what we would have typed in the above examples had our data been on a different time scale:

Daily: pretend your t variable had t=1 corresponding to 15mar1993
```
. gen newt = d(15mar1993) + t - 1
. format newt %td
. tsset newt
```

Weekly: pretend your t variable had t=1 corresponding to 1994w1:
```
. gen newt = w(1994w1) + t - 1
. format newt %tw
. tsset newt
```

Monthly: pretend your t variable had t=1 corresponding to 2004m7:
```
. gen newt = m(2004m7) + t - 1
. format newt %tm
. tsset newt
```

Quarterly: pretend your t variable had t=1 corresponding to 1994q1:
```
. gen newt = q(1994q1) + t - 1
. format newt %tq
. tsset newt
```

Halfyearly: pretend your t variable had t=1 corresponding to 1921h2:
```
. gen newt = h(1921h2) + t - 1
. format newt %th
. tsset newt
```

Yearly: pretend your t variable had t=1 corresponding to 1842:
```
. gen newt = y(1842) + t - 1
. format newt %ty
. tsset newt
```

In each of the above examples, we subtracted one from our time variable in constructing the new time variable `newt` because we assumed that our starting time value was 1. For the quarterly example, if our starting time value had been 5 and that corresponded to 1994q1, we would have typed

```
. generate newt = q(1994q1) + t - 5
```

Had our initial time value been $t = 742$ and this corresponded to 1994q1, we would have typed

```
. generate newt = q(1994q1) + t - 742
```

The %td, %tw, %tm, %tq, %th, and %ty formats can display the date in the form you want; when you type, for instance, %td, you are specify the *default* daily format, which produces dates of the form 15apr2002. If you wanted that to be displayed as "April 15, 2002", %td can do that; see [U] **27.3 Time-series dates**. Similarly, all the other %t formats can be modified to produce the results you want.

❑

▷ Example

Your data might include a time variable that is encoded into a string. Below, each monthly observation is identified by string variable yrmo containing the year and month of the observation, sometimes with punctuation in between:

```
. list yrmo income
```

	yrmo	income
1.	1995 7	1153
2.	1995 8	1181
3.	1995,9	1208
4.	1995 10	1272
5.	1995/11	1236
6.	1995,12	1297
7.	1996-1	1265
8.	1996.2	1230
9.	1996 Mar	1282

The first step is to convert the string to a numeric representation. That is easy using the monthly() function; see [U] **27.3 Time-series dates**.

```
. gen mdate = monthly(yrmo, "ym")
(1 missing value generated)
. list yrmo mdate income
```

	yrmo	mdate	income
1.	1995 7	426	1153
2.	1995 8	427	1181
3.	1995,9	428	1208
	(output omitted)		
9.	1996 Mar	434	1282

Our new variable, mdate, contains the number of months from January, 1960. Now that we have numeric variable mdate, we can tsset the data:

```
. format mdate %tm
. tsset mdate
        time variable:  mdate, 1995m7 to 1996m3
```

In fact, we can combine the two and type

```
. tsset mdate, format(%tm)
        time variable:  mdate, 1995m7 to 1996m3
```

or type

```
. tsset mdate, monthly
        time variable:  mdate, 1995m7 to 1996m3
```

We do not have to bother to format the time variable at all, but formatting makes it display more prettily:

```
. list yrmo mdate income
```

	yrmo	mdate	income
1.	1995 7	1995m7	1153
2.	1995 8	1995m8	1181
3.	1995,9	1995m9	1208
4.	1995 10	1995m10	1272
5.	1995/11	1995m11	1236
6.	1995,12	1995m12	1297
7.	1996-1	1996m1	1265
8.	1996.2	1996m2	1230
9.	1996 Mar	1996m3	1282

◁

❑ Technical Note

In addition to the monthly() function for translating strings to monthly dates, Stata has daily(), weekly(), quarterly(), halfyearly(), and yearly(). Stata also has the yw(), ym(), yq(), and yh() functions to convert from two numeric time variables to a Stata time variable. For example, generate qdate = yq(year,qtr) takes the variable year containing year values and the variable qtr containing quarter values (1–4), and produces the variable qdate containing the number of quarters since 1960q1. See [U] **27.3 Time-series dates**.

❑

▷ Example

Gaps in the time series cause no difficulties:

```
. list yrmo income
```

	yrmo	income
1.	1995 7	1153
2.	1995 8	1181
3.	1995/11	1236
4.	1995,12	1297
5.	1996-1	1265
6.	1996 Mar	1282

```
. gen mdate = monthly(yrmo, "ym")
(1 missing value generated)
. tsset mdate, monthly
        time variable:  mdate, 1995m7 to 1996m1, but with a gap
```

Once the dataset has been tsset, we can use the time-series operators. The D operator specifies first (or higher order) differences:

```
. list mdate income d.income
```

	mdate	income	D.income
1.	1995m7	1153	.
2.	1995m8	1181	28
3.	1995m11	1236	.
4.	1995m12	1297	61
5.	1996m1	1265	-32
6.	1996m3	1282	.

You can use the operators in an expression or varlist context; you do not have to create a new variable to hold D.income. You can use D.income with the list command, or with the regress command, or with any other Stata command that allows time-series varlists.

◁

▷ Example

We stated above that gaps were no problem, and that is true as far as operators are concerned. You might, however, need to fill in the gaps for some analysis, say, by interpolation. This is easy to do with tsfill and ipolate. tsfill will create the missing observations, and then ipolate (see [R] **ipolate**) will fill them in. Staying with the example above, we can fill in the time series by typing

```
. tsfill
. list mdate income ipinc
```

	mdate	income	
1.	1995m7	1153	
2.	1995m8	1181	
3.	1995m9	.	← new
4.	1995m10	.	← new
5.	1995m11	1236	
6.	1995m12	1297	
7.	1996m1	1265	
8.	1996m2	.	← new
9.	1996m3	1282	

We listed the data after tsfill just to show you the role tsfill plays in this. tsfill created the observations. We can now use ipolate to fill them in:

```
. ipolate income mdate, gen(ipinc)
```

(*Continued on next page*)

. list mdate income ipinc

	mdate	income	ipinc
1.	1995m7	1153	1153
2.	1995m8	1181	1181
3.	1995m9	.	1199.333
4.	1995m10	.	1217.667
5.	1995m11	1236	1236
6.	1995m12	1297	1297
7.	1996m1	1265	1265
8.	1996m2	.	1273.5
9.	1996m3	1282	1282

◁

Panel data

▷ Example

Now, let us assume that we have time series on annual income and that we have the series for two groups: individuals who have not completed high school (edlevel = 1) and individuals who have (edlevel = 2).

. list edlevel year income, sep(0)

	edlevel	year	income
1.	1	1988	14500
2.	1	1989	14750
3.	1	1990	14950
4.	1	1991	15100
5.	2	1989	22100
6.	2	1990	22200
7.	2	1992	22800

We declare the data to be a panel by typing

```
. tsset edlevel year, yearly
        panel variable:  edlevel, 1 to 2
         time variable:  year, 1988 to 1992, but with a gap
```

Having tsset the data, we can now use time-series operators. The difference operator, for example, can be used to list annual changes in income:

. list edlevel year income d.income, sep(0)

	edlevel	year	income	D.income
1.	1	1988	14500	.
2.	1	1989	14750	250
3.	1	1990	14950	200
4.	1	1991	15100	150
5.	2	1989	22100	.
6.	2	1990	22200	100
7.	2	1992	22800	.

We see that in addition to producing missing values due to missing times, the difference operator correctly produced a missing value at the start of each panel. Once we have `tsset` our panel data, we can use time-series operators and be assured that they will handle missing time periods and panel changes correctly.

◁

▷ Example

As with nonpanel time series, we can use `tsfill` to fill in gaps in a panel time series. Continuing with our example data,

```
. tsfill
. list edlevel year income
```

	edlevel	year	income	
1.	1	1988	14500	
2.	1	1989	14750	
3.	1	1990	14950	
4.	1	1991	15100	
5.	2	1989	22100	
6.	2	1990	22200	
7.	2	1991	.	← new
8.	2	1992	22800	

We could instead ask `tsfill` to produce fully balanced panels using the `full` option:

```
. tsfill, full
. list edlevel year income, sep(0)
```

	edlevel	year	income	
1.	1	1988	14500	
2.	1	1989	14750	
3.	1	1990	14950	
4.	1	1991	15100	
5.	1	1992	.	← new
6.	2	1988	.	← new
7.	2	1989	22100	
8.	2	1990	22200	
9.	2	1991	.	← new
10.	2	1992	22800	

◁

Saved Results

tsset saves in r():

Scalars
r(tmin)	minimum elapsed time
r(tmax)	maximum elapsed time
r(imin)	minimum panel id
r(imax)	maximum panel id

Macros
r(timevar)	elapsed time variable
r(panvar)	panel variable
r(tmins)	formatted minimum elapsed time
r(tmaxs)	formatted maximum elapsed time
r(tsfmt)	format for the current time variable
r(unit)	daily, weekly, monthly, quarterly, halfyearly, yearly, or generic
r(unit1)	d, w, m, q, h, y, or ""

Methods and Formulas

tsset is implemented as an ado-file.

References

Baum, C. F. 2000. sts17: Compacting time series data. *Stata Technical Bulletin* 57: 44–45. Reprinted in *Stata Technical Bulletin Reprints*, vol 10, pp. 369–370.

Also See

Background:	[U] **14.4.3 Time-series varlists**,
	[U] **15.5.4 Time-series formats**,
	[U] **27.3 Time-series dates**,
	[U] **29.13 Models with time-series data**

Title

> **tssmooth** — Smooth and forecast univariate time-series data

Syntax

Moving average filter with uniform weights

> tssmooth ma ... , $\underline{w}$indow($\#_l\left[\#_c\left[\#_f\right]\right]$) ...

Moving average filter with specified weights

> tssmooth ma ... , $\underline{we}$ights($\left[\textit{numlist}_l\right]$ <$\#_c$> $\left[\textit{numlist}_f\right]$) ...

Single exponential

> tssmooth $\underline{e}$xponential ...

Double exponential

> tssmooth $\underline{d}$exponential ...

Nonseasonal Holt–Winters

> tssmooth $\underline{h}$winters ...

Seasonal Holt–Winters

> tssmooth $\underline{s}$hwinters ...

Nonlinear filter

> tssmooth nl ...

tssmooth is for use with time-series data; see [TS] **tsset**. You must tsset your data before using tssmooth.

See [TS] **tssmooth ma**, [TS] **tssmooth exponential**, [TS] **tssmooth dexponential**, [TS] **tssmooth hwinters**, [TS] **tssmooth shwinters**, and [TS] **tssmooth nl** for details of syntax.

Description

Each of the smoothers in tssmooth generates a new variable by passing the specified expression, usually just a variable name, through the requested filter. The available smoothers can be organized into three groups. The ma smoothers are moving-average filters. The exponential, double exponential, nonseasonal Holt–Winters, and seasonal Holt–Winters smoothers are all based on simple recursions from initial values, and thus form a group of recursive smoothers. The nonlinear smoother constitutes the third group.

All these methods can be used for smoothing, and all of the recursive smoothers can be used for forecasting univariate time series. The single exponential, double exponential, and the two Holt–Winters smoothers can all perform dynamic out-of-sample forecasts. The moving average and nonlinear smoothers are generally used to extract the trend, or signal, from a time series while omitting the high-frequency or noise components. The exponential and double exponential methods are used both as smoothers and as forecasting methods. The Holt–Winters methods are used almost exclusively as forecasting methods.

All of the methods except the nonlinear smoother will automatically work with panel data or missing data. When there is panel data, the data for each panel is treated as if it were itself a time series. In the moving-average smoothers, missing data is handled by placing a weight of zero on missing observations. The recursive smoothers handle missing data by substituting predicted values for any missing observations.

For each of the recursive smoothers, the smoothing parameters may be specified or chosen to minimize the in-sample sum of squared prediction errors. When there is panel data, the optimization is performed separately for each panel.

There are several texts that provide good introductions to the methods available in tssmooth. Chatfield (1996) provides a good discussion of how these methods fit into time-series analysis in general. Abraham and Ledolter (1983); Montgomery, Johnson, and Gardiner (1990); Bowerman and O'Connell (1993); and Chatfield (2001) discuss these methods in the context of modern time-series forecasting methods.

References

Abraham, B. and J. Ledolter. 1983. *Statistical Methods for Forecasting.* New York: John Wiley & Sons.

Bowerman, B. and R. O'Connell. 1993. *Forecasting and Time Series: An Applied Approach.* 3d ed. Pacific Grove, CA: Duxbury Press.

Chatfield, C. 1996. *The Analysis of Time Series: An Introduction.* 5th ed. London: Chapman & Hall.

——. 2001. *Time–Series Forecasting.* London: Chapman & Hall.

Chatfield, C. and M. Yar. 1988. Holt–Winters forecasting: some practical issues. *The Statistician* 37: 129–140.

Montgomery, D. C., L. A. Johnson, and J. S. Gardiner. 1990. *Forecasting and Time Series Analysis.* 2d ed. New York: McGraw–Hill.

Also See

Complementary:	[TS] **tsset**, [TS] **tssmooth dexponential**, [TS] **tssmooth exponential**, [TS] **tssmooth hwinters**, [TS] **tssmooth ma**, [TS] **tssmooth nl**, [TS] **tssmooth shwinters**
Related:	[TS] **arima**, [R] **egen**, [R] **generate**
Background:	[U] **14.4.3 Time-series varlists**

Title

> **tssmooth dexponential** — Double exponential smoothing

Syntax

> tssmooth dexponential $[type]$ *newvarname* = *exp* $[if\ exp]$ $[in\ range]$
>
> $[$, parms$(\#_\alpha)$ samp0$(\#)$ s0$(\#_1\ \#_2)$ forecast$(\#)$ replace $]$

tssmooth dexponential is for use with time-series data or cross-sectional time-series data (panel data). You must tsset your data before using tssmooth; see [TS] **tsset**.

exp may contain time series operators; see [U] **14.4.3 Time-series varlists**.

Description

tssmooth dexponential performs double exponential smoothing on a user-specified expression, which is usually just a variable name, and generates a new variable containing the filtered series.

Options

parms$(\#_\alpha)$ specifies the parameter α for the double exponential smoothers. The parameter α must be in $(0, 1)$.

samp0$(\#)$ specifies the sample to be used to obtain the initial value(s) for the recursions. The integer in samp0() is interpreted as the number of observations to be used. The default is to use one-half of the number of observations. samp0$(\#)$ cannot be specified with s0$(\#_1\ \#_2)$.

s0$(\#_1\ \#_2)$ specifies the initial values for the recursions. By default, the two initial values specified in s0$(\#_1\ \#_2)$ are obtained from the data as discussed in *Methods and Formulas*.

forecast$(\#)$ specifies the number of periods for the out-of-sample prediction. $0 \le \# \le 500$. The default is 0, which is equivalent to not performing an out-of-sample forecast.

replace causes the variable specified in *newvarname* to be replaced, if it already exists.

Remarks

The double exponential smoothing procedure is designed for series that can be locally approximated as

$$\widehat{x}_t = m_t + b_t t$$

where $\widehat{x}_t$ is the smoothed or predicted value of the series x and the terms m_t and b_t change over time. Abraham and Ledolter (1983), Bowerman and O'Connell (1993), and Montgomery et al. (1990) all provide good introductions to double exponential smoothing. Chatfield (2000, 2001) provides helpful discussions of how double exponential smoothing relates to modern time-series methods.

The double exponential method has been used both as a smoother and as a prediction method. Recall from [TS] **tssmooth exponential** that the single exponential smoothed series is given by

$$S_t = \alpha x_t + (1 - \alpha)S_{t-1}$$

where α is the smoothing constant and x_t is the original series. The double exponential smoother is obtained by smoothing the smoothed series,

$$S_t^{[2]} = \alpha S_t + (1 - \alpha)S_{t-1}^{[2]}$$

Values of S_0 and $S_0^{[2]}$ are necessary to begin the process. Following Montgomery et al. (1990), the default method is to obtain S_0 and $S_0^{[2]}$ from a regression of the first N_{pre} values of x_t on $\tilde{t} = (1, \ldots, N_{\mathrm{pre}} - t_0)'$. By default, N_{pre} is equal to one half the number of observations in the sample. N_{pre} can be specified using the samp0() option.

Alternatively, the values of S_0 and $S_0^{[2]}$ can be specified using the option s0().

▷ Example

Suppose that we had some data on the monthly sales of a book, and that we wanted to smooth this series. The graph below illustrates that this series is locally trending over time, so we would not want to use single exponential smoothing.

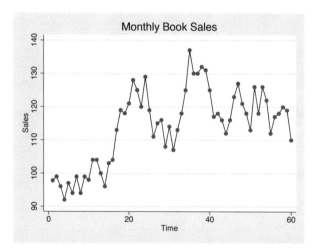

The following illustrates that double exponential smoothing is simply smoothing the smoothed series. Because the starting values are treated as time-zero values, we actually lose two observations when smoothing the smoothed series.

```
. use http://www.stata-press.com/data/r8/sales2

. tssmooth exponential double sm1=sales, p(.7) s0(1031)
exponential coefficient   =      0.7000
sum of squared residuals  =       13923
root mean squared error   =      13.192

. tssmooth exponential double sm2=sm1, p(.7) s0(1031)
exponential coefficient   =      0.7000
sum of squared residuals  =      7698.6
root mean squared error   =      9.8098
```

```
. tssmooth dexponential double sm2b=sales, p(.7) s0(1031 1031)
double exponential coefficient  =        0.7000
sum of squared residuals        =        3724.4
root mean squared error         =        6.8231

. generate sm2c = f2.sm2
(2 missing values generated)

. list sm2 sm2c in 1/10
```

	sm2	sm2c
1.	1031	1031
2.	1031	1028.383
3.	1031	1030.631
4.	1028.3834	1017.810
5.	1030.6306	1022.938
6.	1017.8182	1026.075
7.	1022.938	1041.859
8.	1026.0752	1042.834
9.	1041.8587	1035.957
10.	1042.8341	1030.665

◁

The double exponential method can also be viewed as a forecasting mechanism. The exponential forecast method is a constrained version of the Holt–Winters method implemented in [TS] **tssmooth hwinters** (as discussed by Gardner (1985) and Chatfield (2001)). Chatfield (2001) also notes that the double exponential method arises when the underlying model is an ARIMA(0,2,2) with equal roots.

This method produces predictions $\widehat{x}_t$ for $t = t_1, \ldots, T + \texttt{forecast()}$. These predictions are obtained as a function of the smoothed series and the smoothed-smoothed series. For $t \in [t_0, T]$,

$$\widehat{x}_t = \left(2 + \frac{\alpha}{1-\alpha}\right) S_t - \left(1 + \frac{\alpha}{1-\alpha}\right) S_t^{[2]}$$

where S_t and $S_t^{[2]}$ are as given above.

The out-of-sample predictions are obtained as a function of the constant term, the linear term of the smoothed series at the last observation in the sample, and time. The constant term is $a_T = 2S_T - S_T^{[2]}$, and the linear term is $b_T = \frac{\alpha}{1-\alpha}(S_T - S_T^{[2]})$. The τth step-ahead out-of-sample prediction is given by

$$\widehat{x}_t = a_t + \tau b_T$$

▷ Example

Specifying the `forecast` option will cause Stata to put the double exponential forecast into the new variable instead of the double exponential smoothed series. The code given below uses the smoothed series `sm1` and `sm2` that were generated above to illustrate how the double exponential forecasts are computed.

```
. tssmooth dexponential double f1=sales, p(.7) s0(1031 1031) forecast(4)
double exponential coefficient   =      0.7000
sum of squared residuals         =        20737
root mean squared error          =         16.1
. generate double xhat = (2+ .7/.3)* sm1 - ( 1 + .7/.3)* f.sm2
(5 missing values generated)
. list xhat f1 in 1/10
```

	xhat	f1
1.	1031	1031
2.	1031	1031
3.	1023.524	1023.524
4.	1034.8039	1034.8039
5.	994.0237	994.0237
6.	1032.4463	1032.4463
7.	1031.9015	1031.9015
8.	1071.1709	1071.1709
9.	1044.6454	1044.6454
10.	1023.1855	1023.1855

◁

▷ Example

Generally, when one is forecasting, the smoothing parameter is unknown. tssmooth dexponential will compute the double exponential forecasts of a series and obtain the optimal smoothing parameter by finding the smoothing parameter that minimizes the in-sample sum of squared forecast errors.

```
. tssmooth dexponential f2=sales, forecast(4)
computing optimal double exponential coefficient (0,1)

optimal double exponential coefficient =      0.3631
sum of squared residuals             =    16075.805
root mean squared error              =    14.175598
```

The following graph describes the fit that we obtained by applying the double exponential forecast method to our sales data. Note that the out-of-sample dynamic predictions are not constant, as in the single exponential case.

(Continued on next page)

```
. twoway connected f2 sales t,
> title("Double Exponential Forecast with Optimal alpha")
> ytitle(Sales) xtitle(time)
```

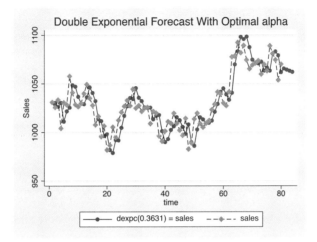

◁

tssmooth dexponential automatically detects panel data from the information provided when the dataset was tsset. The starting values are chosen separately for each series. If the smoothing parameter is chosen to minimize the sum of squared prediction errors, then the optimization is also performed separately on each panel. The saved results will contain the results from the last panel. Missing values at the beginning of the sample are excluded from the sample. After at least one value has been found, missing values are filled in using the one-step predictions from the previous period.

Saved Results

tssmooth saves in r():

Scalars
r(N)	number of observations
r(alpha)	α smoothing parameter
r(rss)	sum of squared errors
r(rmse)	root mean squared error
r(N_pre)	number of observations used in calculating starting values, if starting values calculated
r(s2_0)	initial value for linear term, i.e., $S_0^{[2]}$
r(s1_0)	initial value for constant term, i.e., S_0
r(linear)	final value of linear term
r(constant)	final value of constant term
r(period)	period, if filter is seasonal

Macros
r(exp)	expression specified
r(timevar)	time variables specified in tsset
r(panelvar)	panel variables specified in tsset

Methods and Formulas

tssmooth dexponential is implemented as an ado-file.

A truncated description of the specified double exponential filter is used to label the new variable. See [R] **label** for more information on labels.

An untruncated description of the specified double exponential filter is saved in the characteristic tssmooth for the new variable. See [P] **char** for more information on characteristics.

The updating equations for the smoothing and forecasting versions are as given previously.

The starting values for both the smoothing and forecasting versions of double exponential are obtained using the same method. The method begins with the model

$$x_t = \beta_0 + \beta_1 t$$

where x_t is the series to be smoothed and t is a time variable that has been normalized to equal 1 in the first period included in the sample. The regression coefficient estimates $\widehat{\beta}_0$ and $\widehat{\beta}_1$ are obtained via OLS. The sample is determined by the option samp0(). By default, samp0() includes the first half of the observations. Given the estimates $\widehat{\beta}_0$ and $\widehat{\beta}_1$, the starting values are

$$S_0 = \widehat{\beta}_0 - \{(1 - \alpha)/\alpha\}\widehat{\beta}_1$$
$$S_0^{[2]} = \widehat{\beta}_0 - 2\{(1 - \alpha)/\alpha\}\widehat{\beta}_1$$

References

Abraham, B. and J. Ledolter. 1983. *Statistical Methods for Forecasting*. New York: John Wiley & Sons.

Bowerman, B. and R. O'Connell. 1993. *Forecasting and Time Series: An Applied Approach*. 3d ed. Pacific Grove, CA: Duxbury Press.

Chatfield, C. 1996. *The Analysis of Time Series: An Introduction*. 5th ed. London: Chapman & Hall.

——. 2001. *Time–Series Forecasting*. London: Chapman & Hall.

Chatfield, C. and M. Yar. 1988. Holt–Winters Forecasting: some practical issues. *The Statistician* 37: 129–140.

Gardner, E. S., Jr. 1985. Exponential Smoothing: The State of the Art. *Journal of Forecasting* 4: 1–28.

Montgomery, D. C., L. A. Johnson, and J. S. Gardiner. 1990. *Forecasting and Time Series Analysis*. 2d ed. New York: McGraw–Hill.

Also See

Complementary:	[TS] **tsset**
Related:	[TS] **arima**, [TS] **tssmooth exponential**, [TS] **tssmooth hwinters**,
	[TS] **tssmooth ma**, [TS] **tssmooth nl**, [TS] **tssmooth shwinters**,
	[R] **egen**, [R] **generate**
Background:	[U] **14.4.3 Time-series varlists**,
	[TS] **tssmooth**

Title

tssmooth exponential — Exponential smoothing

Syntax

tssmooth $\underline{e}$xponential $[type]$ *newvarname* = *exp* $[\text{if } exp]$ $[\text{in } range]$

$\Big[$, $\underline{p}$arms($\#_\alpha$) $\underline{samp}$0($\#$) s0($\#$) $\underline{f}$orecast($\#$) replace $\Big]$

tssmooth exponential is for use with time-series data or cross-sectional time-series data (panel data). You must tsset your data before using tssmooth; see [TS] **tsset**.

exp may contain time-series operators; see [U] **14.4.3 Time-series varlists**

Description

tssmooth exponential performs single exponential smoothing on a user-specified expression, which is usually just a variable name, and generates a new variable containing the filtered series.

Options

parms($\#_\alpha$) specifies the parameter α for the exponential smoother. The parameter α must be in $(0, 1)$. If parms($\#_\alpha$) is not specified, the smoothing parameter is chosen to minimize the in-sample sum of squared forecast errors.

samp0($\#$) specifies the sample to be used to obtain the initial value(s) for the recursions. The integer in samp0($\#$) specifies the number of observations to be used. The default is to use one-half of the number of observations. samp0($\#$) may not be specified with s0($\#$).

s0($\#$) specifies the initial value for the recursions. By default, the initial value is set to the mean of *exp* over the first half of the sample. s0($\#$) may not be specified with samp0($\#$).

forecast($\#$) gives the number of periods for the out-of-sample prediction, where $0 \le \# \le 500$. The default value is 0, and is equivalent to not forecasting out-of-sample.

replace causes the variable specified in *newvarname* to be replaced, if it already exists.

Remarks

Exponential smoothing can be viewed either as an adaptive forecasting algorithm, or, equivalently, as a geometrically weighted moving-average filter. Exponential smoothing is most appropriate when used with time-series data that exhibit no linear or higher order trends, but that do exhibit low velocity, aperiodic variation in the mean. Abraham and Ledolter (1983), Bowerman and O'Connell (1993), and Montgomery et al. (1990) all provide good introductions to single exponential smoothing. Chatfield (1996, 2001) provide helpful discussions of how single exponential smoothing relates to modern time-series methods. For example, it can be shown that simple exponential smoothing produces optimal forecasts for several underlying models, including ARIMA(0,1,1) and the random walk plus noise state-space model. (See Chatfield (2001, section 4.3.1) for a discussion.)

The exponential filter with smoothing parameter α creates the series S_t, where

$$S_t = \alpha X_t + (1 - \alpha)S_{t-1} \qquad \text{for } t = 1, \ldots, T$$

and S_0 is the initial value. This is the adaptive forecast updating form of the exponential smoother. This implies that

$$S_t = \alpha \sum_{k=0}^{T-1}(1 - \alpha)^K X_{T-k} + (1 - \alpha)^T S_0$$

which is the weighted moving average representation, with geometrically declining weights. The choice of the smoothing constant α determines how quickly the smoothed series or forecast will adjust to changes in the mean of the unfiltered series. For small values of α, the response will be slow because more weight is placed on the previous estimate of the mean of the unfiltered series, whereas larger values of α will put more emphasis on the most recently observed value of the unfiltered series.

▷ Example

Let's consider some examples using sales data. Here we forecasts sales for 3 periods with a smoothing parameter of .4:

```
. use http://www.stata-press.com/data/r8/sales1
. tssmooth exponential sm1=sales, parms(.4) forecast(3)
exponential coefficient  =      0.4000
sum of squared residuals =        8345
root mean squared error  =      12.919
```

To get an idea of how our forecast compares with the actual data, we graph the series and the forecasted series over time.

```
. twoway connected sm1 sales t, title("Single Exponential Forecast")
> ytitle(Sales) xtitle(time)
```

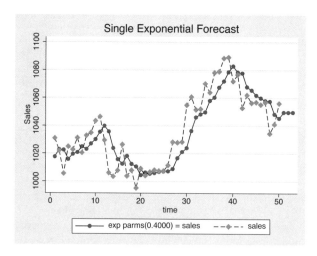

The graph indicates that our forecasted series may not be adjusting rapidly enough to the changes in the actual series. The smoothing parameter α controls the rate at which the forecast adjusts. Smaller values of α cause the forecasts to adjust more slowly. Thus, we suspect that our chosen value of .4 is too small. One way to investigate this suspicion is to ask `tssmooth exponential` to choose the smoothing parameter that minimizes the sum of squared forecast errors.

```
. tssmooth exponential sm2=sales, forecast(3)

computing optimal exponential  coefficient [0,1]

optimal exponential coefficient =        0.7815
sum of squared residuals         =     6727.7056
root mean squared error          =     11.599746
```

The output confirms our conjecture that the value of $\alpha = .4$ was too small. The graph below indicates that the new forecast tracks the series much more closely than the previous forecast.

```
. twoway connected sm2 sales t,
> title ("Single Exponential Forecast With Optimal alpha")
> ytitle(sales) xtitle(time)
```

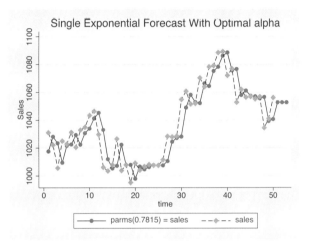

$\triangleleft$

It was noted above that simple exponential forecasts are optimal for an ARIMA $(0,1,1)$ model. (See [TS] **arima** for fitting ARIMA models in Stata.) Chatfield (2001 page 90) gives the following useful derivation that relates the MA coefficient in an ARIMA $(0,1,1)$ model to the smoothing parameter in single exponential smoothing. An ARIMA $(0,1,1)$ is given by

$$x_t - x_{t-1} = \epsilon_t + \theta \epsilon_{t-1}$$

where ϵ_t is an identically and independently distributed white noise error term. Thus, given $\widehat{\theta}$, an estimate of θ, an optimal one-step prediction of $\widehat{x}_{t+1}$ is

$$\widehat{x}_{t+1} = x_t + \widehat{\theta} \epsilon_t$$

Since ϵ_t is not observable, it can be replaced by

$$\widehat{\epsilon_t} = x_t - \widehat{x}_{t-1}$$

yielding

$$\widehat{x}_{t+1} = x_t + \widehat{\theta}(x_t - \widehat{x}_{t-1})$$

Letting $\widehat{\alpha} = 1 + \widehat{\theta}$ and some further rearranging implies that

$$\widehat{x}_{t+1} = (1 + \widehat{\theta})x_t - \widehat{\theta}\widehat{x}_{t-1}$$
$$\widehat{x}_{t+1} = \widehat{\alpha}x_t - (1 - \widehat{\alpha})\widehat{x}_{t-1}$$

▷ Example

Let's compare the estimate of the optimal smoothing parameter of .7815 with the one that we could obtain using [TS] **arima**. Below, we fit an ARIMA(0,1,1) to the sales data, and then back out the estimate of α. The two estimates of α are quite close, given the large estimated standard error of $\widehat{\theta}$.

```
. arima sales, arima(0,1,1)

(setting optimization to BHHH)
Iteration 0:    log likelihood = -189.91037
Iteration 1:    log likelihood = -189.62405
Iteration 2:    log likelihood = -189.60468
Iteration 3:    log likelihood = -189.60352
Iteration 4:    log likelihood = -189.60343
(switching optimization to BFGS)
Iteration 5:    log likelihood = -189.60342

ARIMA regression

Sample:  2 to 50                          Number of obs    =        49
                                          Wald chi2(1)     =      1.41
Log likelihood = -189.6034                Prob > chi2      =    0.2347
```

D.sales	Coef.	OPG Std. Err.	z	P>|z|	[95% Conf. Interval]
sales					
_cons	.5025469	1.382727	0.36	0.716	-2.207548 3.212641
ARMA					
ma					
L1	-.1986561	.1671699	-1.19	0.235	-.5263031 .1289908
/sigma	11.58992	1.240607	9.34	0.000	9.158378 14.02147

```
. display 1 + _b[ARMA:L.ma]
.80134387
```

◁

▷ Example

`tssmooth exponential` automatically detects panel data. Suppose we had sales figures for five companies in long form. Running `tssmooth exponential` on the variable that contains all five series will put the smoothed series and the predictions in a single variable in long form. When the smoothing parameter is chosen to minimize the squared prediction error, an optimal value for the smoothing parameter is chosen separately for each panel.

```
. use http://www.stata-press.com/data/r8/sales_cert

. tsset
      panel variable:  id, 1 to 5
       time variable:  t, 1 to 100

. tssmooth exponential sm5=sales, forecast(3)
```

```
-> id = 1

computing optimal exponential  coefficient [0,1]

optimal exponential coefficient =       0.8702
sum of squared residuals         =    16070.567
root mean squared error          =    12.676974
```

```
-> id = 2

computing optimal exponential  coefficient [0,1]

optimal exponential coefficient =       0.7003
sum of squared residuals         =    20792.393
root mean squared error          =    14.419568
```

```
-> id = 3

computing optimal exponential  coefficient [0,1]

optimal exponential coefficient =       0.6927
sum of squared residuals         =        21629
root mean squared error          =    14.706801
```

```
-> id = 4

computing optimal exponential  coefficient [0,1]

optimal exponential coefficient =       0.3866
sum of squared residuals         =    22321.334
root mean squared error          =    14.940326
```

```
-> id = 5

computing optimal exponential  coefficient [0,1]

optimal exponential coefficient =       0.4540
sum of squared residuals         =    20714.095
root mean squared error          =    14.392392
```

tssmooth exponential computed starting values and chose an optimal α for each panel individually.

◁

Missing values

Missing values in the middle of the data are filled in with the one-step prediction using the previous values. Missing values at the beginning or at the end of the data are treated as if the observations were not there.

▷ Example

Here, the 28th observation is missing. Note that the prediction for the 29th observation is repeated in the new series.

```
. use http://www.stata-press.com/data/r8/sales1
```

```
. tssmooth exponential sm1=sales, parms(.7) forecast(3)
  (output omitted )

. generate sales2=sales if t!=28
(4 missing values generated)

. tssmooth exponential sm3=sales2, parms(.7) forecast(3)
exponential coefficient   =      0.7000
sum of squared residuals  =      6842.4
root mean squared error   =      11.817

. list t sales2 sm3 if t>25 & t < 31
```

	t	sales2	sm3
26.	26	1011.5	1007.5
27.	27	1028.3	1010.3
28.	28	.	1022.9
29.	29	1028.4	1022.9
30.	30	1054.8	1026.75

Since the data for $t = 28$ are missing, the prediction for period 28 has been used in its place. This implies that the updating equation for period 29 is

$$S_{29} = \alpha S_{28} + (1 - \alpha)S_{28} = S_{28}$$

which explains why the prediction for $t = 28$ is repeated.

Because this is a single exponential procedure, the loss of that one observation will not be noticed several periods later.

```
. generate diff = sm3-sm1 if t>28
(28 missing values generated)

. list t diff if t>28 & t < 39
```

	t	diff
29.	29	-3.5
30.	30	-1.050049
31.	31	-.3150635
32.	32	-.0946045
33.	33	-.0283203
34.	34	-.0085449
35.	35	-.0025635
36.	36	-.0008545
37.	37	-.0003662
38.	38	-.0001221

◁

▷ Example

Now consider an example in which there are data missing at the beginning and at the end of the sample.

```
. generate sales3=sales if t>2 & t<49
(7 missing values generated)
```

```
. tssmooth exponential sm4=sales3, parms(.7) forecast(3)
exponential coefficient   =      0.7000
sum of squared residuals  =      6215.3
root mean squared error   =      11.624
. list t sales sales3 sm4 if t<5 | t >45
```

	t	sales	sales3	sm4
1.	1	1031	.	.
2.	2	1022.1	.	.
3.	3	1005.6	1005.6	1016.787
4.	4	1025	1025	1008.956
46.	46	1055.2	1055.2	1057.2
47.	47	1056.8	1056.8	1055.8
48.	48	1034.5	1034.5	1056.5
49.	49	1041.1	.	1041.1
50.	50	1056.1	.	1041.1
51.	51	.	.	1041.1
52.	52	.	.	1041.1
53.	53	.	.	1041.1

The output above illustrates that missing values at the beginning or end of the sample cause the sample to be truncated. The new series begins with nonmissing data and begins predicting immediately after it stops.

One period after the actual data concludes, the exponential forecast becomes a constant. After the actual data end, the forecast at period t is substituted for the missing data. This also illustrates why the forecasted series is a constant.

◁

tssmooth exponential treats an observation excluded from the sample by an if or in statement as if it were missing.

Saved Results

tssmooth saves in r():

Scalars
  r(N)   number of observations
  r(alpha) α smoothing parameter
  r(rss)   sum of squared prediction errors
  r(rmse)   root mean squared error
  r(N_pre)   number of observations used in calculating starting values
  r(s1_0)   initial value for S_t
Macros
  r(exp)   right-hand-side expression

Methods and Formulas

tssmooth exponential is implemented as an ado-file.

The formulas for the derivation of the smoothed series are as given previously. When the value of α is not specified, an optimal value is found that minimizes the mean square forecast error. A method of bisection is employed for finding the solution to this optimization problem.

A truncated description of the specified exponential filter is used to label the new variable. See [R] **label** for more information on labels.

An untruncated description of the specified exponential filter is saved in the characteristic tssmooth for the new variable. See [P] **char** for more information on characteristics.

References

Abraham, B. and J. Ledolter. 1983. *Statistical Methods for Forecasting*. New York: John Wiley & Sons.

Bowerman, B. and R. O'Connell. 1993. *Forecasting and Time Series: An Applied Approach*. 3d ed. Pacific Grove, CA: Duxbury Press.

Chatfield, C. 1996. *The Analysis of Time Series: An Introduction*. 5th ed. London: Chapman & Hall.

——. 2001. *Time–Series Forecasting*. London: Chapman & Hall.

Chatfield, C. and M. Yar. 1988. Holt–Winters Forecasting: some practical issues. *The Statistician* 37: 129–140.

Montgomery, D. C., L. A. Johnson, and J. S. Gardiner. 1990. *Forecasting and Time Series Analysis*. 2d ed. New York: McGraw–Hill.

Also See

Complementary:	[TS] **tsset**
Related:	[TS] **arima**, [TS] **tssmooth dexponential**, [TS] **tssmooth hwinters**, [TS] **tssmooth ma**, [TS] **tssmooth nl**, [TS] **tssmooth shwinters**, [R] **egen**, [R] **generate**
Background:	[U] **14.4.3 Time-series varlists**, [TS] **tssmooth**

Title

tssmooth hwinters — Holt–Winters nonseasonal smoothing

Syntax

tssmooth <u>h</u>winters [*type*] *newvarname* = *exp* [if *exp*] [in *range*]

[, <u>p</u>arms($\#_\alpha$ $\#_\beta$) <u>f</u>orecast(#) <u>s</u>amp0(#) s0($\#_{cons}$ $\#_{lt}$) <u>d</u>iff replace

<u>fr</u>om($\#_\alpha$ $\#_\beta$) nolog <u>nodiff</u>icult *maximize_options*]

tssmooth hwinters is for use with time-series data; see [TS] **tsset**. You must tsset your data before using tssmooth hwinters.

exp may contain time-series operators; see [U] **14.4.3 Time-series varlists**.

Description

tssmooth hwinters performs the Holt–Winters method of smoothing on a user-specified expression, which is usually just a variable name, and generates a new variable containing the forecasted series.

Options

parms($\#_\alpha$ $\#_\beta$) specifies the parameters for the nonseasonal Holt–Winters filter. The Holt–Winters filter has two parameters, α and β. Each parameter must be in the interval $[0, 1]$. If parms() is not specified, then the smoothing parameters will be chosen to minimize the in-sample sum of squared prediction errors.

forecast(#) gives the number of periods for the out-of-sample prediction. $1 \le \# \le 500$.

samp0(#) specifies the sample used to obtain the initial value(s) for the recursions. The integer in samp0() specifies the number of observations to be used. The default is to use one-half of the number of observations. samp0() cannot be specified with s0().

s0($\#_{cons}$ $\#_{lt}$) specifies the initial values for the recursions. $\#_{cons}$ is used to start the recursion on the constant term. $\#_{lt}$ is the initial value for the recursion on the linear term. If initial values are not specified, they will be obtained from a regression of x_t on a constant and a time variable that is normalized to equal one at the beginning of the sample. The estimated constant term from this regression is the starting value for the constant, and the estimated coefficient on the normalized time variable is the starting value for the linear term. s0() may not be specified with samp0() or diff.

diff specifies that the initial value for the recursions on the linear term is to be obtained by averaging the first-difference of the expression to be smoothed, and that the starting value for the constant is $a_0 = x_1 - b_0$, where b_0 is the mean of D.x_t and x_1 is the first observation on the expression to be forecasted. diff may not be specified with s0().

replace causes the variable specified in *newvarname* to be replaced, if it already exists.

from($\#_\alpha$ $\#_\beta$) specifies starting values for the parameters when the parameters are to be chosen so as to minimize the in-sample sum of squared prediction errors. from() may not be specified with parms().

`nolog` suppresses the iteration log when estimating the smoothing parameters.

`nodifficult` specifies that the *maximize_option* `difficult` is not to be used. By default, the *maximize_option* `difficult` is used.

maximize_options control the maximization process; see [R] **maximize**. You will seldom need to specify any of the maximize options, except perhaps `iterate(0)`. If the iteration log shows many "not concave" messages and it is taking many iterations to converge, try providing better starting values using `from()`.

Remarks

Consider a time-series that could be forecasted by

$$\widehat{x}_{t+1} = a_t + b_t t$$

where $\widehat{x}_t$ is the forecast of the original series x_t, a_t is a mean that drifts over time, and b_t is a coefficient on time that also drifts. The Holt–Winters forecasting method is used to forecast series of this form. In fact, as Gardner (1985) has noted, the Holt–Winters method produces optimal forecasts for an ARIMA(0,2,2) model and some local linear models. See [TS] **arima** and the references therein for ARIMA models, and see Harvey (1989) for a discussion of the local linear model and its relationship to the Holt–Winters method. Abraham and Ledolter (1983), Bowerman and O'Connell (1993), and Montgomery et al. (1990) all provide good introductions to the Holt–Winters method. Chatfield (1996, 2001) provides helpful discussions of how this method relates to modern time-series analysis.

The Holt–Winters method can be viewed as an extension of double exponential smoothing with two parameters. These parameters may be explicitly set or chosen so as to minimize the in-sample sum of squared forecast errors. In the latter case, as discussed in *Methods and Formulas*, the smoothing parameters are chosen to minimize the in-sample sum of squared forecast errors plus a penalty term that helps to achieve convergence when one of the parameters is too close to the boundary.

Given the series x_t, the smoothing parameters α and β, and the starting values a_0 and b_0, the updating equations are

$$a_t = \alpha x_t + (1 - \alpha)(a_{t-1} + b_{t-1})$$

$$b_t = \beta(a_t - a_{t-1}) + (1 - \beta)b_{t-1}$$

After computing the series of constant and linear terms, a_t and b_t, respectively, the τ step ahead prediction of x_t is given by

$$\widehat{x}_{t+\tau} = a_t + b_t \tau$$

▷ Example

Below, we illustrate how to use `tssmooth hwinters` with specified smoothing parameters. This example also illustrates that the Holt–Winters method can closely follow a series in which both the mean and the time-coefficient drift over time.

Suppose that we have data on the monthly sales of a book and that we are interested in forecasting this series using the Holt–Winters method.

```
. use http://www.stata-press.com/data/r8/bsales

. tssmooth hwinters hw1=sales, parms(.7 .3) forecast(3)

Specified weights:
                        alpha = 0.7000
                         beta = 0.3000
sum of squared residuals = 2301.046
 root mean squared error = 6.192799


. twoway connected sales hw1 t,
> title("Holt-Winters Forecast with alpha=.7  and beta=.3") ytitle(Sales)
> xtitle(Time)
```

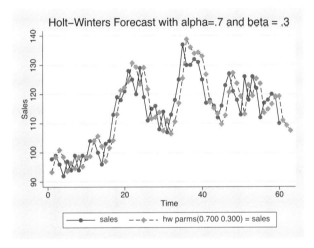

The graph indicates that the forecasts are for linearly decreasing sales. Given a_T and b_T, the out-of-sample predictions are linear functions of time. In this example, the slope appears to be too steep, but this is probably due to our choice of α and β.

◁

▷ Example

The graph in the previous example illustrates that the starting values for the linear and constant series can affect the in-sample fit of the predicted series for the first few observations. The previous example used the default method for obtaining the initial values for the recursion. The output below illustrates that, for some problems, the differenced-based initial values provide a better in-sample fit for the first few observations. However, the differenced-based initial values do not always outperform the regression-based initial values. Furthermore, as shown in the output below, for series of reasonable length, the predictions produced are nearly identical.

```
. tssmooth hwinters hw2=sales, parms(.7 .3) forecast(3) diff

Specified weights:
                        alpha = 0.7000
                         beta = 0.3000
sum of squared residuals = 2261.173
 root mean squared error = 6.13891
```

```
. list hw1 hw2 if _n<6 | _n>57
```

	hw1	hw2
1.	93.31973	97.80807
2.	98.40002	98.11447
3.	100.8845	99.2267
4.	98.50404	96.78276
5.	93.62408	92.2452
58.	116.5771	116.5771
59.	119.2146	119.2146
60.	119.2608	119.2608
61.	111.0299	111.0299
62.	109.2815	109.2815
63.	107.5331	107.5331

When the smoothing parameters are chosen to minimize the in-sample sum of squared forecast errors, changing the initial values can affect the choice of the optimal α and β. When changing the initial values results in different optimal values for α and β, the predictions will also differ.

◁

When the Holt–Winters model fits the data well, finding the optimal smoothing parameters generally proceeds well. When the model fits poorly, it can be difficult to find the α and β that minimize the in-sample sum of squared forecast errors.

▷ Example

In this example, we forecast the book sales data using the α and β that minimize the in-sample squared forecast errors.

```
. tssmooth hwinters hw3=sales, forecast(3)
computing optimal weights

Iteration 0:   penalized RSS = -2632.2073   (not concave)
Iteration 1:   penalized RSS = -1982.8431
Iteration 2:   penalized RSS = -1976.4236
Iteration 3:   penalized RSS = -1975.9175
Iteration 4:   penalized RSS = -1975.9036
Iteration 5:   penalized RSS = -1975.9036

optimal weights:
                          alpha = 0.8209
                           beta = 0.0067
penalized sum of squared residuals = 1975.904
         sum of squared residuals = 1975.904
         root mean squared error = 5.738617
```

The graph below contains the data and the forecast using the optimal α and β. Comparing this graph with the one above illustrates how different choices of α and β can lead to very different forecasts. Instead of linearly decreasing sales, the new forecast is for linearly increasing sales.

```
. twoway connected sales hw3 t,
> title("Holt-Winters Forecast with optimal alpha and beta") ytitle(Sales)
> xtitle(Time)
```

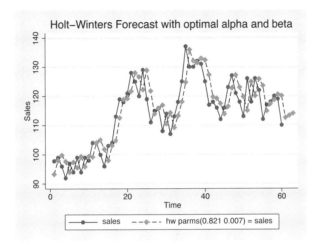

◁

Saved Results

tssmooth saves in r():

Scalars

r(N)	number of observations	r(N_pre)	number of observations used in calculating starting values
r(alpha)	α smoothing parameter		
r(beta)	β smoothing parameter	r(s2_0)	initial value for linear term
r(rss)	sum of squared errors	r(ε1_0)	initial value for constant term
r(prss)	penalized sum of squared errors, if parms() not specified	r(linear)	final value of linear term
		r(constant)	final value of constant term
r(rmse)	root mean squared error		

Macros

r(exp)	expression specified	r(panelvar)	panel variables specified in tsset
r(timevar)	time variables specified in tsset		

Methods and Formulas

tssmooth hwinters is implemented as an ado-file.

A truncated description of the specified Holt–Winters filter is used to label the new variable. See [R] **label** for more information on labels.

An untruncated description of the specified Holt–Winters filter is saved in the characteristic named tssmooth for the new variable. See [P] **char** for more information on characteristics.

Given the series, x_t, the smoothing parameters, α and β, and the starting values, a_0 and b_0, the updating equations are

$$a_t = \alpha x_t + (1 - \alpha)(a_{t-1} + b_{t-1})$$

$$b_t = \beta(a_t - a_{t-1}) + (1 - \beta) b_{t-1}$$

By default, the initial values are found by fitting a linear regression with a time trend. The time variable in this regression is normalized to equal one in the first period included in the sample. By default, one-half of the data is used in this regression, but this sample can be changed using samp0(). a_0 is then set to the estimate of the constant, and b_0 is set to the estimate of the coefficient on the time trend. When the diff option is specified, b_0 is set to the mean of D.x and a_0 to $x_1 - b_0$. Alternatively, s0() can be used to specify the initial values directly.

Sometimes, one or both of the optimal parameters may lie on the boundary of $[0, 1]$. In order to keep the estimates inside $[0, 1]$, tssmooth hwinters parameterizes the objective function in terms of their inverse logits; i.e., in terms of $\frac{\exp(\alpha)}{1+\exp(\alpha)}$ and $\frac{\exp(\beta)}{1+\exp(\beta)}$. When one of these parameters is actually on the boundary, this can complicate the optimization. For this reason, tssmooth hwinters optimizes a penalized sum of squared forecast errors. Let $\widehat{x}_t(\widetilde{\alpha}, \widetilde{\beta})$ be the forecast for the series x_t, given the choices of $\widetilde{\alpha}$ and $\widetilde{\beta}$. Then, the in-sample penalized sum of squared prediction errors is

$$P = \sum_{t=1}^{T} \left[\{x_t - \widehat{x}_t(\widetilde{\alpha}, \widetilde{\beta})\}^2 + I_{|f(\widetilde{\alpha})|>12)}(|f(\widetilde{\alpha})| - 12)^2 + I_{|f(\widetilde{\beta})|>12)}(|f(\widetilde{\beta})| - 12)^2 \right]$$

where $f(x) = \ln\left(\frac{x}{1-x}\right)$. The penalty term is zero unless one of the parameters is very close to the boundary. When one of the parameters is very close the boundary, the penalty term will help to obtain convergence.

Acknowledgment

We would like to thank Nick Cox of the University of Durham for his helpful comments.

References

Abraham, B. and J. Ledolter. 1983. *Statistical Methods for Forecasting*. New York: John Wiley & Sons.

Bowerman, B. and R. O'Connell. 1993. *Forecasting and Time Series: An Applied Approach*. 3d ed. Pacific Grove, CA: Duxbury Press.

Chatfield, C. 1996. *The Analysis of Time Series: An Introduction*. 5th ed. London: Chapman & Hall.

——. 2001. *Time–Series Forecasting*. London: Chapman & Hall.

Chatfield, C. and M. Yar. 1988. Holt–Winters forecasting: some practical issues. *The Statistician* 37: 129–140.

Gardner, E. S., Jr. 1985. Exponential smoothing: the state of the art. *Journal of Forecasting* 4: 1–28.

Montgomery, D. C., L. A. Johnson, and J. S. Gardiner. 1990. *Forecasting and Time Series Analysis*. 2d ed. New York: McGraw–Hill.

Also See

Complementary:	[TS] **tsset**
Related:	[TS] **arima**, [TS] **tssmooth dexponential**, [TS] **tssmooth exponential**,
	[TS] **tssmooth ma**, [TS] **tssmooth nl**, [TS] **tssmooth shwinters**,
	[R] **egen**, [R] **generate**
Background:	[U] **14.4.3 Time-series varlists**,
	[TS] **tssmooth**

Title

> **tssmooth ma** — Moving-average filter

Syntax

Syntax for moving average with uniform weights

> tssmooth ma [*type*] *newvarname* = *exp* [if *exp*] [in *range*] , window($\#_l$[$\#_c$[$\#_f$]])
>
> [replace]

Moving average with specified weights

> tssmooth ma [*type*] *newvarname* = *exp* [if *exp*] [in *range*] ,
>
> weights([*numlist*$_l$] <$\#_c$> [*numlist*$_f$]) [replace]

tssmooth ma is for use with time-series data or cross-sectional time-series data (panel data). You must tsset your data before using tssmooth; see [TS] **tsset**.

exp may contain time-series operators; see [U] **14.4.3 Time-series varlists**.

Description

tssmooth ma applies a moving-average filter to a user-specified expression and generates a new variable that contains the filtered expression.

If the first syntax is chosen, then the window() option is required to specify the span of the filter. tssmooth ma then constructs an uniformly weighted moving average of the expression.

If the second syntax is chosen, then weights() is required. tssmooth ma then applies the user-supplied weights to construct a weighted moving average of the expression.

Options

window($\#_l$ [$\#_c$[$\#_f$]]) describes the span of the uniformly moving average.

The first argument is required and specifies the number of lagged terms to include in the filter. This argument accepts integer values ranging from 0 to one-half of the number of observations in the sample.

The second argument is optional and specifies whether or not the current observation is to be included in the filter. A value of 0 indicates exclusion and a value of 1 indicates inclusion. By default, the current observation is excluded.

The third argument is optional and specifies the number of forward terms to include in the filter. The third argument accepts integer values ranging from 0 to one-half of the number of observations in the sample.

weights([*numlist*$_l$]<$\#_c$>[*numlist*$_f$]) is required for the weighted moving average and describes the span of the moving average as well as the weights that are to be applied to each term in the average.

153

numlist$_l$ is optional and specifies the weights that are to be applied to the lagged terms when computing the moving average.

<#$_c$> is required and specifies the weight that is to be applied to the current term when computing the moving average.

numlist$_f$ is optional and specifies the weights that are to be applied to the forward terms when computing the moving average.

The number of elements in each *numlist* is limited to the one half the number of observations in the sample.

replace specifies that if the variable specified in *newvarname* already exists, it is to be replaced.

Remarks

Moving averages are simple linear filters of the form

$$\widehat{x}_t = \frac{\sum_{i=-l}^{f} w_i x_{t+i}}{\sum_{i=-l}^{f} w_i}$$

where

$\widehat{x}_t$ is the moving average

x_t is the variable or expression to be smoothed

w_i are the weights being applied to the terms in the filter

l is the longest lag in the span of the filter

f is the longest lead in the span of the filter

Moving averages are used primarily as a means of noise-reduction in time-series data. The general applicability of moving averages as a method for signal detection is limited, however, because the moving averages themselves are serially correlated even when the underlying data series is not. Still, Chatfield (1996) discusses moving-average filters and gives a number of specific moving-average filters for extracting certain trends.

In the case of the uniformly weighted moving-average filter, all of the weights, w_i, are 1s. Thus, all that is required is to specify the span of the filter. This is accomplished using the window() option.

▷ Example

Suppose that you have a time series of sales data. Presumably, we will be attempting to separate the data into two components: signal and noise. In order to eliminate the noise, we apply a moving-average filter. In this example, we will use a symmetric moving average with a span of 5. This means that we will average the first two lagged values, the current value, and the first two forward terms of the series, with each term in the average receiving a weight of 1.

```
. use http://www.stata-press.com/data/r8/sales1
. tsset
        time variable:  t, 1 to 50
```

```
. tssmooth ma sm1 = sales, window(2 1 2)
The smoother applied was
        (1/5)*[x(t-2) + x(t-1) + 1*x(t) + x(t+1) + x(t+2)]; x(t)= sales
```

In the case at hand, we want to smooth our series so that there is no autocorrelation in the noise. Below, we compute the noise as the difference between the smoothed series and the series itself. Then, we use **ac**, see [TS] **corrgram**, to check for autocorrelation in the noise.

```
. generate noise = sales-sm1
. ac noise
```

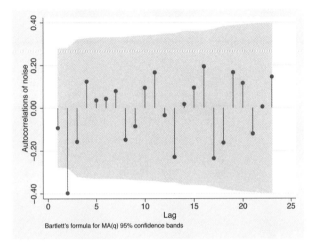

◁

> ## Example

Since there is some evidence of negative second-order autocorrelation, we might consider respecifying our filter. Since the negative autocorrelation may be due to the uniform weighting or the length of the filter, we are going to specify a shorter filter in which the weights decline as the observations become farther away from the current observation.

The weighted moving-average filter requires the user to supply the weights to be applied to each term in the span of the filter. This is accomplished using the **weights()** option. In specifying the weights, one implicitly specifies the span of the filter.

Below, we use the filter

$$\hat{x}_t = (1/9)(1x_{t-2} + 2x_{t-1} + 3x_t + 2x_{t+1} + 1x_{t+2})$$

```
. tssmooth ma sm2 = sales, weights( 1/2 <3> 2/1)
The smoother applied was
        (1/9)*[1*x(t-2) + 2*x(t-1) + 3*x(t) + 2*x(t+1) + 1*x(t+2)]; x(t)= sales
. generate noise2 = sales-sm2
```

Now, we compute the noise and use **ac** to check for autocorrelation.

. ac noise2

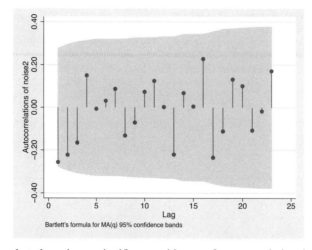

The graph indicates that there is no significant evidence of autocorrelation in the noise from the second filter.

◁

❑ Technical Note

If an observation is missing, tssmooth ma gives this observation a coefficient of zero in both the uniformly and weighted moving-average filters. This simply means that missing values or missing time periods are excluded from the moving average.

Sample restrictions, via if and in, cause the expression smoothed by tssmooth ma to be missing for the excluded observations. Thus, sample restrictions have the same effect as missing values in a variable in the expression that is filtered. Also, gaps in the data that are longer than the span of the filter will generate missing values in the filtered series.

Since the first l observations and the last f observations will be outside the span of the filter, those observations will be set to missing in the moving-average series.

❑

Saved Results

tssmooth saves in r():

Scalars
    r(N)   number of observations in sample
    r(w0)   weight on current observation
    r(wlead$i$)   weight on lead i, if leads specified
    r(wlag$i$)   weight on lag i, if lags specified
Macros
    r(exp)   expression specified
    r(timevar)   time variables specified in tsset
    r(panelvar) panel variables specified in tsset

Methods and Formulas

tssmooth ma is implemented as an ado-file. The formula for moving averages is as previously given.

A truncated description of the specified moving-average filter is used to label the new variable. See [R] **label** for more information on labels.

An untruncated description of the specified moving-average filter is saved in the characteristic tssmooth for the new variable. See [P] **char** for more information on characteristics.

References

Chatfield, C. 1996. *The Analysis of Time Series: An Introduction.* 5th ed. London: Chapman & Hall.

Also See

Complementary:	[TS] **tsset**
Related:	[TS] **arima**, [TS] **tssmooth dexponential**, [TS] **tssmooth exponential**, [TS] **tssmooth hwinters**, [TS] **tssmooth nl**, [TS] **tssmooth shwinters**, [R] **egen**, [R] **generate**
Background:	[U] **14.4.3 Time-series varlists**, [TS] **tssmooth**

Title

> **tssmooth nl** — Nonlinear filter

Syntax

Nonlinear filter

> tssmooth nl $[type]$ *newvarname* = *exp* $[\text{if } exp]$ $[\text{in } range]$,
>
> <u>sm</u>oother(*smoother*$[,\underline{tw}ice$ $]$) replace

where *smoother* is specified as $Sm\big[Sm[\dots]\big]$ and *Sm* is one of

$$\big\{\,1\,|\,2\,|\,3\,|\,4\,|\,5\,|\,6\,|\,7\,|\,8\,|\,9\,\big\}\big[\text{R}\big]$$
$$3\big[\text{R}\big]\text{S}\big[\text{S}\,|\,\text{R}\big]\big[\text{S}\,|\,\text{R}\big]\dots$$
$$\text{E}$$
$$\text{H}$$

Letters may be specified in lowercase if preferred. Examples of *smoother*$[,\texttt{twice}]$ include

3RSSH	3RSSH,twice	4253H	4253H,twice	43RSR2H,twice
3rssh	3rssh,twice	4253h	4253h,twice	43rsr2h,twice

tssmooth nl is for use with time-series data; see [TS] **tsset**. You must tsset your data before using tssmooth nl. *exp* may contain time-series operators; see [U] **14.4.3 Time-series varlists**.

Description

tssmooth nl applies resistant, nonlinear smoothers to a user-specified expression, which is usually just a variable name, and generates a new variable containing the smoothed series.

Options

smooth(*smoother*$[,\texttt{twice}]$) specifies the choice of nonlinear smoother.

replace causes the variable specified in *newvarname* to be replaced, if it already exists.

Remarks

tssmooth nl works as a front-end to smooth. See [R] **smooth** for details.

Saved Results

tssmooth saves in r():

Scalars
    r(N)   number of observations

Macros
    r(method)   nl
    r(smoother)   specified smoother
    r(timevar)   time variables specified in tsset
    r(panelvar)   panel variables specified in tsset

Methods and Formulas

tssmooth nl is implemented as an ado-file. The methods are documented in [R] **smooth**.

A truncated description of the specified nonlinear filter is used to label the new variable. See [R] **label** for more information on labels.

An untruncated description of the specified nonlinear filter is saved in the characteristic tssmooth for the new variable. See [P] **char** for more information on characteristics.

Also See

Complementary:	[TS] **tsset**
Related:	[TS] **tssmooth dexponential**, [TS] **tssmooth exponential**, [TS] **tssmooth hwinters**, [TS] **tssmooth ma**, [TS] **tssmooth shwinters**, [R] **egen**, [R] **generate**
Background:	[U] **14.4.3 Time-series varlists**, [TS] **tssmooth**

Title

> **tssmooth shwinters** — Holt–Winters seasonal smoothing

Syntax

> tssmooth $\underline{\text{sh}}$winters $\begin{bmatrix} type \end{bmatrix}$ newvarname = exp $\begin{bmatrix} \text{if } exp \end{bmatrix}$ $\begin{bmatrix} \text{in } range \end{bmatrix}$
>
> $\Big[$, $\underline{\text{pa}}$rms($\#_\alpha$ $\#_\beta$ $\#_\gamma$) $\underline{\text{sa}}$mp0(#) s0($\#_{\text{cons}}$ $\#_{\text{lt}}$) $\underline{\text{f}}$orecast(#) $\underline{\text{fr}}$om($\#_\alpha$ $\#_\beta$ $\#_\gamma$)
>
> $\underline{\text{ad}}$ditive $\underline{\text{pe}}$riod(#) sn0_0(varname) sn0_v(newvarname) snt_v(newvarname)
>
> $\underline{\text{n}}$ormalize $\underline{\text{alt}}$starts replace nolog $\underline{\text{nodif}}$ficult maximize_options $\Big]$

tssmooth shwinters is for use with time-series data; see [TS] **tsset**. You must tsset your data before using tssmooth shwinters.

exp may contain time-series operators; see [U] **14.4.3 Time-series varlists**.

Description

tssmooth shwinters performs the seasonal Holt–Winters method on a user-specified expression, which is usually just a variable name, and generates a new variable containing the forecasted series.

Options

parms($\#_\alpha$ $\#_\beta$ $\#_\gamma$) specifies the values for the 3 parameters α, β, and γ. The domain for each of these 3 parameters is $[0, 1]$. If parms($\#_\alpha$ $\#_\beta$ $\#_\gamma$) is not specified, then the smoothing parameters are chosen to minimize the in-sample sum of squared prediction errors. parms() may not be specified with from().

samp0(#) specifies the sample to be used to obtain the initial value(s) for the recursions. The integer specifies the number of seasons to be used. The default is to use one-half of the number of observations. samp0() may not be specified with s0().

s0($\#_{\text{cons}}$ $\#_{\text{lt}}$) specifies the initial values for the recursions. The first number is the initial value for the recursion on the constant term. The second number is the initial value for the recursion on the linear term. s0() may not be specified with samp0() or altstarts.

forecast(#) specifies number of periods for the out-of-sample prediction. $1 \le \# \le 500$.

from($\#_\alpha$ $\#_\beta$ $\#_\gamma$) specifies starting values for the three parameters when they are to be chosen to minimize the in-sample sum of squared prediction errors.

additive specifies that the additive seasonal Holt–Winters method is to be used instead of the default multiplicative seasonal Holt–Winters method.

period(#) specifies the periodicity of the seasonality. By default, tssmooth shwinters uses the information stored by [TS] **tsset** to determine the period of seasonality. For example, if you have tsset your data to be quarterly, the period is automatically determined to be 4. If the dataset was tsset without specifying the period, then tsset does not contain any information about the seasonality of the data and period() is required. period() accepts integer values ranging from 2 to one-half the number of observations in the sample.

sn0_0(*varname*) specifies the name of the variable that holds the initial values to be used for the seasonal terms. sn0_0() may not be specified with sn0_v().

sn0_v(*newvarname*) specifies the name of the new variable that will hold the initial values for the seasonal term. sn0_v() may not be specified with sn0_0().

snt_v(*newvarname*) specifies the name of the variable that holds the estimated seasonal terms for the last year in the sample.

normalize specifies that the seasonal values are to be normalized. In the multiplicative model, they are normalized to sum to one. In the additive model, the seasonal values are normalized to sum to zero.

altstarts specifies that an alternative method for computing the starting values for the constant, the linear, and the seasonal terms is to be used. The default and the alternative methods are described in *Methods and Formulas*. altstarts may not be specified with s0().

replace specifies that if the variable specified in *newvarname* already exists, it is to be replaced.

nolog suppresses the iteration log when choosing the smoothing parameters to minimize the in-sample sum of forecast squared errors.

nodifficult specifies that the *maximize_option* difficult is not to be used. By default, the *maximize_option* difficult is used.

maximize_options control the maximization process; see [R] **maximize**. You will seldom need to specify any of the maximize options, except for iterate(0), and possibly difficult. If the iteration log shows many "not concave" messages and it is taking many iterations to converge, try providing better starting values using from($#_\alpha$ $#_\beta$ $#_\gamma$).

Remarks

The seasonal Holt–Winters methods are used to forecast univariate series that have a seasonal component. If the amplitude of the seasonal component grows with the series, then the Holt–Winters multiplicative method should be used. If the amplitude of the seasonal component is not growing with the series, then the Holt–Winters additive method should be used. Abraham and Ledolter (1983), Bowerman and O'Connell (1993), and Montgomery et al. (1990) provide good introductions to the Holt–Winters methods in the context of recursive univariate forecasting methods. Chatfield (1996, 2001) provides introductions in the broader context of modern time-series analysis.

Like the other recursive methods in tssmooth, tssmooth shwinters uses the information stored by tsset to automatically detect panel data. When applied to panel data, each series is smoothed separately and the starting values are computed separately for each panel. If the smoothing parameters are chosen to minimize the in-sample sum of squared forecast errors, the optimization is performed separately on each panel.

When there are missing values at the beginning of the series, the sample begins with the first nonmissing observation. When there are missing values after the first nonmissing observation, they are filled in with forecasted values.

Holt–Winters seasonal multiplicative method

This method is used to forecast seasonal time series in which the amplitude of the seasonal component grows with the series. Chatfield (2001) notes that there are some nonlinear state-space models whose optimal prediction equations correspond to the multiplicative Holt–Winters method. This procedure is best applied to data that could be described by

$$x_{t+j} = (\mu_t + \beta j)S_{t+j} + \epsilon_{t+j}$$

where x_t is the series, μ_t is the time-varying mean at time t, β is a parameter, S_t is the seasonal component at time t, and ϵ_t is an idiosyncratic error. See the *Methods and Formulas* section for the updating equations.

▷ Example

We have quarterly data on turkey sales by a new producer in the 1990s. The data have a strong seasonal component and an upward trend. We use the multiplicative Holt–Winters method to forecast sales for the year 2000. Since we have already `tsset` our data to the quarterly format, we do not need to specify the `period()` option.

```
. use http://www.stata-press.com/data/r8/turksales

. tssmooth shwinters shw1 = sales, forecast(4)
computing optimal weights

Iteration 0:   penalized RSS = -189.34609   (not concave)
Iteration 1:   penalized RSS = -108.68038
Iteration 2:   penalized RSS = -107.50224
Iteration 3:   penalized RSS = -106.25535
Iteration 4:   penalized RSS = -106.14102
Iteration 5:   penalized RSS = -106.14093
Iteration 6:   penalized RSS = -106.14093

optimal weights:
                           alpha = 0.1310
                            beta = 0.1428
                           gamma = 0.2999
penalized sum of squared residuals = 106.1409
         sum of squared residuals = 106.1409
        root mean squared error = 1.628964
```

The graph below describes the fit and the forecast that was obtained.

```
. twoway connected sales shw1 t, title("Multiplicative Holt-Winters forecast")
> xtitle(Time) ytitle(Sales)
```

(Continued on next page)

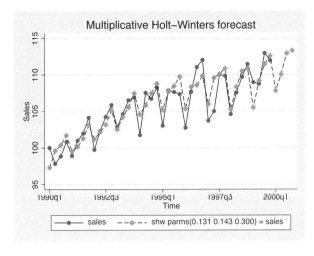

◁

Holt–Winters seasonal additive method

This method is similar to the previous one, but the seasonal effect is assumed to be additive rather than multiplicative. This method is used to forecast series that can be described by the equation

$$x_{t+j} = (\mu_t + \beta j) + S_{t+j} + \epsilon_{t+j}$$

See *Methods and Formulas* for the updating equations.

▷ Example

In this example, we have data on tourism in Hawaii during the 1990's. The data are in millions of visitor days per month. We are interested in predicting the first six months of 2000. Below, we use `tssmooth shwinters` to forecast the series using the smoothing parameters that minimize the in-sample penalized squared forecast errors. We use the `snt_v()` option to save the last year's seasonal terms in the new variable `seas`.

```
. use http://www.stata-press.com/data/r8/htourism
. tssmooth shwinters shwa = mvdays, forecast(6) snt_v(seas) normalize additive
computing optimal weights
Iteration 0:   penalized RSS = -9.7722243  (not concave)
Iteration 1:   penalized RSS = -6.5493745
Iteration 2:   penalized RSS = -6.4306839
Iteration 3:   penalized RSS = -6.3964306
Iteration 4:   penalized RSS = -6.3873126
Iteration 5:   penalized RSS = -6.3854755
Iteration 6:   penalized RSS = -6.3850352
Iteration 7:   penalized RSS = -6.3849915
Iteration 8:   penalized RSS = -6.3849849
Iteration 9:   penalized RSS =  -6.384982
Iteration 10:  penalized RSS = -6.3849806
Iteration 11:  penalized RSS = -6.3849799  (backed up)
```

```
optimal weights:
                                 alpha = 0.5150
                                  beta = 0.0000
                                 gamma = 0.4395
       penalized sum of squared residuals = 6.38498
                 sum of squared residuals = 6.38498
                 root mean squared error = .230669
```

The graph below describes the fit and the forecast. The fact that there is no apparent time trend in the model explains why the optimal $\beta = 0$.

```
. twoway connected mvdays shwa t, title("Multiplicative Holt-Winters forecast")
> xtitle(Time) ytitle(Sales) legend(span)
```

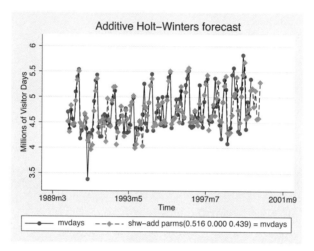

As a check on whether the estimated seasonal components are intuitively sound, we list the last year's seasonal components.

```
. list t seas if seas < .
```

	t	seas
109.	1999m1	.139266
110.	1999m2	-.2216016
111.	1999m3	.09418608
112.	1999m4	-.46097168
113.	1999m5	-.44980284
114.	1999m6	.25432897
115.	1999m7	.64315406
116.	1999m8	.54739016
117.	1999m9	-.40815178
118.	1999m10	-.15800782
119.	1999m11	-.24677337
120.	1999m12	.26698382

The output indicates that the signs of the estimated seasonal components are intuitively reasonable.

◁

Saved Results

tssmooth saves in r():

Scalars

r(N)	number of observations	r(N_pre)	number of seasons used
r(alpha)	α smoothing parameter		in calculating starting values
r(beta)	β smoothing parameter	r(s2_0)	initial value for linear term
r(gamma)	γ smoothing parameter	r(s1_0)	initial value for constant term
r(prss)	penalized sum of squared errors	r(linear)	final value of linear term
r(rss)	sum of squared errors	r(constant)	final value of constant term
r(rmse)	root mean squared error	r(period)	period, if filter is seasonal

Macros

r(method)	shwinters, additive or	r(exp)	expression specified
	shwinters, multiplicative	r(timevar)	time variables specified in tsset
r(normalize	normalize, if specified	r(panelvar)	panel variables specified in tsset

Methods and Formulas

tssmooth shwinters is implemented as an ado-file.

A truncated description of the specified seasonal Holt–Winters filter is used to label the new variable. See [R] **label** for more information on labels.

An untruncated description of the specified seasonal Holt–Winters filter is saved in the characteristic named tssmooth for the new variable. See [P] **char** for more information on characteristics.

When the parms() option is not specified, the smoothing parameters are chosen to minimize the in-sample sum of penalized squared forecast errors. Sometimes, one or more of the three optimal parameters will lie on the boundary $[0, 1]$. In order to keep the estimates inside $[0, 1]$, tssmooth shwinters parameterizes the objective function in terms of their inverse logits; i.e., in terms of $\frac{\exp(\alpha)}{1+\exp(\alpha)}$, $\frac{\exp(\beta)}{1+\exp(\beta)}$, and $\frac{\exp(\gamma)}{1+\exp(\gamma)}$. When one of these parameters is actually on the boundary, this can complicate the optimization. For this reason, tssmooth shwinters optimizes a penalized sum of squared forecast errors. Let $\widehat{x}_t(\widetilde{\alpha}, \widetilde{\beta}, \widetilde{\gamma})$ be the forecast for the series x_t given the choices of $\widetilde{\alpha}$, $\widetilde{\beta}$, and $\widetilde{\gamma}$. Then the in-sample penalized sum of squared prediction errors is

$$P = \sum_{t=1}^{T} \left[\{x_t - \widehat{x}_t(\widetilde{\alpha}, \widetilde{\beta})\}^2 + I_{|f(\widetilde{\alpha})|>12)}(|f(\widetilde{\alpha})| - 12)^2 + I_{|f(\widetilde{\beta})|>12)}(|f(\widetilde{\beta})| - 12)^2 \right.$$
$$\left. + I_{|f(\widetilde{\gamma})|>12)}(|f(\widetilde{\gamma})| - 12)^2 \right]$$

where $f(x) = \ln\left(\frac{x}{1-x}\right)$. The penalty term is zero unless one of the parameters is very close to the boundary. When one of the parameters is very close to the boundary, the penalty term will help to obtain convergence.

Holt–Winters seasonal multiplicative procedure

Like the other recursive methods in `tssmooth`, there are three aspects to implementing the Holt–Winters seasonal multiplicative procedure: the forecasting equation, the initial values, and the updating equations. Unlike the previous methods, the data are now assumed to be seasonal with period L.

Given the estimates $a(t)$, $b(t)$, and $s(t + \tau - L)$, a τ step-ahead point forecast of x_t, denoted by $\widehat{y}_{t+\tau}$, is

$$\widehat{y}_{t+\tau} = \{a(t) + b(t)\tau\}\, s(t + \tau - L)$$

Given the smoothing parameters α, β, and γ, the updating equations are

$$a(t) = \alpha \frac{x_t}{s(t - L)} + (1 - \alpha)\{a(t - 1) + b(t - 1)\}$$

$$b(t) = \beta\{a(t) - a(t - 1)\} + (1 - \beta)\, b(t - 1)$$

and

$$s(t) = \gamma \left\{ \frac{x_t}{a(t)} \right\} + (1 - \gamma)s(t - L)$$

To impose the restriction that the seasonal terms sum to 1 over each year, specify the `normalize` option.

The updating equations require the $L + 2$ initial values $a(0), b(0), s(1 - L), s(2 - L), \ldots s(0)$. There are two methods for calculating the initial values. Both methods calculate the initial values using the first m years, each of which contains L seasons. By default, m is set to the number of seasons in one-half the sample.

The initial value of the trend component, $b(0)$, can be estimated by

$$b(0) = \frac{\overline{x}_m - \overline{x}_1}{(m - 1)L}$$

where $\overline{x}_m$ is the average level of x_t in year m and $\overline{x}_1$ is the average level of x_t in the first year. The initial value for the linear term, $a(0)$, is then calculated as

$$a(0) = \overline{x}_1 - \frac{L}{2} b(0)$$

In order to calculate the initial values for the seasons $1, 2, \ldots, L$, we first calculate the deviation adjusted values,

$$S(t) = \frac{x_t}{\overline{x}_i - \left\{ \frac{(L+1)}{2} - j \right\} b(0)}$$

where i is the year that corresponds to time t, j is the season that corresponds to time t, and $\overline{x}_i$ is the average level of x_t in year i.

Next, for each season $l = 1, 2, \ldots, L$, define $\overline{s}_l$ to be the average S_t over the years. That is,

$$\bar{s}_l = \frac{1}{m} \sum_{k=0}^{m-1} S_{l+kL} \qquad \text{for } l = 1, 2, \ldots, L$$

Then, the initial seasonal estimates are

$$s_{0l} = \bar{s}_l \left(\frac{L}{\sum_{l=1}^{L} \bar{s}_l} \right) \qquad \text{for } l = 1, 2, \ldots, L$$

and these values are used to fill in $s(1 - L), \ldots, s(0)$.

If the `altstarts` option is specified, then the starting values are computed based on a regression with seasonal indicator variables. Specifically, the series x_t is regressed on a time variable normalized to equal one in the first period in the sample and a constant. Then, $b(0)$ is set to the estimated coefficient on the time variable and $a(0)$ is set to the estimated constant term. To calculate the seasonal starting values, x_t is regressed on a set of L seasonal dummy variables. The lth seasonal starting value is set to $\left(\frac{1}{\mu}\right)\widehat{\beta}_l$, where μ is the of mean x_t and $\widehat{\beta}_l$ is estimated coefficient on the lth seasonal dummy variable. The sample used in both regressions and the mean computation is restricted to include the first `samp0()` years. By default, `samp0()` includes half of the data.

❏ Technical Note

If there are missing values in the first few years, a small value of m can cause the starting value methods for seasonal term to fail. In this case, users should either specify a larger values of m using `samp0()` or directly specify the seasonal starting values using the `snt0_0()` option.

❏

Holt–Winters seasonal additive procedure

This procedure is similar to the previous one except that the data are assumed to be described by

$$x_t = (\beta_0 + \beta_1 t) + s_t + \epsilon_t$$

As in the multiplicative case, there are three smoothing parameters, α, β, and γ, which can either be set or be chosen to minimize the in-sample sum of squared forecast errors.

The updating equations are

$$a(t) = \alpha \left\{ x_t - s(t - L) \right\} + (1 - \alpha) \left\{ a(t - 1) + b(t_1) \right\}$$

$$b(t) = \beta \left\{ a(t) - a(t - 1) \right\} + (1 - \beta)b(t - 1)$$

and

$$s(t) = \gamma \left\{ x_t - a(t) \right\} + (1 - \gamma)s(t - L)$$

To impose the restriction that the seasonal terms sum to 0 over each year, specify the `normalize` option.

A τ step-ahead forecast, denoted by $\widehat{y}_{t+\tau}$, is given by

$$\widehat{x}_{t+\tau} = a(t) + b(t)\tau + s(t + \tau - L)$$

As in the multiplicative case, there are two methods for setting the initial values.

The default method is to obtain the initial values for $a(0), b(0), s(1 - L), \ldots, s(0)$ from the regression

$$x_t = a(0) + b(0)t + \beta_{s,1-L}D_1 + \beta_{s,2-L}D_2 + \cdots + \beta_{s,0}D_L + e_t$$

where the $D_1, \ldots, D_L$ are dummy variables with

$$D_i = \left\{ \begin{array}{ll} 1 & \text{if } t \text{ corresponds to season } i \\ 0 & \text{Otherwise} \end{array} \right\}$$

When the `altstarts` option is specified, an alternative method is used where the x_t series is regressed on a time variable that has been normalized to equal one in the first period in the sample and a constant term. $b(0)$ is set to the estimated coefficient on the time variable, and $a(0)$ is set to the estimated constant term. Then, the demeaned series $\widetilde{x}_t = x_t - \mu$ is created where μ is the mean of the x_t. The $\widetilde{x}_t$ are regressed on L seasonal dummy variables. The lth seasonal starting value is then set to β_l, where β_l is the estimated coefficient on the lth seasonal dummy variable. The sample in both the regression and the mean calculation is restricted to include the first `samp0` years, where, by default, `samp0()` includes half of the data.

Acknowledgment

We would like to thank Nick Cox of the University of Durham for his helpful comments.

References

Abraham, B. and J. Ledolter. 1983. *Statistical Methods for forecasting*. New York: John Wiley & Sons.

Bowerman, B. and R. O'Connell. 1993. *Forecasting and Time Series: An Applied Approach*. 3d ed. Pacific Grove, CA: Duxbury Press.

Chatfield, C. 1996. *The Analysis of Time Series: An Introduction*. 5th ed. London: Chapman & Hall.

——. 2001. *Time–Series Forecasting*. London: Chapman & Hall.

Chatfield, C. and M. Yar. 1988. Holt–Winters Forecasting: some practical issues. *The Statistician* 37: 129–140.

Montgomery, D. C., L. A. Johnson, and J. S. Gardiner. 1990. *Forecasting and Time Series Analysis*. 2d ed. New York: McGraw–Hill.

Also See

Complementary:	[TS] **tsset**
Related:	[TS] **arima**, [TS] **tssmooth dexponential**, [TS] **tssmooth exponential**,
	[TS] **tssmooth hwinters**, [TS] **tssmooth ma**, [TS] **tssmooth nl**,
	[R] **egen**, [R] **generate**
Background:	[U] **14.4.3 Time-series varlists**,
	[TS] **tssmooth**

Title

> **var intro** — An introduction to vector autoregression models

Description

Stata has a suite of commands for fitting, forecasting, interpreting, and performing inference on vector autoregressions (VARs) and structural vector autoregressions (SVARs). The suite includes several commands for estimating and interpreting impulse–response functions (IRFs) and forecast-error variance decompositions (FEVDs). The table below lists all the commands, along with a brief description of what they do.

Fitting a VAR or SVAR

var	[TS] **var**	Vector autoregression models
svar	[TS] **var svar**	Structural vector autoregression models
varbasic	[TS] **varbasic**	Fit a simple VAR and graph impulse–response functions

Model diagnostics and inference

varstable	[TS] **varstable**	Check stability condition of var or svar estimates
varsoc	[TS] **varsoc**	Obtain lag-order selection statistics for a set of VARs
varwle	[TS] **varwle**	Obtain Wald lag exclusion statistics after var or svar
vargranger	[TS] **vargranger**	Perform pairwise Granger causality tests after var or svar
varlmar	[TS] **varlmar**	Obtain LM statistics for residual autocorrelation after var or svar
varnorm	[TS] **varnorm**	Tests for normally distributed disturbances after var or svar

Forecasting after fitting a VAR or SVAR

varfcast clear	[TS] **varfcast clear**	Drop variables containing previous forecasts from varfcast
varfcast compute	[TS] **varfcast compute**	Compute dynamic forecasts of dependent variables after var or svar
varfcast graph	[TS] **varfcast compute**	Graph forecasts of dependent variables after var or svar

Obtaining IRFs and FEVDs

varirf create	[TS] **varirf create**	Obtain impulse–response functions and FEVDs

(Continued on next page)

Analyze and present IRFs and FEVDs

varirf table	[TS] **varirf table**	Create tables of impulse–response functions and FEVDs
varirf ctable	[TS] **varirf ctable**	Make combined tables of impulse–response functions and FEVDs
varirf graph	[TS] **varirf graph**	Graph impulse–response functions and FEVDs
varirf cgraph	[TS] **varirf cgraph**	Make combined graphs of impulse–response and FEVDs
varirf ograph	[TS] **varirf ograph**	Graph overlaid impulse–response functions and FEVDs

Manage IRF and FEVD results

varirf add	[TS] **varirf add**	Add VARIRF results from one VARIRF file to another
varirf describe	[TS] **varirf describe**	Describe a VARIRF file
varirf dir	[TS] **varirf dir**	List the VARIRF files in a directory
varirf drop	[TS] **varirf drop**	Drop VARIRF results from the active VARIRF file
varirf erase	[TS] **varirf erase**	Erase a VARIRF file
varirf rename	[TS] **varirf rename**	Rename a VARIRF result in a VARIRF file
varirf set	[TS] **varirf set**	Set active VARIRF file

Here, we briefly discuss vector autoregressions, structural vector autoregressions, and the commands in Stata for these models. For further details on the commands, see their manual entries as listed above. For general introductions to vector autoregressions, see Hamilton (1994), Lütkepohl (1993), Sims (1980), and Stock and Watson (2001). See Amisano and Giannini (1997) for a detailed introduction to structural vector autoregressions.

A pth order vector autoregression (VAR(p)) models K variables as linear functions of p of their own lags, p lags of each of the other $K-1$ variables, a K dimensional disturbance term, and, sometimes, additional exogenous variables. While the elements of the K-dimensional disturbance vector may be contemporaneously correlated, they are assumed to be serially uncorrelated.

VAR(p)s are widely used for forecasting and for investigating how shocks affect the dynamics of a system of related variables. VARs have also been used for performing inference about which variables are useful in predicting other variables.

In the absence of any contemporaneous exogenous variables, the disturbance variance–covariance matrix contains all the information about contemporaneous correlations among the variables. VARs are sometimes classified into three types, where the types differ in how they account for this contemporaneous correlation. (See Stock and Watson (2001) for one derivation of this taxonomy.) In a reduced form VAR, aside from estimating the variance–covariance matrix of the disturbance, no attempt is made to account for contemporaneous correlations. In a recursive VAR, the K variables are assumed to form a recursive dynamic structural equation model in which the first variable is only a function of lagged variables, the second is a function of contemporaneous values of the first variable and lagged values, and so on. In a structural VAR, theory is used to place restrictions on the contemporaneous correlations that are not necessarily recursive.

Stata has three commands for fitting VARs: var, svar, and varbasic. Reduced form VARs can be fit using var or varbasic. var allows for constraints to be imposed on the coefficients. svar fits short-run and long-run structural VAR models. varbasic allows you to quickly fit a simple VAR without constraints and graph the impulse–response functions.

Since one is interested in fitting a VAR of the correct order, varsoc offers several methods for choosing the order p of the VAR to fit. After fitting a VAR, and before proceeding with inference, interpretation, or forecasting, it is important to check that the VAR fits the data. varlmar can be used to check for autocorrelation in the disturbances. `varwle` performs Wald tests to determine if certain lags can be excluded. `varnorm` tests the null hypothesis that the disturbances are normally distributed. `varstable` checks the eigenvalue condition for stability. Stability is necessary for meaningful interpretation of the impulse response functions and forecast-error variance decompositions.

Impulse response functions (IRFs) describe how the K endogenous variables react over time to a one-time shock to one of the K disturbances. Since the disturbances may be contemporaneously correlated, these functions do not answer the question, "How does variable i react to a one-time increase in the innovation to variable j after s periods, holding everything else constant?" In order to answer questions of this type, one must start with orthogonalized innovations, so that the "holding everything else constant" assumption is reasonable. Recursive VARs use a Cholesky decomposition to orthogonalize the disturbances, and thereby obtain structurally interpretable impulse–response functions. Structural VARs use theory to impose sufficient restrictions, which need not be recursive, to decompose the contemporaneous correlations into orthogonal components.

Forecast error variance decompositions (FEVDs) are another tool for interpreting how the orthogonalized innovations affect the K variables over time. The FEVD from j to i gives the fraction of the s-step forecast-error variance of variable i that can be attributed to the jth orthogonalized innovation.

`varirf create` can estimate impulse–response functions, (Cholesky) orthogonalized impulse–response functions, and structural impulse–response functions and their standard errors. It can also estimate Cholesky and structural forecast-error variance decompositions. Stata has `varirf graph`, `varirf cgraph`, `varirf ograph`, `varirf table`, and `varirf ctable` to easily make graphs and tables of these estimates. In addition, Stata has several other commands to manage IRF and FEVD results. See [TS] **varirf** for a description of these commands.

`varfcast compute` computes dynamic forecasts and their standard errors from VARs. The command `varfcast graph` produces graphs of the forecasts that are generated using `varfcast compute`.

VARs allow researchers to investigate whether one variable is useful in predicting another variable. A variable x is said to Granger cause a variable y if, given the past values of y, past values of x are useful for predicting y. The Stata command `vargranger` performs Wald tests to investigate Granger causality between the variables in a VAR.

References

Amisano, G. and C. Giannini. 1997. *Topics in Structural VAR Econometrics*. 2d ed. Heidelberg: Springer–Verlag.

Hamilton, J. D. 1994. *Time Series Analysis*. Princeton: Princeton University Press.

Lütkepohl, H. 1993. *Introduction to Multiple Time Series Analysis*. 2d ed. New York: Springer.

Sims, C. A. 1980. Macroeconomics and reality. *Econometrica* 48(1): 1–48.

Stock, J. H. and M. W. Watson. 2001. Vector autoregressions. *Journal of Economic Perspectives*. 15(4): 101–115.

(Continued on next page)

Also See

Complementary:	[TS] **var**, [TS] **var svar**, [TS] **varfcast**, [TS] **varfcast clear**,
	[TS] **varfcast compute**, [TS] **varfcast graph**, [TS] **vargranger**,
	[TS] **varirf**, [TS] **varirf add**, [TS] **varirf cgraph**, [TS] **varirf create**,
	[TS] **varirf ctable**, [TS] **varirf describe**, [TS] **varirf dir**, [TS] **varirf drop**,
	[TS] **varirf erase**, [TS] **varirf graph**, [TS] **varirf ograph**,
	[TS] **varirf rename**, [TS] **varirf set**, [TS] **varirf table**, [TS] **varlmar**,
	[TS] **varnorm**, [TS] **varsoc**, [TS] **varstable**, [TS] **varwle**
Related:	[TS] **arima**,
	[R] **regress**, [R] **sureg**
Background:	[U] **14.4.3 Time-series varlists**,
	[TS] **varirf**

Title

> **var** — Vector autoregression models

Syntax

var *depvarlist* [if *exp*] [in *range*] [, <u>lags</u>(*numlist*) <u>ex</u>og(*varlist*)

    <u>c</u>onstraints(*numlist*) <u>noc</u>onstant dfk <u>lut</u>stats nobigf <u>nois</u>ure <u>iter</u>ate(#)

    <u>tol</u>erance(#) nolog <u>sma</u>ll <u>l</u>evel(#)]

var is for use with time-series data; see [TS] **tsset**. You must tsset your data before using var.

by ... : may be used with var; see [R] **by**.

depvarlist and *varlist* may contain time-series operators; see [U] **14.4.3 Time-series varlists**.

var shares the features of all estimation commands; see [U] **23 Estimation and post-estimation commands**.

Syntax for predict

predict [*type*] *newvarname* [if *exp*] [in *range*] [, <u>e</u>quation(*eqno* | *eqname*)

    xb <u>r</u>esiduals stdp]

These statistics are available both in and out of sample; type predict ... if e(sample) ... if wanted only for the estimation sample.

Other post-estimation commands

See the following entries for information on post-estimation commands that can be used after var:

[TS]	**varstable**	Check stability condition of var or svar estimates
[TS]	**varsoc**	Obtain lag-order selection statistics for a set of VARs
[TS]	**varwle**	Obtain Wald lag exclusion statistics after var or svar
[TS]	**vargranger**	Perform pairwise Granger causality tests after var or svar
[TS]	**varlmar**	Obtain LM statistics for residual autocorrelation after var or svar
[TS]	**varnorm**	Test for normally distributed disturbances after var or svar
[TS]	**varfcast**	Dynamic forecasts of dependent variables after var or svar
[TS]	**varirf**	An introduction to the varirf commands

Description

var fits vector autoregressive (VAR(p)) models. The lag structure of the VAR(p) need not be complete, and the model may contain exogenous variables. Linear constraints may be placed on any of the coefficients in the VAR(p), but var does not allow constraints on Σ, the error variance–covariance matrix; see [TS] **var svar** for an estimator that does allow one to impose structure on the error variance–covariance matrix. See Lütkepohl (1993) and Hamilton (1994) for general treatments of VAR(p) models. Stock and Watson (2001) also provides a very good introduction to VAR(p) models.

Options

lags(*numlist*) specifies the lags to be included in the model. The default is lags(1 2). Note that this option takes a *numlist* and not simply an integer for the maximum lag. For example, lags(2) would include only the second lag in the model, whereas lags(1/2) would include both the first and second lags in the model. See [U] **14.1.8 numlist** and [U] **16.8 Time-series operators** for further discussion of numlists and lags.

exog(*varlist*) specifies a list of exogenous variables to be included in the VAR(p).

constraints(*numlist*) specifies the constraint number(s) of the linear constraint(s) to be applied during estimation. Constraints are specified using the constraint command; see [R] **constraint**. Since VAR(p) is a multiple-equation estimator, constraints must specify the equation name for all but the first equation.

noconstant suppresses the constant terms (intercepts) in the model.

dfk specifies that a small-sample degrees-of-freedom adjustment is to be used when estimating Σ, the error variance–covariance matrix. Specifically, $1/(T - \overline{m})$ is used instead of the large sample $1/T$, where $\overline{m}$ is the average number of parameters in the functional form for $\mathbf{y}_t$ over the K equations.

lutstats specifies that the Lütkepohl (1993) versions of the lag-order selection statistics should be reported. See the *Methods and Formulas* section in [TS] **varsoc** for a discussion of these statistics.

nobigf causes var to not compute the estimated parameter vector that incorporates coefficients that have been implicitly constrained to be zero; i.e., when some lags have been omitted from a model. Consider a model with lags(1 3). This is the same as a model with lags(1/3) in which all the coefficients on the second lag are constrained to zero. In the lags(1 3) specification, e(b) and e(V) will only include estimates for the coefficients on lags 1 and 3; i.e., those that are explicitly included in the model. However, by default, var saves in e(bf), the vector of estimated parameters that corresponds to the lags(1/3) model specification with explicit constraints applied. e(bf) is used for computing asymptotic standard errors in the post-estimation commands varirf create and varfcast. Therefore, specifying nobigf implies that the asymptotic standard errors will not be available from the varirf create and varfcast post-estimation routines.

noisure specifies that the estimates in the presence of constraints are to be obtained via one-step seemingly unrelated regression. By default, var obtains estimates in the presence of constraints via iterated seemingly unrelated regression. When the constraints() option is not specified, the estimates are obtained via OLS and noisure has no effect. For this reason, noisure can only be specified when the constraints() option is specified.

iterate(#) specifies an integer that sets the maximum number of iterations when the estimates are obtained via iterated seemingly unrelated regression. By default, the limit is 1600. When the constraints() option is not specified, the estimates are obtained via OLS and iterate() has no effect. For this reason, iterate() can only be specified when the constraints() option is specified. Similarly, iterate() cannot be combined with noisure.

tolerance(#) specifies a number that must be greater than zero and less than 1 for the convergence tolerance of the iterated seemingly unrelated regression algorithm. By default, the tolerance is 1e-6. When the constraints() option is not specified, the estimates are obtained via OLS and tolerance() has no effect. For this reason, tolerance() can only be specified when the constraints() option is specified. Similarly, tolerance() cannot be combined with noisure.

nolog suppresses the log from the iterated seemingly unrelated regression algorithm. By default, the iteration log is displayed when the coefficients are estimated via iterated seemingly unrelated regression. When the constraints() option is not specified, the estimates are obtained via OLS and nolog has no effect. For this reason, nolog can only be specified when the constraints() option is specified. Similarly, nolog cannot be combined with noisure.

small causes var to report small-sample t and F statistics instead of the large-sample normal and chi-squared statistics.

level(#) specifies the confidence level, in percent, for confidence intervals. The default is level(95) or as set by set level; see [U] **23.6 Specifying the width of confidence intervals**.

Options for predict

equation(*eqno* | *eqname*) specifies to which equation you are referring.

equation() is filled in with one *eqno* or *eqname* for options xb, stdp, and residuals. equation(#1) would mean the calculation is to be made for the first equation, equation(#2) would mean the second, and so on. Alternatively, you could refer to the equation by its name. equation(income) would refer to the equation named income and equation(hours) to the equation named hours.

If you do not specify equation(), the results are as if you specified equation(#1).

xb, the default, calculates the fitted values for the specified equation.

residuals calculates the residuals.

stdp calculates the standard error of the linear prediction for the specified equation.

For more information on using predict after multiple-equation estimation commands, see [R] **predict**.

Remarks

Remarks are presented under the headings

> *Introduction to vector autoregressions*
> *Fitting models with some lags excluded*
> *Fitting models with exogenous variables*
> *Fitting models with constraints on the coefficients*
> *Model selection and hypothesis testing*
> *Forecasting*

Introduction to vector autoregressions

var estimates the parameters in vector autoregressive (VAR(p)) models. A VAR(p) is a model in which K variables are specified as linear functions of p of their own lags, p lags of the other $K - 1$ variables, and possibly additional exogenous variables. Algebraically, a pth-order vector autoregressive model with exogenous variables $\mathbf{x}_t$ is given by

$$\mathbf{y}_t = \mathbf{v} + \mathbf{A}_1 \mathbf{y}_{t-1} + \cdots + \mathbf{A}_p \mathbf{y}_{t-p} + \mathbf{B}\mathbf{x}_t + \mathbf{u}_t \qquad t \in \{-\infty, \infty\} \tag{1}$$

where

$\mathbf{y}_t = (y_{1t}, \ldots, y_{Kt})'$ is a $K \times 1$ random vector,
the $\mathbf{A}_i$ are fixed $K \times K$ matrices of parameters,
$\mathbf{x}_t$ is an $M \times 1$ vector of exogenous variables,
$\mathbf{B}$ is a $K \times M$ matrix of coefficients,
$\mathbf{v}$ is a $K \times 1$ vector of fixed parameters, and
$\mathbf{u}_t$ is assumed to be white noise; that is,

$$E(\mathbf{u}_t) = \mathbf{0},$$
$$E(\mathbf{u}_t \mathbf{u}_t') = \boldsymbol{\Sigma}, \text{ and}$$
$$E(\mathbf{u}_t \mathbf{u}_s') = \mathbf{0} \text{ for } t \neq s$$

There are $K \times K \times p + K \times (M + 1)$ parameters in the functional form for $\mathbf{y}_t$, and there are $\{K \times (K + 1)\}/2$ parameters in the covariance matrix $\boldsymbol{\Sigma}$. One way of reducing the number of parameters is to specify an incomplete VAR(p), in which some of the $\mathbf{A}_i$ matrices are set to zero. Any of the $\mathbf{A}_i$, $i = 1, \ldots, p$ may be set to a zero matrix by simply omitting this lag from the *numlist* specified in the lags() option. Another way of reducing the number of parameters to estimate is by specifying linear constraints on any of the coefficients in the VAR(p). var does not allow constraints on $\boldsymbol{\Sigma}$, the error variance–covariance matrix; see [TS] **var svar** for an estimator that does.

A VAR(p) can be viewed as the reduced form of a system of dynamic simultaneous equations. Consider the system

$$\mathbf{W}_0 \mathbf{y}_t = \mathbf{a} + \mathbf{W}_1 \mathbf{y}_{t-1} + \cdots + \mathbf{W}_p \mathbf{y}_{t-p} + \mathbf{W}_x \mathbf{x}_t + \mathbf{e}_t \qquad (2)$$

where $\mathbf{a}$ is a $K \times 1$ vector of parameters, each $\mathbf{W}_i$, $i = 1, \ldots, p$, is a $K \times K$ matrix of parameters, and $\mathbf{e}_t$ is a $K \times 1$ disturbance vector. In the traditional dynamic simultaneous equations approach, sufficient restrictions are placed on the $\mathbf{W}_i$ to obtain identification. Assuming that $\mathbf{W}_0$ is nonsingular, (2) can be rewritten as

$$\mathbf{y}_t = \mathbf{W}_0^{-1}\mathbf{a} + \mathbf{W}_0^{-1}\mathbf{W}_1 \mathbf{y}_{t-1} + \cdots + \mathbf{W}_0^{-1}\mathbf{W}_p \mathbf{y}_{t-p} + \mathbf{W}_0^{-1}\mathbf{W}_x \mathbf{x}_t + \mathbf{W}_0^{-1}\mathbf{e}_t \qquad (3)$$

which is a VAR(p). Just define

$$\mathbf{v} = \mathbf{W}_0^{-1}\mathbf{a}$$
$$\mathbf{A}_i = \mathbf{W}_0^{-1}\mathbf{W}_i$$
$$\mathbf{B} = \mathbf{W}_0^{-1}\mathbf{W}_x$$
$$\mathbf{u}_t = \mathbf{W}_0^{-1}\mathbf{e}_t$$

The cross-equation error variance–covariance matrix $\boldsymbol{\Sigma}$ contains all the information about contemporaneous correlations in a VAR(p), and may be the VAR(p)'s greatest strength and its greatest weakness. Since no questionable *a priori* assumptions are imposed, fitting a VAR(p) allows the dataset to speak for itself. However, without imposing some restrictions on the structure of $\boldsymbol{\Sigma}$, no causal interpretation of the results is possible.

At the cost of additional technical assumptions, we can derive another representation of the VAR(p) in (1) that provides some intuition. To simplify the notation, consider the case without exogenous variables. If a VAR(p) without exogenous variables is stable (see [TS] **varstable**), we can rewrite the variables in $\mathbf{y}_t$ as

$$\mathbf{y}_t = \boldsymbol{\mu} + \sum_{i=0}^{\infty} \boldsymbol{\Phi}_i \mathbf{u}_{t-i} \qquad (4)$$

where $\boldsymbol{\mu}$ is the $K \times 1$ time-invariant mean of the process and the $\boldsymbol{\Phi}_i$ are $K \times K$ matrices of parameters. Equation (4) states that the process by which the variables in $\mathbf{y}_t$ fluctuate about their time-invariant means, $\boldsymbol{\mu}$, is completely determined by the parameters in $\boldsymbol{\Phi}_i$ and the (infinite) past history of independent and identically distributed (i.i.d.) shocks, or innovations, $\mathbf{u}_{t-1}, \mathbf{u}_{t-2}, \ldots$. Equation (4) is known as the vector moving-average representation of the VAR(p). The moving-average coefficients $\boldsymbol{\Phi}_i$ are also known as the simple impulse–response functions at horizon i. The precise relationship between the $\mathbf{A}_i$ and the $\boldsymbol{\Phi}_i$ is derived in the *Methods and Formulas* section of [TS] **varirf create**.

The distributions and joint distributions of the $\mathbf{y}_t$ are uniquely determined by the distributions of $\mathbf{x}_t$, $\mathbf{u}_t$, and the parameters $\mathbf{v}$, $\mathbf{B}$, and $\mathbf{A}_i$. In order to estimate the parameters in a VAR(p), certain restrictions on the temporal properties of the joint distribution of the $\mathbf{y}_t$ and $\mathbf{x}_t$ are required. Specifically, the variables in $\mathbf{y}_t$ and $\mathbf{x}_t$ are required to be covariance stationary, meaning that their first two moments exist and are time invariant. More explicitly, a random vector $\mathbf{z}_t$ is covariance stationary if

1. $E[\mathbf{z}_t]$ is finite and independent of t;

2. $\mathrm{Var}[\mathbf{z}_t]$ is finite and independent of t; and

3. $\mathrm{Cov}[\mathbf{z}_t, \mathbf{z}_s]$ is a finite function of $|t - s|$ but neither of t nor s.

If the $\mathbf{u}_t$ form a zero mean, i.i.d. vector process and $\mathbf{y}_t$ and $\mathbf{x}_t$ are covariance stationary and uncorrelated with the $\mathbf{u}_t$, then consistent and efficient estimates of $\mathbf{B}$, the $\mathbf{A}_i$, and $\mathbf{v}$ are obtained via seemingly unrelated regression, yielding estimators that are asymptotically normally distributed. When the equations for the variables $\mathbf{y}_t$ have the same set of regressors, equation-by-equation OLS estimates are the conditional maximum likelihood estimates.

Much of the interest in VAR(p) models is focused on the forecasts, impulse–response functions, and the forecast-error variance decompositions, all of which are functions of the estimated parameters. Estimation of these functions is straightforward, but their asymptotic standard errors are usually obtained by assuming that $\mathbf{u}_t$ forms a zero mean, i.i.d. Gaussian vector process. Also, some of the specification tests for VAR(p)s have been derived using the LR principle and the stronger Gaussian assumption.

In general, only stable $\mathbf{y}_t$ processes are of interest. A method for investigating the stability of a process is presented in [TS] **varstable**. While stability implies covariance stationarity, the converse is not true, as there are some processes that are not stable but are still covariance stationary.

▷ Example

Let's fit a simple VAR(2) on the data used by Lütkepohl (1993). The VAR(2) is given by

$$\mathbf{y}_t = \mathbf{v} + \mathbf{A}_1 \mathbf{y}_{t-1} + \mathbf{A}_2 \mathbf{y}_{t-2} + \boldsymbol{\epsilon}_t$$

where $\mathbf{y}_t$, $\mathbf{v}$, and $\boldsymbol{\epsilon}_t$ are all 3×1 vectors and the $\mathbf{A}_1$ and $\mathbf{A}_2$ are 3×3 matrices. Following Lütkepohl (1993), the three variables are the first difference of natural log of investment, dlinvestment, the first difference of the natural log of income, dlincome, and the first difference of the natural log of consumption, dlconsumption. Actually, we use the first-difference of the natural log of each variable, because the log series is believed to be first-difference stationary.

In this example, the sample is restricted to the one used in Lütkepohl (1993, 72).

```
. use http://www.stata-press.com/data/r8/lutkepohl
(Quarterly SA West German macro data, Bil DM, from Lutkepohl 1993 Table E.1)

. tsset
        time variable:  qtr, 1960q1 to 1982q4
```

```
. var dlinvestment dlincome dlconsumption if qtr <= q(1978q4),
> lags(1/2) lutstats dfk
Vector autoregression
Sample: 1960q4   1978q4
```

Equation	Obs	Parms	RMSE	R-sq	chi2	P
dlinvestment	73	7	.046148	0.1286	9.736909	0.1362
dlincome	73	7	.011719	0.1142	8.508289	0.2032
dlconsumpt~n	73	7	.009445	0.2513	22.15096	0.0011

Model lag order selection statistics (lutstats)

FPE	AIC	HQIC	SBIC	LL	Det(Sigma_ml)
2.183e-11	-24.631632	-24.406561	-24.066861	606.30704	1.226e-11

	Coef.	Std. Err.	z	P>|z|	[95% Conf. Interval]	
dlinvestment						
dlinvestment						
L1	-.3196318	.1254564	-2.55	0.011	-.5655218	-.0737419
L2	-.1605508	.1249066	-1.29	0.199	-.4053633	.0842616
dlincome						
L1	.1459851	.5456664	0.27	0.789	-.9235013	1.215472
L2	.1146009	.5345709	0.21	0.830	-.9331388	1.162341
dlconsumpt~n						
L1	.9612288	.6643086	1.45	0.148	-.3407922	2.26325
L2	.9344001	.6650949	1.40	0.160	-.369162	2.237962
_cons	-.0167221	.0172264	-0.97	0.332	-.0504852	.0170409
dlincome						
dlinvestment						
L1	.0439309	.0318592	1.38	0.168	-.018512	.1063739
L2	.0500302	.0317196	1.58	0.115	-.0121391	.1121995
dlincome						
L1	-.1527311	.1385702	-1.10	0.270	-.4243237	.1188615
L2	.0191634	.1357525	0.14	0.888	-.2469067	.2852334
dlconsumpt~n						
L1	.2884992	.168699	1.71	0.087	-.0421448	.6191431
L2	-.0102	.1688987	-0.06	0.952	-.3412354	.3208353
_cons	.0157672	.0043746	3.60	0.000	.0071932	.0243412
dlconsumpt~n						
dlinvestment						
L1	-.002423	.0256763	-0.09	0.925	-.0527476	.0479016
L2	.0338806	.0255638	1.33	0.185	-.0162235	.0839847
dlincome						
L1	.2248134	.1116778	2.01	0.044	.005929	.4436978
L2	.3549135	.1094069	3.24	0.001	.1404798	.5693471
dlconsumpt~n						
L1	-.2639695	.1359595	-1.94	0.052	-.5304451	.0025062
L2	-.0222264	.1361204	-0.16	0.870	-.2890175	.2445646
_cons	.0129258	.0035256	3.67	0.000	.0060157	.0198358

The lag() option takes a *numlist* of lags. To specify a model that includes the first and second lags, type

```
. var y1 y2 y3, lags(1/2)
```

not

```
. var y1 y2 y3, lags(2)
```

because the latter specification would fit a model that included only the second lag.

The output has two parts: a header and the standard Stata output table for the coefficients, standard errors, and confidence intervals. The header contains summary statistics for each of the equations in the VAR(p) and statistics used in selecting the lag order of the VAR(p). Although there are standard formulas for all of the lag-order statistics, Lütkepohl (1993) gives different versions of the three information criteria, which drop the constant term from the likelihood. To obtain the Lütkepohl (1993) versions, we specified the `lutstats` option. The formulas for both the standard and Lütkepohl versions of these statistics are given in the *Methods and Formulas* section of [TS] **varsoc**.

The `dfk` option specifies that the small-sample divisor $1/(T - \overline{m})$ is to be used in estimating Σ instead of the ML divisor $1/T$. Here, $\overline{m}$ is set to the average number of parameters included in each of the K equations. All of the lag-order statistics are computed using the ML estimator of Σ. Thus, specifying `dfk` will not change the computed lag-order statistics, but it will change the estimated variance–covariance matrix. Also, when `dfk` is specified, a `dfk` adjusted log-likelihood is computed and saved in `e(ll_dfk)`.

◁

Fitting models with some lags excluded

Suppose that you wanted to fit a model that only had a fourth lag; that is,

$$\mathbf{y}_t = \mathbf{v} + \mathbf{A}_4\mathbf{y}_{t-4} + \mathbf{u}_t$$

Of course, this is equivalent to fitting the more general model

$$\mathbf{y}_t = \mathbf{v} + \mathbf{A}_1\mathbf{y}_{t-1} + \mathbf{A}_2\mathbf{y}_{t-2} + \mathbf{A}_3\mathbf{y}_{t-3} + \mathbf{A}_4\mathbf{y}_{t-4} + \mathbf{u}_t$$

with $\mathbf{A}_i, i = 1, 2, 3$, constrained to be $\mathbf{0}$. When you fit a model with some lags excluded (the *excluded lag parameterization*), `var` will estimate the coefficients included in the specification, e.g., $\mathbf{A}_4$, and, as usual, save these estimates in `e(b)`. However, by default, `var` will also create the full vector of estimated parameters for the constrained estimation problem and save it in `e(bf)`, since the full vector of parameter estimates (with the constraints applied, of course) is needed to obtain the asymptotic standard errors in the post-estimation routines. Because it is not too difficult to write down models for which the full vector of parameter estimates is too large for Stata, option `nobigf` specifies that `var` not compute the larger constrained versions. Of course, this means that the asymptotic standard errors of the post-estimation functions will not be available, although bootstrapping is still possible. Building `e(bf)` can be time consuming, so if you do not need these constrained matrices and speed is an issue, specifying `nobigf` may be to your advantage.

Fitting models with exogenous variables

▷ Example

Use the `exog()` option to include exogenous variables in the VAR(p).

```
var dlincome dlconsumption if qtr <= q(1978q4), lags(1/2)
> dfk exog(dlinvestment)
Vector autoregression
Sample:  1960q4   1978q4
```

Equation	Obs	Parms	RMSE	R-sq	chi2	P
dlincome	73	6	.011917	0.0702	5.059587	0.4087
dlconsumpt~n	73	6	.009197	0.2794	25.97262	0.0001

Model lag order selection statistics

FPE	AIC	HQIC	SBIC	LL	Det(Sigma_ml)
9.637e-09	-12.782638	-12.632591	-12.406124	478.56629	6.932e-09

	Coef.	Std. Err.	z	P>|z|	[95% Conf. Interval]	
dlincome						
dlincome						
L1	-.1343345	.1391074	-0.97	0.334	-.4069801	.1383111
L2	.0120331	.1380346	0.09	0.931	-.2585097	.2825759
dlconsumpt~n						
L1	.3235342	.1652769	1.96	0.050	-.0004027	.647471
L2	.0754177	.1648624	0.46	0.647	-.2477066	.398542
dlinvestment	.0151546	.0302319	0.50	0.616	-.0440987	.074408
_cons	.0145136	.0043815	3.31	0.001	.0059259	.0231012
dlconsumpt~n						
dlincome						
L1	.2425719	.1073561	2.26	0.024	.0321578	.452986
L2	.3487949	.1065281	3.27	0.001	.1400036	.5575862
dlconsumpt~n						
L1	-.3119629	.1275524	-2.45	0.014	-.5619611	-.0619648
L2	-.0128502	.1272325	-0.10	0.920	-.2622213	.2365209
dlinvestment	.0503616	.0233314	2.16	0.031	.0046329	.0960904
_cons	.0131013	.0033814	3.87	0.000	.0064738	.0197288

All the post-estimation commands for analyzing VAR(p)s work when exogenous variables are included in a model, but the asymptotic standard errors for the h-step-ahead forecasts are not available.

◁

Fitting models with constraints on the coefficients

var permits model specifications that include constraints on the coefficient. var does not allow for constraints on $\widehat{\Sigma}$; see [TS] **var svar** for more on this topic.

▷ Example

In the first example, we fit a full VAR(2) to a three-equation model. Several coefficients were safely insignificant, the coefficients in the equation for dlinvestment were jointly insignificant, as were the coefficients in the equation for dlincome. In this example, we constrain the coefficient on L2.dlincome in the equation for dlinvestment and the coefficient on L2.dlconsumption in the equation for dlincome to zero.

```
. constraint define 1 [dlinvestment]L2.dlincome = 0
. constraint define 2 [dlincome]L2.dlconsumption = 0
```

```
. var dlinvestment dlincome dlconsumption  if qtr <= q(1978q4),
> lags(1/2) lutstats constraints(1 2) dfk
Estimating VAR coefficients

Iteration 1:   tolerance =  .00737681
Iteration 2:   tolerance =  3.998e-06
Iteration 3:   tolerance =  2.730e-09

Vector autoregression

Constraints:
 ( 1)   [dlinvestment]L2.dlincome = 0
 ( 2)   [dlincome]L2.dlconsumption = 0
Sample:  1960q4   1978q4
```

Equation	Obs	Parms	RMSE	R-sq	chi2	P
dlinvestment	73	6	.043895	0.1280	9.842338	0.0798
dlincome	73	6	.011143	0.1141	8.584446	0.1268
dlconsumpt~n	73	7	.008981	0.2512	22.86958	0.0008

Model lag order selection statistics (lutstats)

FPE	AIC	HQIC	SBIC	LL	Det(Sigma_ml)
1.772e-14	-31.692539	-31.467468	-31.127768	606.28044	1.052e-14

	Coef.	Std. Err.	z	P>|z|	[95% Conf. Interval]	
dlinvestment						
dlinvestment						
L1	-.320713	.1247512	-2.57	0.010	-.5652208	-.0762051
L2	-.1607084	.124261	-1.29	0.196	-.4042555	.0828386
dlincome						
L1	.1195448	.5295669	0.23	0.821	-.9183873	1.157477
L2	-2.55e-17	1.18e-16	-0.22	0.829	-2.57e-16	2.06e-16
dlconsumpt~n						
L1	1.009281	.623501	1.62	0.106	-.2127586	2.231321
L2	1.008079	.5713486	1.76	0.078	-.1117438	2.127902
_cons	-.0162102	.016893	-0.96	0.337	-.0493199	.0168995
dlincome						
dlinvestment						
L1	.0435712	.0309078	1.41	0.159	-.017007	.1041495
L2	.0496788	.0306455	1.62	0.105	-.0103852	.1097428
dlincome						
L1	-.1555119	.1315854	-1.18	0.237	-.4134146	.1023908
L2	.0122353	.1165811	0.10	0.916	-.2162595	.2407301
dlconsumpt~n						
L1	.29286	.1568345	1.87	0.062	-.01453	.6002501
L2	1.78e-19	8.28e-19	0.22	0.829	-1.45e-18	1.80e-18
_cons	.015689	.003819	4.11	0.000	.0082039	.0231741
dlconsumpt~n						
dlinvestment						
L1	-.0026229	.0253538	-0.10	0.918	-.0523154	.0470696
L2	.0337245	.0252113	1.34	0.181	-.0156888	.0831378
dlincome						
L1	.2224798	.1094349	2.03	0.042	.0079912	.4369683
L2	.3469758	.1006026	3.45	0.001	.1497984	.5441532
dlconsumpt~n						
L1	-.2600227	.1321622	-1.97	0.049	-.519056	-.0009895
L2	-.0146825	.1117618	-0.13	0.895	-.2337315	.2043666
_cons	.0129149	.003376	3.83	0.000	.0062981	.0195317

None of the free parameter estimates changed by much. While the coefficients in the equation dlinvestment are now significant at the 10% level, the coefficients in the equation for dlincome remain jointly insignificant.

◁

Model selection and hypothesis testing

See the following sections for information on model selection and hypothesis testing after var.

[TS] **vargranger**	Perform pairwise Granger causality tests after var or svar
[TS] **varlmar**	Obtain LM statistics for residual autocorrelation after var or svar
[TS] **varnorm**	Test for normally distributed disturbances after var or svar
[TS] **varsoc**	Obtain lag-order selection statistics for a set of VARs
[TS] **varstable**	Check stability condition of var or svar estimates
[TS] **varwle**	Obtain Wald lag exclusion statistics after var or svar

Forecasting

There are two types of forecasts available after fitting a VAR(p): a one-step-ahead forecast and a dynamic h-step-ahead forecast.

The one-step-ahead forecast produces a prediction of the value of an endogenous variable in the current period using the estimated coefficients, the past values of the endogenous variables, and any exogenous variables. Keep in mind that if you include contemporaneous values of exogenous variables in your model, you must have observations on the exogenous variables that are contemporaneous with the period in which the prediction is being made in order to compute the prediction. In Stata terms, these one-step-ahead predictions are just the standard linear predictions available after any estimation command. Thus, predict, xb eq(*eqno* | *eqname*) produces one-step-ahead forecasts for the specified equation. predict, stdp eq(*eqno* | *eqname*) produces the standard error of the linear prediction for the specified equation. Note that the standard error of the linear prediction is not the same as the standard error of the forecast. The latter includes an estimate of the variability due to innovations, while the former does not.

The dynamic h-step-ahead forecast begins by using the estimated coefficients, the lagged values of the endogenous variables, and any exogenous variables, to predict one-step ahead for each endogenous variable. Then, the one-step-ahead forecast is used to produce two-step-ahead forecasts for each endogenous variable. The process continues for h periods. Since each step uses the predictions of the previous steps, these forecasts are known as dynamic forecasts. See [TS] **varfcast** for information on obtaining dynamic forecasts and their standard errors.

(Continued on next page)

Saved Results

var saves in e():

Scalars
e(N)	number of observations
e(T)	number of observations
e(neqs)	number of equations
e(df_eq)	average number of parameters in an equation
e(tparms)	number of parameters in all equations
e(k_#)	number of parameters in equation #
e(obs_#)	number of observations on equation #
e(df_m)	degrees of freedom in model
e(df_m#)	model degrees of freedom for equation #
e(df_r)	residual degrees of freedom, if small specified.
e(rmse_#)	root mean square for equation #
e(r2_#)	R-squared for equation #
e(ll)	log likelihood
e(ll_#)	log likelihood for equation #
e(ll_dfk)	dfk adjusted log-likelihood (dfk only)
e(F_#)	F statistic for equation # (small only)
e(aic)	Akaike information criteria
e(sbic)	Schwartz–Bayesian information criteria
e(hqic)	Hannan–Quinn information criteria
e(fpe)	final prediction error
e(mlag)	highest lag in VAR
e(tmax)	last time period in sample
e(tmin)	first time period in sample
e(N_gaps)	number of gaps in sample
e(detsig)	determinant of e(Sigma)
e(detsig_ml)	determinant of $\widehat{\Sigma}_{ml}$

Macros
e(cmd)	var
e(depvar)	name(s) of dependent variable(s)
e(exog)	names of exogenous variables, if specified
e(eqnames)	names of equations
e(lags)	lags in model
e(constraints)	constraints, if constraints specified
e(small)	small, if specified
e(timevar)	name of timevar
e(fsfmt)	format of timevar
e(dfk)	dfk, if specified
e(predict)	program used to implement predict

Matrices
e(b)	coefficient vector
e(V)	variance–covariance matrix of the estimators
e(bf)	constrained coefficient vector
e(G)	Gamma matrix, see *Methods and Formulas*
e(Sigma)	$\widehat{\Sigma}$ matrix

Functions
e(sample)	marks estimation sample

Methods and Formulas

`var` is implemented as an ado-file.

When there are no constraints placed on the coefficients, the VAR(p) is a seemingly unrelated regression model with the same explanatory variables in each equation. As discussed in Lütkepohl (1993) and Greene (2003), performing linear regression on each equation produces the maximum likelihood estimates of the coefficients. The estimated coefficients can then be used to calculate the residuals which in turn are used to estimate the cross-equation error variance–covariance matrix Σ.

Following the notation of Lütkepohl (1993), we write the VAR(p) with exogenous variables as

$$\mathbf{y}_t = \mathbf{A}\mathbf{Y}_{t-1} + \mathbf{B}_0\mathbf{x}_t + \mathbf{u}_t \tag{5}$$

where

$\mathbf{y}_t$ is the $K \times 1$ vector of endogenous variables,

$\mathbf{A}$ is a $K \times Kp$ matrix of coefficients,

$\mathbf{B}_0$ is a $K \times M$ matrix of coefficients,

$\mathbf{x}_t$ is the $M \times 1$ vector of exogenous variables,

$\mathbf{u}_t$ is the $K \times 1$ vector of white noise innovations,

and $\mathbf{Y}_t$ is the $Kp \times 1$ matrix given by $\mathbf{Y}_t = \begin{pmatrix} \mathbf{y}_t \\ \vdots \\ \mathbf{y}_{t-p+1} \end{pmatrix}$

While (5) is easier to read, the formulas are much easier to manipulate if it is instead written as

$$\mathbf{Y} = \mathbf{B}\mathbf{Z} + \mathbf{U}$$

where

$$
\begin{aligned}
\mathbf{Y} &= (\mathbf{y}_1, \dots, \mathbf{y}_T) & &\mathbf{Y} \text{ is } Kp \times T \\
\mathbf{B} &= (\mathbf{A}, \mathbf{B}_0) & &\mathbf{B} \text{ is } K \times (Kp + M) \\
\mathbf{Z} &= \begin{pmatrix} \mathbf{Y}_0 \dots, \mathbf{Y}_{T-1} \\ \mathbf{x}_1 \dots, \mathbf{x}_T \end{pmatrix} & &\mathbf{Z} \text{ is } (Kp + M) \times T \\
\mathbf{U} &= (\mathbf{u}_1, \dots, \mathbf{u}_T) & &\mathbf{U} \text{ is } K \times T
\end{aligned}
$$

If there are intercept terms in the model, then they are included in $\mathbf{x}_t$. If there are no exogenous variables and no intercept terms in the model, then $\mathbf{x}_t$ is empty.

The coefficients are estimated by iterated seemingly unrelated regression. Since the estimation is actually performed by `reg3`, the methods are documented in [R] **reg3**. See [P] **matrix constraint** for more on estimation with constraints.

Let $\widehat{\mathbf{U}}$ be the matrix of residuals, that are obtained via $\mathbf{Y} - \widehat{\mathbf{B}}\mathbf{Z}$, where $\widehat{\mathbf{B}}$ is the matrix of estimated coefficients. Then, the estimator of Σ is

$$\widehat{\Sigma} = \frac{1}{T}\widehat{\mathbf{U}}'\widehat{\mathbf{U}}$$

By default, the maximum likelihood divisor of $\widetilde{T} = T$ is used. When dfk is specified, a small-sample degrees-of-freedom adjustment is used in estimating $\mathbf{\Sigma}$. Specifically, $1/(T - \overline{m})$; i.e., $\widetilde{T} = T - \overline{m}$, where $\overline{m}$ is the average number of parameters in the functional form for $\mathbf{y}_t$ over the K equations.

small specifies that Wald tests after are assumed to have F or t distributions instead of chi-squared or standard normal distributions. The standard errors from each equation are computed using the degrees of freedom for the equation.

The "gamma" matrix saved in e(G), referred to in *Saved Results*, is the $(1 + Kp) \times (1 + Kp)$ matrix given by

$$\frac{1}{T} \sum_{t=1}^{T} (1, \mathbf{Y}_t')(1, \mathbf{Y}_t')'$$

The formulas for the lag-order selection criteria and the log likelihood are discussed in [TS] **varsoc**.

Formulas for predict

predict with the xb option provides the one-step forecast. If exogenous variables are specified, then the forecast is conditional on the exogenous $\mathbf{x}_t$ variables. Specifying the residuals option calculates the errors of the one-step forecasts. Specifying the stdp option calculates the standard errors of the one-step forecasts.

Acknowledgment

We would like to thank Christopher Baum, Boston College, for his helpful comments.

References

Greene, W. H. 2003. *Econometric Analysis*. 5th ed. Upper Saddle River, NJ: Prentice–Hall.

Hamilton, J. D. 1994. *Time Series Analysis*. Princeton: Princeton University Press.

Lütkepohl, H. 1993. *Introduction to Multiple Time Series Analysis*. 2d ed. New York: Springer.

Stock, J. H. and M. W. Watson. 2001. Vector autoregressions. *Journal of Economic Perspectives* 15(4): 101–115.

Also See

Complementary:	[TS] **tsset**,
	[R] **adjust**, [R] **lincom**, [R] **linktest**, [R] **lrtest**, [R] **mfx**, [R] **nlcom**,
	[R] **predict**, [R] **predictnl**, [R] **test**, [R] **testnl**, [R] **vce**
Related:	[TS] **arch**, [TS] **arima**, [TS] **var svar**, [TS] **varbasic**
	[R] **reg3**, [R] **regress**, [R] **sureg**
Background:	[U] **16.5 Accessing coefficients and standard errors**,
	[U] **23 Estimation and post-estimation commands**,
	[TS] **var intro**

Title

> **var svar** — Structural vector autoregression models

Syntax

Short-run constraints

svar *depvarlist* [if *exp*] [in *range*] ,

    { <u>aco</u>nstraints(*numlist$_a$*) <u>ae</u>q(*matrix$_{aeq}$*) <u>acns</u>(*matrix$_{acns}$*)

    <u>bco</u>nstraints(*numlist$_b$*) <u>be</u>q(*matrix$_{beq}$*) <u>bcns</u>(*matrix$_{bcns}$*) }

    [<u>lags</u>(*numlist*) <u>ex</u>og(*varlist$_{exog}$*) <u>varc</u>onstraints(*numlist$_v$*) <u>noconst</u>ant

    dfk <u>noiden</u>check <u>lut</u>stats <u>nobigf</u> <u>noisure</u> <u>isiter</u>ate(#) <u>istol</u>erance(#)

    <u>noislog</u> <u>small</u> <u>full</u> var <u>level</u>(#) *maximize_options*]

Long-run constraints

svar *depvarlist* [if *exp*] [in *range*] ,

    { <u>lrc</u>onstraints(*numlist$_{lr}$*) <u>lre</u>q(*matrix$_{lreq}$*) <u>lrcns</u>(*matrix$_{lrcns}$*) }

    [<u>lags</u>(*numlist*) <u>ex</u>og(*varlist$_{exog}$*) <u>varc</u>onstraints(*numlist$_v$*) <u>noconst</u>ant

    dfk <u>noiden</u>check <u>lut</u>stats <u>nobigf</u> <u>noisure</u> <u>isiter</u>ate(#) <u>istol</u>erance(#)

    <u>noislog</u> <u>small</u> <u>full</u> var <u>level</u>(#) *maximize_options*]

svar is for use with time-series data; see [TS] **tsset**. You must tsset your data before using svar.

by ...: may be used with svar; see [R] **by**.

depvarlist and *varlist$_{exog}$* may contain time-series operators; see [U] **14.4.3 Time-series varlists**.

svar shares the features of all estimation commands; see [U] **23 Estimation and post-estimation commands**.

Syntax for predict

    predict [*type*] *newvarname* [if *exp*] [in *range*] [, <u>eq</u>uation(*eqno | eqname*)

    xb <u>r</u>esiduals stdp]

These statistics are available both in and out of sample; type predict ... if e(sample) ... if wanted only for the estimation sample.

Description

svar estimates the parameters of a structural vector autoregression (SVAR) and the parameters of the underlying vector autoregression (VAR).

For good introductions to SVARs, see Amisano and Giannini (1997) and Hamilton (1994). For good introductions to VARs, see Lütkepohl (1993), Hamilton (1994), and Stock and Watson (2001).

Options

aconstraints(*numlist_a*), aeq(*matrix*$_{aeq}$), acns(*matrix*$_{acns}$)

bconstraints(*numlist_b*), beq(*matrix*$_{beq}$), bcns(*matrix*$_{bcns}$)

These options specify the short-run constraints in an SVAR. To specify a short-run SVAR model, at least one of these options must be specified. The first list of options specifies constraints on the parameters of the **A** matrix; the second list specifies constraints on the parameters of the **B** matrix (see *Methods and Formulas*). If at least one option is selected from the first list, and none are selected from the second list, svar sets **B** to the identity matrix. Similarly, if at least one option is selected from the second list, and none are selected from the first list, svar sets **A** to the identity matrix.

None of these options may be specified with any of the options that define long-run constraints.

aconstraints(*numlist_a*) specifies a *numlist* of previously defined Stata constraints that are to be applied to **A** during estimation.

aeq(*matrix*$_{aeq}$) specifies a matrix that defines a set of equality constraints. This matrix must be square with dimension equal to the number of equations in the underlying VAR. The elements of this matrix must be either *missing* or real numbers. A missing value in the i, j element of this matrix specifies that the i, j element of **A** is a free parameter. A real number in the i, j element of this matrix constrains the i, j element of **A** to this real number. For example,

$$\mathbf{A} = \begin{bmatrix} 1 & 0 \\ . & 1.5 \end{bmatrix}$$

specifies that $\mathbf{A}[1, 1] = 1$, $\mathbf{A}[1, 2] = 0$, $\mathbf{A}[2, 2] = 1.5$, and that $\mathbf{A}[2, 1]$ is a free parameter.

acns(*matrix*$_{acns}$) specifies a matrix which defines a set of exclusion or cross-parameter equality constraints on **A**. This matrix must be square with dimension equal to the number of equations in the underlying VAR. Each element of this matrix must be either *missing*, 0, or a strictly positive integer. A missing value in the i, j element of this matrix specifies that no constraint is placed on this element of **A**. A zero in the i, j element of this matrix constrains the i, j element of **A** to be zero. Any strictly positive integers must be in two or more elements of this matrix. A strictly positive integer in the i, j element of this matrix constrains the i, j element of **A** to be equal to all the other elements of **A** that correspond to elements in this matrix that contain the same integer. For example, consider the matrix

$$\mathbf{A} = \begin{bmatrix} . & 1 \\ 1 & 0 \end{bmatrix}$$

Specifying, acns(A) in a two equation SVAR would constrain $\mathbf{A}[2, 1] = \mathbf{A}[1, 2]$ and $\mathbf{A}[2, 2] = 0$ while leaving $\mathbf{A}[1, 1]$ free.

bconstraints(*numlist_a*) specifies a *numlist* of previously defined Stata constraints that are to be applied to **B** during estimation.

beq(*matrix*$_{beq}$) specifies a matrix that defines a set of equality constraints. This matrix must be square with dimension equal to the number of equations in the underlying VAR. The elements of this matrix must be either *missing* or real numbers. The syntax of implied constraints is analogous to the one described in aeq(), except that what is stated there applies to **B** rather than **A**.

bcns(*matrix*_{bcns}) specifies a matrix that defines a set of exclusion or cross-parameter equality
constraints on **B**. This matrix must be square with dimension equal to the number of equations
in the underlying VAR. Each element of this matrix must be either *missing*, 0, or a strictly
positive integer. The format of the implied constraints is the same as the one described in the
acns() option above.

lrconstraints(*numlist*_{lr}), lreq(*matrix*_{lreq}), lrcns(*matrix*_{lrcns})

These options specify the long-run constraints in an SVAR. To specify a long-run SVAR model,
at least one of these options must be specified. The list of options specifies constraints on the
parameters of the long-run **C** matrix (see *Methods and Formulas* for the definition of **C**). None
of these options may be specified with any of the options that define short-run constraints.

lrconstraints(*numlist*_{lr}) specifies a *numlist* of previously defined Stata constraints that are
to be applied to **C** during estimation.

lreq(*matrix*_{lreq}) specifies a matrix that defines a set of equality constraints on the elements
of **C**. This matrix must be square with dimension equal to the number of equations in the
underlying VAR. The elements of this matrix must be either *missing* or real numbers. The syntax
of implied constraints is analogous to the one described in option aeq() above.

lrcns(*matrix*_{lrcns}) specifies a matrix that defines a set of exclusion or cross-parameter equality
constraints on **C**. This matrix must be square with dimension equal to the number of equations
in the underlying VAR. Each element of this matrix must be either *missing*, 0, or a strictly
positive integer. The syntax of the implied constraints is the same as the one described for the
acns() option above.

lags(*numlist*) specifies the lags to be included in the underlying VAR model. Note that this option
takes a *numlist* and not simply an integer for the maximum lag. For instance, lags(2) would
include only the second lag in the model, while lags(1/2) would include both the first and
second lags in the model. See [U] **14.1.8 numlist** and [U] **16.8 Time-series operators** for further
discussion of *numlist*s and lags.

exog(*varlist*) specifies a list of exogenous variables to be included in the underlying VAR.

varconstraints(*numlist*) specifies a list of constraints that should be applied to coefficients in
the underlying VAR. Since VAR estimates on multiple equations, the constraints must specify the
equation name for all but the first equation.

noconstant specifies that the constant (intercept) terms should be removed from the underlying VAR
model.

dfk specifies that a small-sample degrees of freedom adjustment should be used in estimating the
covariance matrix of the VAR disturbances, Σ. Specifically, $1/(T - \overline{m})$ is used instead of the large
sample $1/T$, where $\overline{m}$ is the average number of parameters in the functional form for $\mathbf{y}_t$, over
the K equations.

noidencheck specifies that the Amisano and Giannini (1997) check for local identification should not
be performed. As noted by Amisano and Giannini (1997), this check is local to the starting values
used. Because of this dependence on the starting values, one may wish to suppress this check by
specifying the noidencheck option. However, great care should be taken before specifying this
option. Models that are not structurally identified can still converge, thereby producing meaningless
results that only appear to have meaning.

lutstats specifies that the Lütkepohl versions of the lag order selection statistics should be reported.
See the *Methods and Formulas* section in [TS] **varsoc** for a discussion of these statistics.

nobigf causes var not to compute the estimated parameter vector that incorporates coefficients that
have been implicitly set to zero. When some lags are omitted from a model, the coefficients on

those omitted lags have been implicitly set to zero. Consider a model with lags(1 3). This is the same as a model with lags(1/3) in which all the coefficients on the second lag are constrained to equal zero. In this sense, omitting lags from a model implies constraints. In the lags(1 3) specification, e(b_var) and e(V_var) will only include estimates for the coefficients on lags 1 and 3. However, by default, the vector of estimated parameters that correspond to lags(1/3) with constraints applied is saved in e(bf_var).

Specifying nobigf suppresses the creation of this expanded vector. This can prove useful in large problems where matsize limitations could be exceeded otherwise. A consequence of nobigf is that standard errors will not be available to those using the post-estimation routines varirf create and varfcast.

noisure specifies that, in the presence of varconstraints(), the VAR coefficients are to be estimated via one-step seemingly unrelated regression. By default, svar estimates the coefficients in the VAR (in the presence of varconstraints()) via iterated seemingly unrelated regression. When the varconstraints() option is not specified, the VAR coefficient estimates are obtained via OLS, a non-iterative procedure. As a result, noisure may only be specified with varconstraints().

isiterate(#) sets the maximum number of iterations for the iterated seemingly unrelated regression algorithm. The default limit is 1600. When the varconstraints() option is not specified, the VAR coefficients are estimated via OLS, a non-iterative procedure. As a result, isiterate() may only be specified with varconstraints(). Similarly, isiterate() may not be combined with noisure.

istolerance(#) specifies the convergence tolerance of the iterated seemingly unrelated regression algorithm. The default tolerance is 1e-6. When the varconstraints() option is not specified, the VAR coefficients are estimated via OLS, a non-iterative procedure. As a result, istolerance() may only be specified with varconstraints(). Similarly, istolerance() may not be combined with noisure.

noislog prevents svar from displaying the iteration log from the iterated seemingly unrelated regression algorithm. When the varconstraints() option is not specified, the estimates are obtained via OLS, an non-iterative procedure. As a result, noislog may only be specified with varconstraints(). Similarly, noislog may not be combined with noisure.

small causes svar and var to calculate and report small sample t and F statistics instead of the large sample normal and chi-squared statistics.

full specifies that all the parameters should appear in the output table. By default, any constrained parameters are omitted from the output table. full may be specified at estimation or upon replay.

var specifies that the output from var should also be displayed. By default, the underlying VAR(p) is fit quietly.

level(#) specifies the confidence level, in percent, for confidence intervals. The default is level(95) or as set by set level; see [U] **23.6 Specifying the width of confidence intervals**.

maximize_options control the maximization process; see [R] **maximize**. You will seldom need to specify any of the maximize options except for iterate(0) and possibly difficult. If the iteration log shows many "not concave" messages and it is taking many iterations to converge, try the difficult option to see if that helps it to converge in fewer steps, or try providing better starting values using the from() option.

Options for predict

equation(*eqno* | *eqname*) specifies to which equation you are referring.

> equation() is filled in with one *eqno* or *eqname* for options xb, stdf, and residuals. equation(#1) would mean the calculation is to be made for the first equation, equation(#2) would mean the second, and so on. Alternatively, you could refer to the equation by its name. equation(income) would refer to the equation named income and equation(hours) to the equation named hours.

If you do not specify equation(), the results are as if you specified equation(#1).

xb, the default, calculates the fitted values for the specified equation.

residuals calculates the residuals.

stdp calculates the standard error of the linear prediction for the specified equation.

For more information on using predict after multiple-equation estimation commands, see [R] **predict**.

Remarks

This entry assumes that you have already read [TS] **var**; if not, please do.

One way of describing an SVAR is to illustrate how it links the VAR framework and the dynamic structural simultaneous equation framework. Consider a dynamic structural simultaneous-equation model,

$$\boldsymbol{\Gamma}_0 \mathbf{y}_t = \mathbf{c} + \boldsymbol{\Gamma}_1 \mathbf{y}_{t-1} + \cdots + \boldsymbol{\Gamma}_p \mathbf{y}_{t-p} + \boldsymbol{\Gamma}_x \mathbf{x}_t + \boldsymbol{\xi}_t \tag{1}$$

where

> $t = 1, \ldots, T$
> $\mathbf{y}_t$ is the tth observation on a $K \times 1$ vector of dependent variables,
> $\mathbf{c}$ is a $K \times 1$ vector of parameters,
> $\boldsymbol{\Gamma}_i \quad i = 0, \ldots, p$ are $K \times K$ matrices of parameters,
> $\boldsymbol{\Gamma}_x$ is a $K \times m$ matrix of parameters,
> $\mathbf{x}_t$ is $m \times 1$ matrix of exogenous variables, and
> $\boldsymbol{\xi}_t$ is a $K \times 1$ vector of unobservable disturbances with $\boldsymbol{\xi}_t \sim N(\mathbf{0}, \boldsymbol{\Omega})$ over t
> $\quad E(\boldsymbol{\xi}_t) = \mathbf{0}$
> $\quad E(\boldsymbol{\xi}_t \boldsymbol{\xi}_t') = \boldsymbol{\Omega}$
> $\quad E(\boldsymbol{\xi}_t \boldsymbol{\xi}_s') = \mathbf{0}_K$ for $t \neq s$

The dynamic simultaneous structural equation framework assumes that some variables can be treated as exogenous and others as endogenous. These assumptions place restrictions on the parameters in $\mathbf{c}$, $\boldsymbol{\Gamma}_x$, and the $\boldsymbol{\Gamma}_i$, in order to obtain identification.

Assuming that $\boldsymbol{\Gamma}_0$ is invertible, (1) can be rewritten as

$$\mathbf{y}_t = \mathbf{v} + \mathbf{A}_1 \mathbf{y}_{t-1} + \cdots + \mathbf{A}_p \mathbf{y}_{t-p} + \mathbf{A}_x \mathbf{x}_t + \mathbf{u}_t \tag{2}$$

where $\mathbf{v} = \boldsymbol{\Gamma}_0^{-1} \mathbf{c}$, $\mathbf{A}_i = \boldsymbol{\Gamma}_0^{-1} \boldsymbol{\Gamma}_i \; i = 1, \ldots, p$, $\mathbf{A}_x = \boldsymbol{\Gamma}_0^{-1} \boldsymbol{\Gamma}_x$, and $\mathbf{u}_t = \boldsymbol{\Gamma}_0^{-1} \boldsymbol{\xi}_t$. Equation (2) defines a VAR with exogenous variables in which $\mathbf{u}_t \sim N(\mathbf{0}, \boldsymbol{\Sigma})$ with $E(\mathbf{u}_t \mathbf{u}_s') = \mathbf{0}_K$ for $t \neq s$, and $\boldsymbol{\Sigma} = \boldsymbol{\Gamma}_0^{-1} \boldsymbol{\Omega} \boldsymbol{\Gamma}_0'^{-1}$. This simple manipulation reveals that a VAR is a reduced form of a dynamic simultaneous equations model in which all of the contemporaneous correlation is modeled in $\boldsymbol{\Sigma}$. The $\mathbf{u}_t$ are often referred to as the *innovations*.

The dynamic simultaneous equation framework and the VAR/SVAR framework are frequently used to investigate different types of questions. In the dynamic simultaneous equations framework, interest is generally focused on how changes in the exogenous variables affect the endogenous variables. In contrast, in the VAR/SVAR framework interest is generally focused on how the innovations to one endogenous variable affect other endogenous variables.

To see how the innovations in one endogenous variable affects another, we need to rewrite (2). To simplify the discussion, let's consider a VAR(p) that does not include exogenous variables,

$$\mathbf{y}_t = \mathbf{v} + \mathbf{A}_1 \mathbf{y}_{t-1} + \cdots + \mathbf{A}_p \mathbf{y}_{t-p} + \mathbf{u}_t \tag{3}$$

As discussed in [TS] **varstable**, if the VAR(p) in (3) is stable, it can be inverted and rewritten as an infinite order moving average process; i.e., we can write $\mathbf{y}_t$ as the sum of an infinite series of serially uncorrelated innovations. Specifically, if the VAR(p) in (3) is stable, it can rewritten as

$$\mathbf{y}_t = \boldsymbol{\mu} + \sum_{s=0}^{\infty} \boldsymbol{\Phi}_s \mathbf{u}_{t-s} \tag{4}$$

where $\boldsymbol{\mu}$ is the $K \times 1$ time-invariant mean of $\mathbf{y}_t$ and

$$\boldsymbol{\Phi}_s = \begin{cases} \mathbf{I}_K & \text{if } s = 0 \\ \sum_{j=1}^{s} \boldsymbol{\Phi}_{s-j} \mathbf{A}_j & \text{if } s = 1, 2, \ldots \end{cases}$$

The $\boldsymbol{\Phi}_s$ are known as the simple impulse–response functions at horizon s. The i, j element of $\boldsymbol{\Phi}_s$ gives the effect of a one-time unit increase in an innovation to variable j on variable i after s periods holding everything else constant. As discussed in [TS] **varirf create**, the problem is that since $E[\mathbf{u}_t \mathbf{u}_t']$ is not restricted to being a diagonal matrix, an increase in an innovation to one variable provides information about the innovations to other variables. This implies that no causal interpretation of the simple impulse–response functions is possible.

However, suppose that we had a matrix $\mathbf{P}$ such that $\boldsymbol{\Sigma} = \mathbf{P}\mathbf{P}'$. It can then be shown that the variables in $\mathbf{P}^{-1}\mathbf{u}_t$ have zero mean and that $E\{\mathbf{P}^{-1}\mathbf{u}_t(\mathbf{P}^{-1}\mathbf{u}_t)'\} = \mathbf{I}_K$. We could then rewrite (4) as

$$\begin{aligned} \mathbf{y}_t &= \boldsymbol{\mu} + \sum_{s=0}^{\infty} \boldsymbol{\Phi}_s \mathbf{P}\mathbf{P}^{-1} \mathbf{u}_{t-s} \\ &= \boldsymbol{\mu} + \sum_{s=0}^{\infty} \boldsymbol{\Theta}_s \mathbf{P}^{-1} \mathbf{u}_{t-s} \\ &= \boldsymbol{\mu} + \sum_{s=0}^{\infty} \boldsymbol{\Theta}_s \mathbf{w}_{t-s} \end{aligned} \tag{5}$$

where $\boldsymbol{\Theta}_s = \boldsymbol{\Phi}_s \mathbf{P}$ and $\mathbf{w}_t = \mathbf{P}^{-1}\mathbf{u}_t$. If we had such a $\mathbf{P}$, then the $\mathbf{w}_k$ would be mutually orthogonal, and no information would be lost in the *ceteris paribus* assumption. The $\boldsymbol{\Theta}_s$ would then have the causal interpretation that we seek.

SVAR models provide a framework for estimation of, and inference about, a broad class of $\mathbf{P}$ matrices. As described in [TS] **varirf create**, the estimated $\mathbf{P}$ matrices can then be used to estimate structural impulse–response functions and structural forecast-error variance decompositions. There are short-run and long-run SVAR models. Short-run SVAR models identify a $\mathbf{P}$ matrix by placing restrictions on the contemporaneous correlations between the variables. Long-run SVAR models identify a $\mathbf{P}$ matrix by placing restrictions on the long-run accumulated effects of the innovations.

Short-run SVAR models

A short-run SVAR model without exogenous variables can be written as

$$\mathbf{A}(\mathbf{1}_K - \mathbf{A}_1 L - \mathbf{A}_2 L^2 - \cdots - \mathbf{A}_p L^p)\mathbf{y}_t = \mathbf{A}\boldsymbol{\epsilon}_t = \mathbf{B}\mathbf{e}_t \tag{6}$$

where L is the lag operator, $\mathbf{A}$, $\mathbf{B}$, and $\mathbf{A}_i$ $i = 1, \ldots, p$ are $K \times K$ matrices of parameters, $\boldsymbol{\epsilon}_t$ is a $K \times 1$ vector of innovations with $\boldsymbol{\epsilon}_t \sim N(\mathbf{0}, \boldsymbol{\Sigma})$ and $E[\boldsymbol{\epsilon}_t \boldsymbol{\epsilon}_s'] = \mathbf{0}_K$ for all $s \neq t$, and $\mathbf{e}_t$ is a $K \times 1$ vector of orthogonalized disturbances; i.e., $\mathbf{e}_t \sim N(\mathbf{0}, \mathbf{I}_K)$ and $E[\mathbf{e}_t \mathbf{e}_s'] = \mathbf{0}_K$ for all $s \neq t$. These transformations of the innovations allow us to analyze the dynamics of the system in terms of a change to an element of $\mathbf{e}_t$.

In a short-run SVAR model, identification is obtained by placing restrictions on $\mathbf{A}$ and $\mathbf{B}$, which are assumed to be nonsingular.

Among the implications of (6) is that $\mathbf{P}_{sr} = \mathbf{A}^{-1}\mathbf{B}$, where $\mathbf{P}_{sr}$ is the $\mathbf{P}$ matrix identified by a particular short-run SVAR model. To see this point, note that the latter equality in (6) implies that

$$\mathbf{A}\boldsymbol{\epsilon}_t \boldsymbol{\epsilon}_t' \mathbf{A}' = \mathbf{B}\mathbf{e}_t \mathbf{e}_t' \mathbf{B}'$$

Taking the expectation of both sides yields

$$\boldsymbol{\Sigma} = \mathbf{P}_{sr} \mathbf{P}'_{sr}$$

Assuming that the underlying VAR is stable, (see [TS] **varstable** for a discussion of stability), one can invert the autoregressive representation of the model in (6) to an infinite-order moving average representation of the form

$$\mathbf{y}_t = \boldsymbol{\mu} + \sum_{s=0}^{\infty} \boldsymbol{\Theta}_s^{sr} \mathbf{e}_{t-s} \tag{7}$$

whereby $\mathbf{y}_t$ is expressed in terms of the mutually orthogonal, unit-variance structural innovations $\mathbf{e}_t$. The $\boldsymbol{\Theta}_s^{sr}$ contain the structural impulse–response functions at horizon s.

In a short-run SVAR model, the $\mathbf{A}$ and $\mathbf{B}$ matrices model all the information about contemporaneous correlations. The $\mathbf{B}$ matrix also scales the innovations $\mathbf{u}_t$ to have unit variance. This allows the structural impulse–response functions constructed from (7) to be interpreted as the effect on variable i of a one-time unit increase in the structural innovation to variable j, after s periods.

❑ Technical Note

Another useful implication of (6) is that the free parameters in $\mathbf{A}$ and $\mathbf{B}$ can be estimated from a concentrated log-likelihood.

Since $\boldsymbol{\Sigma}$ can be estimated from the underlying VAR(p), the concentrated log-likelihood for the short-run SVAR model is

$$L(\mathbf{A}, \mathbf{B}) = -\frac{NK}{2} \ln(2\pi) + \frac{N}{2} \ln(|\mathbf{W}|^2) - \frac{N}{2} \text{tr}(\mathbf{W}'\mathbf{W}\widehat{\boldsymbol{\Sigma}})$$

where $N = T$ is the number of observations in the underlying VAR, K is the number of the equations in the underlying VAR, and $\mathbf{W} = \mathbf{B}^{-1}\mathbf{A}$.

As discussed in [TS] **var**, specifying the dfk option causes the small sample adjusted estimator of Σ to be used instead of the default method that uses a divisor of $1/T$. Hence, specifying the dfk option will alter the resulting estimate of Σ. Since the concentrated log-likelihood depends on $\widehat{\Sigma}$, specifying dfk will cause the estimates of the parameters in $\mathbf{A}$ and $\mathbf{B}$ to change.

❑

$\mathbf{P}_{sr}$ identifies the structural impulse–response functions, and $\mathbf{P}_{sr}$ itself is identified by the restrictions placed on the parameters in $\mathbf{A}$ and $\mathbf{B}$. $\mathbf{P}_{sr}$ defines a transformation of Σ that identifies the impulse–response functions. Since there are only $K(K+1)/2$ free parameters in Σ, only $K(K+1)/2$ parameters may be estimated in an identified $\mathbf{P}_{sr}$. Since there are $2K^2$ total parameters in $\mathbf{A}$ and $\mathbf{B}$, the order condition for identification requires that there must be at least $2K^2 - K(K+1)/2$ restrictions placed on those parameters. Just as in the simultaneous equations framework, this order condition is necessary but not sufficient. Amisano and Giannini (1997) derived a method to check that an SVAR model is locally identified in the neighborhood of some specified values for $\mathbf{A}$ and $\mathbf{B}$. As discussed in *Methods and Formulas*, by default, svar performs the Amisano and Giannini (1997) check for identification before estimation. If the check for identification fails, then svar exits with an error message.

❑ Technical Note

As noted by Amisano and Giannini (1997), this check for identification is local to the starting values used. Because of this dependence on the starting values, one may wish to suppress this check by specifying the noidencheck option; however, great care should be taken before doing so. Models that are not structurally identified can still converge, thereby producing meaningless results that only appear to have meaning. Modelers should only use the noidencheck option if they are certain that the model is identified and that the identification check is failing because of the starting values. Finally, it is a good idea to check local identification of the converged results from any model that was fit without previously checking identification. This can be done by specifying the converged results as starting values and re-fitting the same model.

❑

▷ Example

Following Sims (1980), the Cholesky decomposition is one method of identifying the impulse–response functions in a VAR. Thus, the Cholesky decomposition identification method corresponds to an SVAR. There are several sets of constraints on $\mathbf{A}$ and $\mathbf{B}$ that are easily manipulated back to the Cholesky decomposition, and the following example illustrates this point.

One way of imposing the Cholesky restrictions is to assume an SVAR model of the form

$$\widetilde{\mathbf{A}}(\mathbf{I}_K - \mathbf{A}_1 - \mathbf{A}_2 L^2 - \cdots \mathbf{A}_p L^p)\mathbf{y}_t = \widetilde{\mathbf{B}}\mathbf{e}_t$$

where $\widetilde{\mathbf{A}}$ is a lower triangular matrix with ones on the diagonal and $\widetilde{\mathbf{B}}$ is a diagonal matrix. Since the $\mathbf{P}$ matrix for this model is $\mathbf{P}_{sr} = \widetilde{\mathbf{A}}^{-1}\widetilde{\mathbf{B}}$, its estimate, $\widehat{\mathbf{P}}_{sr}$, obtained by plugging in estimates of $\widetilde{\mathbf{A}}$ and $\widetilde{\mathbf{B}}$, should be equal to the Cholesky decomposition of $\widehat{\Sigma}$.

Let's consider a simple three-equation model that uses the German macroeconomic data discussed in Lütkepohl (1993). (These are the same data used in [TS] **var**). In this example, $\mathbf{y}_t = ($dlinvestment, dlincome, dlconsumption$)$, where dlinvestment the first difference of the log of investment, dlincome is the first difference of the log of income, and dlconsumption is the first difference of the log of consumption. Since the first difference of the natural log of a variable can be treated as an approximation of the percentage change in that variable, we shall simply refer to these variables as percentage changes in investment, income and consumption, respectively.

One way of imposing the Cholesky restrictions on this system is by applying equality constraints using the constraint matrices

$$
\mathbf{A} = \begin{bmatrix} 1 & 0 & 0 \\ . & 1 & 0 \\ . & . & 1 \end{bmatrix} \quad \text{and} \quad \mathbf{B} = \begin{bmatrix} . & 0 & 0 \\ 0 & . & 0 \\ 0 & 0 & . \end{bmatrix}
$$

In these structural restrictions, we assume that the percentage change in investment is not contemporaneously affected by the percentage changes in either income nor consumption. We also assume that the percentage change of income is affected by contemporaneous changes in investment but not consumption. Finally, we assume that percentage changes in consumption are affected by contemporaneous changes in both investment and income.

The following fits an SVAR model with these constraints.

```
. use http://www.stata-press.com/data/r8/lutkepohl
(Quarterly SA West German macro data, Bil DM, from Lutkepohl 1993 Table E.1)
. mat A = (1,0,0\.,1,0\.,.,1)
. mat B = (.,0,0\0,.,0\0,0,.)
. svar dlinvestment dlincome dlconsumption if qtr <= q(1978q4), aeq(A) beq(B)
Estimating short-run parameters
Iteration 0:   log likelihood = -230.04647
  (output omitted)
Structural vector autoregression

Constraints:
 ( 1)   [a_1_1]_cons = 1
 ( 2)   [a_1_2]_cons = 0
 ( 3)   [a_1_3]_cons = 0
 ( 4)   [a_2_2]_cons = 1
 ( 5)   [a_2_3]_cons = 0
 ( 6)   [a_3_3]_cons = 1
 ( 7)   [b_1_2]_cons = 0
 ( 8)   [b_1_3]_cons = 0
 ( 9)   [b_2_1]_cons = 0
 (10)   [b_2_3]_cons = 0
 (11)   [b_3_1]_cons = 0
 (12)   [b_3_2]_cons = 0
```

Sample: 1960q4 1978q4	Number of obs	=	73
	Log likelihood	=	606.30704

Just-identified model

Equation	Obs	Parms	RMSE	R-sq	chi2	P
dlinvestment	73	7	.046148	0.1286	10.76961	0.0958
dlincome	73	7	.011719	0.1142	9.410683	0.1518
dlconsumpt~n	73	7	.009445	0.2513	24.50031	0.0004

(Continued on next page)

```
VAR Model lag order selection statistics
```

FPE	AIC	HQIC	SBIC	LL	Det(Sigma_ml)
2.183e-11	-16.035809	-15.773227	-15.37691	606.30704	1.226e-11

	Coef.	Std. Err.	z	P>|z|	[95% Conf. Interval]
a_2_1					
_cons	-.0336288	.0294605	-1.14	0.254	-.0913702 .0241126
a_3_1					
_cons	-.0435846	.0194408	-2.24	0.025	-.0816879 -.0054812
a_3_2					
_cons	-.424774	.0765548	-5.55	0.000	-.5748187 -.2747293
b_1_1					
_cons	.0438796	.0036315	12.08	0.000	.036762 .0509972
b_2_2					
_cons	.0110449	.0009141	12.08	0.000	.0092534 .0128365
b_3_3					
_cons	.0072243	.0005979	12.08	0.000	.0060525 .0083962

The SVAR output has five parts: an iteration log, a display of the constraints imposed, an initial header with sample and SVAR log-likelihood information, a header with statistics from the underlying VAR, and a table displaying the estimates of the parameters from the **A** and **B** matrices. From the output above, we can see that the equality constraint matrices supplied to svar did indeed impose the intended constraints and that the SVAR header informs us that the model we fit is just-identified. Interpretation of the VAR header information is discussed in [TS] **var**. The need for identification restrictions implies that there many unestimated parameters in **A** and **B**. By default, to make the output table more readable, only the estimated parameters appear in the table. However, the saved matrices, e(b) and e(V), contain the information on all the parameters, and a complete output table is available by specifying the full option. Stata's replay facility allows you to simply type svar, full after a previously fitted SVAR to produce a complete output table. With respect to the coefficients estimated in this example, note that the estimates of a_2_1, a_3_1, and a_3_2 are all negative. Since the off diagonal elements of the **A** matrix contain the negative of the actual contemporaneous effects, the estimated effects are positive, as expected.

While a complete discussion of the saved results is available below, here we should note that svar saves the estimates in their matrix form in e(A) and e(B), respectively. This allows us to easily compute the estimated Cholesky decomposition.

```
. mat Aest = e(A)
. mat Best = e(B)
. mat chol_est = inv(Aest)*Best
. mat list chol_est

chol_est[3,3]
                    dlinvestment        dlincome   dlconsumption
   dlinvestment        .04387957               0               0
       dlincome        .00147562       .01104494               0
  dlconsumption        .00253928        .0046916       .00722432
```

The estimated Σ is the same for var and svar, and svar saves off this estimate in e(Sigma). The output below illustrates computation of the Cholesky decomposition of $\widehat{\Sigma}$. Note that it is the same as the one computed from the SVAR estimates.

```
. mat sig_var = e(Sigma)

. mat chol_var = cholesky(sig_var)

. mat list chol_var

chol_var[3,3]
                    dlinvestment       dlincome   dlconsumption
    dlinvestment       .04387957              0               0
        dlincome       .00147562       .01104494              0
   dlconsumption       .00253928        .0046916       .00722432
```

◁

❏ Technical Note

At this point, one might wonder why we bother with obtaining parameter estimates via nonlinear estimation if they can simply be obtained by a transform of the estimates produced by var. When the model is just-identified, as in the previous example, the SVAR parameter estimates can be computed via a transform of the VAR estimates. However, when the model is over-identified, such is not the case.

❏

▷ Example

As noted above, an identified SVAR imposes sufficient restrictions to provide a structural interpretation to the structural impulse–response functions. This is a common reason for fitting an SVAR. In the previous example, we estimated the parameters of an SVAR that was equivalent to a Cholesky decomposition of the estimated Σ. This equivalence implies that we should expect the orthogonalized impulse–response functions and the structural impulse–response functions to be identical. Below, we use varirf create (see [TS] **varirf create**) to estimate the impulse–response functions and varirf ctable, (see [TS] **varirf ctable**), to create a table containing the estimated structural impulse response functions, the estimated orthogonalized impulse–response functions, and their standard errors.

```
. varirf create chol1, set(irfs1, new)
current varirf data file is irfs1.vrf
file irfs1.vrf saved

. varirf ctable (chol1 dlincome dlconsumption oirf sirf), noci std
```

| | (1) | (1) | (1) | (1) |
step	oirf	S.E.	sirf	S.E.
0	.004692	.00093	.004692	.00093
1	.001245	.001036	.001245	.001036
2	.003397	.00106	.003397	.00106
3	-.000658	.000758	-.000658	.000758
4	.00086	.000667	.00086	.000667
5	.000312	.000382	.000312	.000382
6	.00002	.000321	.00002	.000321
7	.000147	.000166	.000147	.000166

(1) irfname = chol1, impulse = dlincome, and response = dlconsumption

The table shows that the estimated orthogonalized impulse–response functions are the same as the estimated structural impulse–response functions, and that the estimates of their corresponding standard errors are also the same.

◁

▷ Example

The Cholesky decomposition example above fit a just-identified model. This example considers an overidentified model. In the previous example, the a_2_1 parameter was not significant. Suppose that we had a theory in which changes in our measure of investment only affect changes in income with a lag. This would place one extra restriction on the Cholesky structure assumed in the previous example. This would correspond to an SVAR of the form

$$\mathbf{A}(\mathbf{I}_3 - \mathbf{A}_1 - \mathbf{A}_2 L^2)\mathbf{y}_t = \mathbf{B}\mathbf{e}_t$$

where

$$\mathbf{A} = \begin{bmatrix} 1 & 0 & 0 \\ 0 & 1 & 0 \\ . & . & 1 \end{bmatrix} \quad \text{and} \quad \mathbf{B} = \begin{bmatrix} . & 0 & 0 \\ 0 & . & 0 \\ 0 & 0 & . \end{bmatrix}$$

The output below contains the command and results that we obtained from fitting this model on the Lütkepohl data.

```
. mat B = (.,0,0\0,.,0\0,0,.)
. mat A = (1,0,0\0,1,0\.,.,1)
. svar dlinvestment dlincome dlconsumption if qtr <= q(1978q4), aeq(A) beq(B)
Estimating short-run parameters
Iteration 0:   log likelihood =  -229.9865
 (output omitted )
Structural vector autoregression
Constraints:
 ( 1)   [a_1_1]_cons = 1
 ( 2)   [a_1_2]_cons = 0
 ( 3)   [a_1_3]_cons = 0
 ( 4)   [a_2_1]_cons = 0
 ( 5)   [a_2_2]_cons = 1
 ( 6)   [a_2_3]_cons = 0
 ( 7)   [a_3_3]_cons = 1
 ( 8)   [b_1_2]_cons = 0
 ( 9)   [b_1_3]_cons = 0
 (10)   [b_2_1]_cons = 0
 (11)   [b_2_3]_cons = 0
 (12)   [b_3_1]_cons = 0
 (13)   [b_3_2]_cons = 0
Sample:  1960q4   1978q4                Number of obs    =        73
                                        Log likelihood   = 605.66129
LR test of overidentifying restrictions LR chi2(  1)     = 1.2915054
                                        Prob > chi2      =    0.2558
```

Equation	Obs	Parms	RMSE	R-sq	chi2	P
dlinvestment	73	7	.046148	0.1286	10.76961	0.0958
dlincome	73	7	.011719	0.1142	9.410683	0.1518
dlconsumpt~n	73	7	.009445	0.2513	24.50031	0.0004

(Continued on next page)

```
VAR Model lag order selection statistics
```

FPE	AIC	HQIC	SBIC	LL	Det(Sigma_ml)
2.183e-11	-16.035809	-15.773227	-15.37691	606.30704	1.226e-11

	Coef.	Std. Err.	z	P>|z|	[95% Conf. Interval]
a_3_1						
_cons	-.0435911	.0192696	-2.26	0.024	-.0813589	-.0058233
a_3_2						
_cons	-.4247741	.0758806	-5.60	0.000	-.5734973	-.2760508
b_1_1						
_cons	.0438796	.0036315	12.08	0.000	.036762	.0509972
b_2_2						
_cons	.0111431	.0009222	12.08	0.000	.0093356	.0129506
b_3_3						
_cons	.0072243	.0005979	12.08	0.000	.0060525	.0083962

Note that the SVAR header in this example contains a test of the overidentifying restrictions. As discussed in Amisano and Giannini (1997), since this test of the over-identifying restrictions assumes that the identifying restrictions are valid, the null hypothesis of this test is that the identifying and the overidentifying restrictions are jointly valid.

In the case at hand, we cannot reject this null hypothesis at any of the conventional levels.

◁

▷ Example

In the previous example, we proceeded with our SVAR estimation without any concern for the underlying VAR, since we were only trying to highlight how to use the svar command. Of course, if we were actually analyzing these data, we would want to pay close attention to the form and fit of the underlying VAR. All of the diagnostic post-estimation commands for VARs, work after svar. (See [TS] **var** for a complete list.) Since these post-estimation commands are demonstrated elsewhere, we will proceed with an example demonstrating how to fit an SVAR model in which additional constraints are imposed on the coefficients of the underlying VAR.

Let's begin by taking a quick look at the underlying VAR for the SVARs that we have used in the previous two examples.

```
. var dlinvestment dlincome dlconsumption if qtr <= q(1978q4)
Vector autoregression
Sample:  1960q4   1978q4
```

Equation	Obs	Parms	RMSE	R-sq	chi2	P
dlinvestment	73	7	.046148	0.1286	10.76961	0.0958
dlincome	73	7	.011719	0.1142	9.410683	0.1518
dlconsumpt~n	73	7	.009445	0.2513	24.50031	0.0004

Model lag order selection statistics

FPE	AIC	HQIC	SBIC	LL	Det(Sigma_ml)
2.183e-11	-16.035809	-15.773227	-15.37691	606.30704	1.226e-11

	Coef.	Std. Err.	z	P>|z|	[95% Conf. Interval]	
dlinvestment						
dlinvestment						
L1	-.3196318	.1192898	-2.68	0.007	-.5534355	-.0858282
L2	-.1605508	.118767	-1.35	0.176	-.39333	.0722283
dlincome						
L1	.1459851	.5188451	0.28	0.778	-.8709326	1.162903
L2	.1146009	.508295	0.23	0.822	-.881639	1.110841
dlconsumpt~n						
L1	.9612288	.6316557	1.52	0.128	-.2767936	2.199251
L2	.9344001	.6324034	1.48	0.140	-.3050877	2.173888
_cons	-.0167221	.0163796	-1.02	0.307	-.0488257	.0153814
dlincome						
dlinvestment						
L1	.0439309	.0302933	1.45	0.147	-.0154427	.1033046
L2	.0500302	.0301605	1.66	0.097	-.0090833	.1091437
dlincome						
L1	-.1527311	.131759	-1.16	0.246	-.4109741	.1055118
L2	.0191634	.1290799	0.15	0.882	-.2338285	.2721552
dlconsumpt~n						
L1	.2884992	.1604069	1.80	0.072	-.0258926	.6028909
L2	-.0102	.1605968	-0.06	0.949	-.3249639	.3045639
_cons	.0157672	.0041596	3.79	0.000	.0076146	.0239198
dlconsumpt~n						
dlinvestment						
L1	-.002423	.0244142	-0.10	0.921	-.050274	.045428
L2	.0338806	.0243072	1.39	0.163	-.0137607	.0815219
dlincome						
L1	.2248134	.1061884	2.12	0.034	.0166879	.4329389
L2	.3549135	.1040292	3.41	0.001	.1510199	.558807
dlconsumpt~n						
L1	-.2639695	.1292766	-2.04	0.041	-.517347	-.010592
L2	-.0222264	.1294296	-0.17	0.864	-.2759039	.231451
_cons	.0129258	.0033523	3.86	0.000	.0063554	.0194962

The equation-level model tests reported in the header indicate that we cannot reject the null hypotheses that all of the coefficients in the first equation are zero, and similarly for the second equation. In order to address this problem, we use a combination of theory and the p-values from the output above to place some exclusion restrictions on the underlying VAR(2). Specifically, in the equation for the percentage change of investment, we constrain the coefficients on L2.dlinvestment, L.dlincome, L2.dlincome and L2.dlconsumption to be zero. In the equation for dlincome, we constrain the coefficients on L2.dlinvestment, L2.dlincome, and L2.dlconsumption to be zero. Finally, in the equation for dlconsumption, we constrain L.dlinvestment and L2.dlconsumption to be zero. We then refit the SVAR from the previous example.

```
. constraint define 1 [dlinvestment]L2.dlinvestment = 0
. constraint define 2 [dlinvestment]L.dlincome = 0
. constraint define 3 [dlinvestment]L2.dlincome = 0
. constraint define 4 [dlinvestment]L2.dlconsumption = 0
. constraint define 5 [dlincome]L2.dlinvestment = 0
```

```
. constraint define 6 [dlincome]L2.dlincome = 0

. constraint define 7 [dlincome]L2.dlconsumption = 0

. constraint define 8 [dlconsumption]L.dlinvestment = 0

. constraint define 9 [dlconsumption]L2.dlconsumption = 0

. svar dlinvestment dlincome dlconsumption if qtr <= q(1978q4), aeq(A) beq(B)
> varconstraints(1/9) noislog

Estimating short-run parameters

Iteration 0:   log likelihood = -229.99297
  (output omitted)

Structural vector autoregression

Constraints:
 ( 1)   [a_1_1]_cons = 1
 ( 2)   [a_1_2]_cons = 0
 ( 3)   [a_1_3]_cons = 0
 ( 4)   [a_2_1]_cons = 0
 ( 5)   [a_2_2]_cons = 1
 ( 6)   [a_2_3]_cons = 0
 ( 7)   [a_3_3]_cons = 1
 ( 8)   [b_1_2]_cons = 0
 ( 9)   [b_1_3]_cons = 0
 (10)   [b_2_1]_cons = 0
 (11)   [b_2_3]_cons = 0
 (12)   [b_3_1]_cons = 0
 (13)   [b_3_2]_cons = 0
```

```
Sample: 1960q4   1978q4                  Number of obs    =          73
                                         Log likelihood   =   601.85907
LR test of overidentifying restrictions  LR chi2(  1)     =   .84481283
                                         Prob > chi2      =      0.3580
```

Equation	Obs	Parms	RMSE	R-sq	chi2	P
dlinvestment	73	3	.045185	0.0759	6.036828	0.0489
dlincome	73	4	.011372	0.0774	7.326193	0.0622
dlconsumpt~n	73	5	.009023	0.2441	27.16134	0.0000

VAR Model lag order selection statistics

FPE	AIC	HQIC	SBIC	LL	Det(Sigma_ml)
1.902e-11	-15.92552	-15.662937	-15.266621	602.28148	1.369e-11

	Coef.	Std. Err.	z	P>|z|	[95% Conf. Interval]
a_3_1						
_cons	-.0418708	.0187579	-2.23	0.026	-.0786356	-.0051061
a_3_2						
_cons	-.4255808	.0745298	-5.71	0.000	-.5716565	-.2795051
b_1_1						
_cons	.0451851	.0037395	12.08	0.000	.0378557	.0525145
b_2_2						
_cons	.0113723	.0009412	12.08	0.000	.0095276	.013217
b_3_3						
_cons	.0072417	.0005993	12.08	0.000	.006067	.0084164

If we had displayed the underlying VAR(2) results using the `var` option, we would see that most of the unconstrained coefficients are now significant at the 10% level, and none of the equation level model statistics reject the null at the 10% level. The `svar` output reveals that while the p-value of the overidentification test rose by a small amount, the coefficient on `a_3_1` is still insignificant at the 1% level, but not at the 5% level.

◁

Before moving on to models with long-run constraints, we should note some limitations. It is not possible to place constraints on the elements of **A** in terms of the elements of **B**, or vice versa. This limitation is imposed by the form of the check for identification derived by Amisano and Giannini (1997). As noted in *Methods and Formulas*, this test requires separate constraint matrices for the parameters in **A** and **B**. Another limitation is that it is not possible to mix short-run and long-run constraints.

Long-run restrictions

Recall that a general short-run SVAR has the form

$$\mathbf{A}(\mathbf{I}_K - \mathbf{A}_1 L - \mathbf{A}_2 L^2 - \cdots - \mathbf{A}_p L^p)\mathbf{y}_t = \mathbf{B}\mathbf{e}_t$$

To simplify the notation, let $\bar{\mathbf{A}} = (\mathbf{I}_K - \mathbf{A}_1 L - \mathbf{A}_2 L^2 - \cdots - \mathbf{A}_p L^p)$. Since the model is assumed to be *stable*, $\bar{\mathbf{A}}^{-1}$, the matrix of estimated long-run, or accumulated, effects of the reduced form VAR shocks, is well defined. ([TS] **varstable** provides a brief discussion of stability; Lutkepohl (1993, Chapter 2) derives the applicable results.) Constraining **A** to be an identity matrix allows us to rewrite this equation as

$$\mathbf{y}_t = \bar{\mathbf{A}}^{-1}\mathbf{B}\mathbf{e}_t$$

Since **A** is set to an identity matrix, $\boldsymbol{\Sigma} = \mathbf{B}\mathbf{B}'$. Thus, $\mathbf{C} = \bar{\mathbf{A}}^{-1}\mathbf{B}$ is the matrix of long-run responses to the orthogonalized shocks, and

$$\mathbf{y}_t = \mathbf{C}\mathbf{e}_t$$

In long-run models, the constraints are placed on the elements of **C**, and the free parameters are estimated. These constraints are frequently exclusion restrictions. For instance, constraining $\mathbf{C}[1, 2]$ to be zero can be interpreted as setting the long-run response of variable 1 to the structural shocks driving variable 2 to be zero.

▷ Example

Suppose that we had a theory which stated that unexpected changes to the money supply should have no long-run effect on changes in output, and that, similarly, unexpected changes in output should have no long-run effect on changes in the money supply. The **C** matrix implied by this conjecture is

$$\mathbf{C} = \begin{bmatrix} \cdot & 0 \\ 0 & \cdot \end{bmatrix}$$

Our theory has placed one overidentifying restriction on the model, and this overidentifying restriction implies that we will obtain a test of the null hypothesis that the restrictions are valid.

```
. use http://www.stata-press.com/data/r8/m1gdp

. mat lr = (.,0\0,.)

. svar d.ln_m1 d.ln_gdp, lreq(lr)
Estimating long-run parameters

Iteration 0:   log likelihood = -27.958026
 (output omitted )

Structural vector autoregression

Constraints:
 ( 1)  [c_1_2]_cons = 0
 ( 2)  [c_2_1]_cons = 0

Sample: 1959q4   2002q2                  Number of obs     =        171
                                         Log likelihood    =  1151.6143
LR test of overidentifying restrictions  LR chi2( 1)       =  .13675517
                                         Prob > chi2       =     0.7115
```

Equation	Obs	Parms	RMSE	R-sq	chi2	P
D.ln_m1	171	5	.008509	0.4732	153.5779	0.0000
D.ln_gdp	171	5	.008448	0.1140	22.00553	0.0002

```
VAR Model lag order selection statistics
```

FPE	AIC	HQIC	SBIC	LL	Det(Sigma_ml)
5.444e-09	-13.353014	-13.278467	-13.169291	1151.6827	4.843e-09

	Coef.	Std. Err.	z	P>|z|	[95% Conf. Interval]	
c_1_1						
_cons	.0301007	.0016277	18.49	0.000	.0269106	.0332909
c_2_2						
_cons	.0129691	.0007013	18.49	0.000	.0115946	.0143436

We have assumed that the underlying VAR has 2 lags; four of the five selection order criteria computed by varsoc (see [TS] **varsoc**) recommended this choice. The header indicates that the equations in the VAR(2) fit well and that the null hypothesis that the restrictions on the model are valid cannot be rejected.

◁

Just as in the short-run model, the $\mathbf{P}_{lr}$ matrix identifies the structural impulse–response functions. $\mathbf{P}_{lr} = \mathbf{C}$ is identified by the restrictions placed on the parameters in $\mathbf{C}$. There are K^2 parameters in $\mathbf{C}$, and the order condition for identification requires that there be at least $K^2 - K(K+1)/2$ restrictions placed on those parameters. As in the short-run model, this order condition is necessary but not sufficient, so the Amisano and Giannini (1997) check for local identification is performed by default.

(Continued on next page)

Saved Results

svar saves in e():

Scalars

e(neqs)	number of equations
e(neqs_var)	number of equations in underlying VAR
e(df_eq_var)	average number of parameters in an equation
e(k_var)	number of coefficients in VAR
e(obs_#_var)	number of observations on equation #
e(r2_#_var)	R-squared for equation #
e(df_m#_var)	model degrees of freedom for equation #
e(df_r_var)	if small, residual degrees of freedom
e(chi2_#_var)	χ^2 statistic for equation #
e(ll_var)	log likelihood from var
e(sbic_var)	Schwartz–Bayesian information criteria
e(aic_var)	Akaike information criteria
e(mlag_var)	highest lag in VAR
e(tmin)	first time period in sample
e(detsig_var)	determinant of e(Sigma)
e(df_m_var)	degrees of freedom in model
e(chi2_oid)	overidentification test
e(ic_ml)	iteration count from ml
e(N)	number of observations
e(T_var)	number of observations
e(tparms_var)	number of parameters in all equations
e(k_#_var)	number of coefficients in equation #
e(rmse_#_var)	root mean square for equation #
e(ll_#_var)	log likelihood for equation #
e(df_r#_var)	residual degrees of freedom for equation # (small only)
e(F_#_var)	F statistic for equation # (small only)
e(ll)	log likelihood from svar
e(hqic_var)	Hannan–Quinn information criteria
e(fpe_var)	final prediction error
e(tmax)	last time period in sample
e(N_gaps_var)	number of gaps in sample
e(detsig_ml_var)	determinant of $\widehat{\Sigma}_m l$
e(oid_df)	number of overidentification restrictions
e(rc_ml)	return code from ml
e(N_cns)	number of constraints

(Continued on next page)

Macros

e(cmd)	svar
e(exog_var)	names of exogenous variables, if specified
e(depvar_var)	names of dependent variable(s)
e(small)	small, if specified
e(tsfmt)	format of timevar
e(nocons_var)	noconstant, if noconstant specified
e(dfk_var)	alternate divisor (dfk), if specified
e(lrmodel)	long-run model, if specified
e(endog_var)	names of endogenous variables
e(eqnames_var)	names of equations
e(lags_var)	lags in model
e(lutstats_var)	lutstats, if specified
e(timevar)	name of timevar
e(constraints_var)	constraints_var, if constraints on VAR
e(vcetype)	EIM
e(predict)	program used to implement predict

Matrices

e(b)	coefficient vector
e(bf_var)	full coefficient vector with zeros in dropped lags
e(G_var)	G matrix saved by var, see [TS] **var** *Methods and formulas*
e(Sigma)	$\widehat{\boldsymbol{\Sigma}}$ matrix
e(B)	estimated B matrix
e(A1)	estimated $\bar{\mathbf{A}}$ matrix, if long-run model
e(beq)	beq(*matrix*), if specified
e(bcns)	bcns(*matrix*), if specified
e(lrcns)	lrcns(*matrix*), if specified
e(V)	variance–covariance matrix of the estimators
e(Cns_var)	constraint matrix from var, if varconstraints() specified
e(A)	estimated A matrix, if short-run model
e(C)	estimated C matrix, if long-run model
e(aeq)	aeq(*matrix*), if specified
e(acns)	acns(*matrix*), if specified
e(lreq)	lreq(*matrix*), if specified

Functions

e(sample)	marks estimation sample

Methods and Formulas

svar is implemented as an ado-file.

As derived in the text, the log-likelihood function for models with short-run constraints is

$$L(\mathbf{A}, \mathbf{B}) = -\frac{NK}{2}\ln(2\pi) + \frac{N}{2}\ln(|\mathbf{W}|^2) - \frac{N}{2}\mathrm{tr}(\mathbf{W}'\mathbf{W}\widehat{\boldsymbol{\Sigma}})$$

where $\mathbf{W} = \mathbf{B}^{-1}\mathbf{A}$.

When there are long-run constraints, since $\mathbf{C} = \bar{\mathbf{A}}^{-1}\mathbf{B}$ and $\mathbf{A} = \mathbf{I}_K$, $\mathbf{W} = \mathbf{B}^{-1} = \mathbf{C}^{-1}\bar{\mathbf{A}}^{-1} = (\bar{\mathbf{A}}\mathbf{C})^{-1}$. Substituting the last term for $\mathbf{W}$ in the short-run log-likelihood produces the long-run log-likelihood,

$$L(\mathbf{C}) = -\frac{NK}{2}\ln(2\pi) + \frac{N}{2}\ln(|\widetilde{\mathbf{W}}|^2) - \frac{N}{2}\mathrm{tr}(\widetilde{\mathbf{W}}'\widetilde{\mathbf{W}}\widehat{\mathbf{\Sigma}})$$

where $\widetilde{\mathbf{W}} = (\bar{\mathbf{A}}\mathbf{C})^{-1}$.

For both the short-run and the long-run models, the maximization is performed via the method of scoring. See Harvey (1999) for a discussion of this method.

Using results from Amisano and Giannini (1997), the score vector for the short-run model is

$$\frac{\partial L(\mathbf{A}, \mathbf{B})}{\partial[\mathrm{vec}(\mathbf{A}), \mathrm{vec}(\mathbf{B})]} = N\left[\{\mathrm{vec}(\mathbf{W}'^{-1})\}' - \{\mathrm{vec}(\mathbf{W})\}'(\widehat{\mathbf{\Sigma}} \otimes \mathbf{I}_K)\right] \times$$
$$\left[(\mathbf{I}_K \otimes \mathbf{B}^{-1}), -(\mathbf{A}'\mathbf{B}'^{-1} \otimes \mathbf{B}^{-1})\right]$$

and the expected information matrix is

$$I\left[\mathrm{vec}(\mathbf{A}), \mathrm{vec}(\mathbf{B})\right] = N\begin{bmatrix}(\mathbf{W}^{-1} \otimes \mathbf{B}'^{-1}) \\ -(\mathbf{I}_K \otimes \mathbf{B}'^{-1})\end{bmatrix}(\mathbf{I}_{K^2} + \oplus)\left[(\mathbf{W}'^{-1} \otimes \mathbf{B}^{-1}), -(\mathbf{I}_K \otimes \mathbf{B}^{-1})\right]$$

where $\oplus$ is the commutation matrix defined on pages 46–48 of Magnus and Neudecker (1999).

Using results from Amisano and Giannini (1997), one can derive the score vector and the expected information matrix for the case with long-run restrictions. The score vector is

$$\frac{\partial L(\mathbf{C})}{\partial \mathrm{vec}(\mathbf{C})} = N\left[\{\mathrm{vec}(\mathbf{W}'^{-1})\}' - \{\mathrm{vec}(\mathbf{W})\}'(\widehat{\mathbf{\Sigma}} \otimes \mathbf{I}_K)\right]\left[-(\bar{\mathbf{A}}'^{-1}\mathbf{C}'^{-1} \otimes \mathbf{C}^{-1})\right]$$

and the expected information matrix is

$$I\left[\mathrm{vec}(\mathbf{C})\right] = N(\mathbf{I}_K \otimes \mathbf{C}'^{-1})(\mathbf{I}_{K^2} + \oplus)(\mathbf{I}_K \otimes \mathbf{C}'^{-1})$$

Checking for identification

This section describes the methods used to check for identification of models with short-run or long-run constraints. Both methods depend on the starting values. By default, `svar` uses starting values constructed by taking a vector of appropriate dimension and applying the constraints. By default, if there are m parameters in the model, then the jth element of the $1 \times m$ vector is $1 + m/100$. `svar` also allows the user to provide starting values.

For the short-run case, the model is identified if the matrix

$$\mathbf{V}_{sr}^* = \begin{bmatrix} \mathbf{N}_K & \mathbf{N}_K \\ \mathbf{N}_K & \mathbf{N}_K \\ \mathbf{R}_a(\mathbf{W}' \otimes \mathbf{B}) & \mathbf{0}_{K^2} \\ \mathbf{0}_{K^2} & \mathbf{R}_a(\mathbf{I}_K \otimes \mathbf{B}) \end{bmatrix}$$

has full column rank of $2K^2$, where $\mathbf{N}_K = (1/2)(\mathbf{I}_{K^2} + \oplus)$, $\mathbf{R}_a$ is the constraint matrix for the parameters in $\mathbf{A}$; i.e., $\mathbf{R}_a \text{vec}(\mathbf{A}) = \mathbf{r}_a$, and $\mathbf{R}_b$ is the constraint matrix for the parameters in $\mathbf{B}$; i.e., $\mathbf{R}_b \text{vec}(\mathbf{B}) = \mathbf{r}_b$.

For the long-run case, using results from the $\mathbf{C}$ model in Amisano and Giannini (1997), it can be shown that the model is identified if the matrix

$$\mathbf{V}_{lr}^* = \left[\begin{array}{c} (\mathbf{I} \otimes \mathbf{C}'^{-1})(2\mathbf{N}_K)(\mathbf{I} \otimes \mathbf{C}^{-1}) \\ \mathbf{R}_c \end{array} \right]$$

has full column rank of K^2, where $\mathbf{R}_c$ is the constraint matrix for the parameters in $\mathbf{C}$; i.e. $\mathbf{R}_c \text{vec}(\mathbf{C}) = \mathbf{r}_c$.

The test of the overidentifying restrictions is computed as

$$LR = 2(LL_{\text{var}} - LL_{\text{svar}})$$

where LR is the value of the test statistic against the null hypothesis that the overidentifying restrictions are valid, LL_{var} is the log likelihood from the underlying VAR(p) model and LL_{svar} is the log likelihood from the SVAR model. The test statistic is asymptotically distributed as $\chi^2(q)$, where q is the number of overidentifying restrictions. Amisano and Giannini (1993, 38–39) emphasize that since this test of the validity of the overidentifying restrictions is an omnibus test, it can be interpreted as a test of the null hypothesis that all of the restrictions are valid.

Since constraints might not be independent by either construction or to due the data, the number of restrictions is not necessarily equal to the number of constraints. The rank of e(V) gives the number of parameters that were independently estimated after applying the constraints. The maximum number of parameters that can be estimated in an identified short-run or long-run SVAR is $K(K + 1)/2$. This implies that the number of overidentifying restrictions, q, is equal to $K(K + 1)/2$ minus the rank of e(V).

The number of overidentifying restrictions is also linked to the order condition for each model. In a short-run SVAR model, there are $2K^2$ parameters. Since no more than $K(K + 1)/2$ parameters may be estimated, the order condition for a short-run SVAR model is that there be at least $2K^2 - K(K + 1)/2$ restrictions placed on the model. Similarly, there are K^2 parameters in long-run SVAR model. Since no more than $K(K + 1)/2$ parameters may be estimated, the order condition for a long-run SVAR model is that there be at least $K^2 - K(K + 1)/2$ restrictions placed on the model.

Acknowledgment

We would like to thank Gianni Amisano, Università di Brescia, for his helpful comments.

References

Amisano, G. and C. Giannini. 1997. *Topics in Structural VAR Econometrics*. 2d ed. Heidelberg: Springer.

Hamilton, J. D. 1994. *Time Series Analysis*. Princeton: Princeton University Press.

Harvey, A. C. 1999. *The Econometric Analysis of Time Series*. 2d ed. Cambridge, MA: The MIT Press.

Lütkepohl, H. 1993. *Introduction to Multiple Time Series Analysis*. 2d ed. New York: Springer.

Magnus, J. R. and H. Neudecker. 1999. *Matrix Differential Calculus with Applications in Statistics and Econometrics*. rev. ed. New York: John Wiley & Sons.

Sims, C. A. 1980. Macroeconomics and reality. *Econometrica* 48: 1–48.

Stock, J. H. and M. W. Watson. 2001. Vector autoregressions. *Journal of Economic Perspectives* 15(4): 101–115.

Also See

Complementary:	[TS] **tsset**,
	[R] **adjust**, [R] **lincom**, [R] **lrtest**, [R] **nlcom**, [R] **predict**,
	[R] **predictnl**, [R] **test**, [R] **testnl**, [R] **vce**
Related:	[TS] **arch**, [TS] **arima**, [TS] **var**, [TS] **varbasic**,
	[R] **reg3**, [R] **regress**, [R] **sureg**
Background:	[U] **16.5 Accessing coefficients and standard errors**,
	[U] **23 Estimation and post-estimation commands**,
	[TS] **var intro**

Title

> **varbasic** — Fit a simple VAR and graph impulse–response functions

Syntax

> varbasic *depvarlist* [if *exp*] [in *range*] [, <u>l</u>ags(*numlist*) <u>s</u>tep(*#*) irf <u>f</u>evd
>
> <u>nog</u>raph]

varbasic is for use with time-series data; see [TS] **tsset**. You must tsset your data before using varbasic.
depvarlist may contain time-series operators; see [U] **14.4.3 Time-series varlists**.
varbasic shares the features of all estimation commands; see [U] **23 Estimation and post-estimation commands**.

Syntax for predict

> predict [*type*] *newvarname* [if *exp*] [in *range*] [, <u>e</u>quation(*eqno* | *eqname*)
>
> xb <u>r</u>esiduals stdp]

These statistics are available both in and out of sample; type predict ... if e(sample) ... if wanted only for the estimation sample.

Description

varbasic fits a basic vector autoregressive (VAR(*p*)) model and graphs the impulse–response functions (IRF), the orthogonalized impulse–response functions (OIRF), or the forecast-error variance decompositions (FEVD).

Options

lags(*numlist*) specifies the lags to be included in the model. The default is lags(1 2). Note that this option takes a numlist and not simply an integer for the maximum lag. For instance, lags(2) would include only the second lag in the model, whereas lags(1/2) would include both the first and second lags in the model. See [U] **14.1.8 numlist** and [U] **16.8 Time-series operators** for further discussion of numlists and lags.

step(*#*) specifies the forecast horizon for estimating the IRFs, OIRFs, and FEVDs. The default is 8 periods.

irf causes varbasic to produce a matrix graph of the IRFs instead of a matrix graph of the OIRFs, which is produced by default. irf may not be combined with fevd.

fevd causes varbasic to produce a matrix graph of the FEVDs instead of a matrix graph of the OIRFs, which is produced by default. fevd may not be combined with irf.

nograph specifies that no graph is to be produced. The IRFs, OIRFs and FEVDs are still estimated and saved in the VARIRF file _varbasic. nograph may not be specified with either irf, or with fevd.

Options for predict

equation(*eqno* | *eqname*) specifies to which equation you are referring.

> equation() is filled in with one *eqno* or *eqname* for options xb, stdp, and residuals. equation(#1) would mean the calculation is to be made for the first equation, equation(#2) would mean the second, and so on. Alternatively, you could refer to the equation by its name. equation(income) would refer to the equation named income and equation(hours) to the equation named hours.

> If you do not specify equation(), the results are as if you specified equation(#1).

xb, the default, calculates the fitted values—for the specified equation.

residuals calculates the residuals.

stdp calculates the standard error of the linear prediction for the specified equation.

Remarks

varbasic makes it easy to fit simple VARs and graph the IRFs, the OIRFs, or the FEVDs. See [TS] **var** and [TS] **var svar** for fitting more advanced VAR models and structural vector autoregressive models (SVAR). All the post-estimation commands discussed in [TS] **var intro** work after varbasic.

This entry does not discuss the methods for fitting a VAR(p) nor the methods surrounding the IRFs, OIRFs, and FEVDs. See [TS] **var** and [TS] **varirf create** and the references therein for more on these methods. This entry illustrates how one can use varbasic to easily obtain results. It also illustrates how varbasic serves as an entry point to further analysis.

▷ Example

Suppose that we wanted to fit a three variable VAR(2) to the German macro data used by Lütkepohl (1993). The three variables are the first difference of natural log of investment, dlinvestment, the first difference of the natural log of income, dlincome, and the first difference of the natural log of consumption, dlconsumption. In addition to fitting the VAR, we want to see the OIRFs. Below, we use varbasic to fit a VAR(2) model on the data from the second quarter of 1961 through the fourth quarter of 1978. By default, varbasic produces graphs of the OIRFs.

(Continued on next page)

```
. use http://www.stata-press.com/data/r8/lutkepohl
(Quarterly SA West German macro data, Bil DM, from Lutkepohl 1993 Table E.1)
. varbasic dlinvestment dlincome dlconsumption if qtr<=q(1978q4)
Vector autoregression
Sample:  1960q4    1978q4
```

Equation	Obs	Parms	RMSE	R-sq	chi2	P
dlinvestment	73	7	.046148	0.1286	10.76961	0.0958
dlincome	73	7	.011719	0.1142	9.410683	0.1518
dlconsumpt~n	73	7	.009445	0.2513	24.50031	0.0004

Model lag order selection statistics

FPE	AIC	HQIC	SBIC	LL	Det(Sigma_ml)
2.183e-11	-16.035809	-15.773227	-15.37691	606.30704	1.226e-11

	Coef.	Std. Err.	z	P>|z|	[95% Conf. Interval]
dlinvestment						
dlinvestment						
L1	-.3196318	.1192898	-2.68	0.007	-.5534355	-.0858282
L2	-.1605508	.118767	-1.35	0.176	-.39333	.0722283
dlincome						
L1	.1459851	.5188451	0.28	0.778	-.8709326	1.162903
L2	.1146009	.508295	0.23	0.822	-.881639	1.110841
dlconsumpt~n						
L1	.9612288	.6316557	1.52	0.128	-.2767936	2.199251
L2	.9344001	.6324034	1.48	0.140	-.3050877	2.173888
_cons	-.0167221	.0163796	-1.02	0.307	-.0488257	.0153814
dlincome						
dlinvestment						
L1	.0439309	.0302933	1.45	0.147	-.0154427	.1033046
L2	.0500302	.0301605	1.66	0.097	-.0090833	.1091437
dlincome						
L1	-.1527311	.131759	-1.16	0.246	-.4109741	.1055118
L2	.0191634	.1290799	0.15	0.882	-.2338285	.2721552
dlconsumpt~n						
L1	.2884992	.1604069	1.80	0.072	-.0258926	.6028909
L2	-.0102	.1605968	-0.06	0.949	-.3249639	.3045639
_cons	.0157672	.0041596	3.79	0.000	.0076146	.0239198
dlconsumpt~n						
dlinvestment						
L1	-.002423	.0244142	-0.10	0.921	-.050274	.045428
L2	.0338806	.0243072	1.39	0.163	-.0137607	.0815219
dlincome						
L1	.2248134	.1061884	2.12	0.034	.0166879	.4329389
L2	.3549135	.1040292	3.41	0.001	.1510199	.558807
dlconsumpt~n						
L1	-.2639695	.1292766	-2.04	0.041	-.517347	-.010592
L2	-.0222264	.1294296	-0.17	0.864	-.2759039	.231451
_cons	.0129258	.0033523	3.86	0.000	.0063554	.0194962

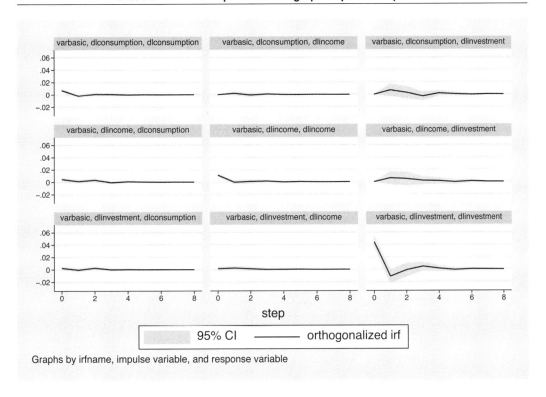

Graphs by irfname, impulse variable, and response variable

Since we are also interested in looking at the FEVDs, we can use `varirf graph` to obtain the graphs. While the details are available in [TS] **varirf** and [TS] **varirf graph**, the command below produces what we want after the call to `varbasic`.

(Continued on next page)

. varirf graph fevd, lstep(1)

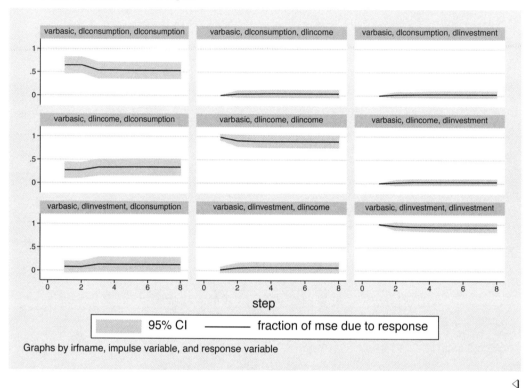

Graphs by irfname, impulse variable, and response variable

◁

❏ **Technical Note**

Stata stores the estimated IRFs OIRFs, and FEVDs in a VARIRF file called _varbasic.vrf in the current working directory. varbasic will replace any _varbasic.vrf that already exists. Finally, varbasic makes _varbasic the active VARIRF file. This means that the graph and table commands varirf graph, varirf cgraph, varirf ograph, varirf table, and varirf ctable will all display results that correspond to the VAR fit by varbasic.

❏

▷ **Example**

All the post-estimation commands discussed in [TS] **var intro** work after varbasic. Suppose that we are interested in testing the hypothesis that there is no autocorrelation in the VAR disturbances. Continuing the example above, we now use varlmar to test this hypothesis.

. varlmar

H0: no autocorrelation at lag order j

j	chi2	df	p
1	4.0531	9	0.90788
2	4.9264	9	0.84068

Since we cannot reject the null hypothesis of no autocorrelation in the residuals, this test does not indicate that there is any model misspecification.

◁

Saved Results

See *Saved Results* in [TS] **var**.

Methods and Formulas

varbasic is implemented as an ado-file.

varbasic uses var and varirf graph to obtain its results. See [TS] **var** and [TS] **varirf graph** for a discussion of how those commands obtain their results.

References

Lütkepohl, H. 1993. *Introduction to Multiple Time Series Analysis.* 2d ed. New York: Springer.

Also See

Complementary:	[TS] **tsset**, [TS] **varirf**
	[R] **adjust**, [R] **lincom**, [R] **linktest**, [R] **lrtest**, [R] **mfx**, [R] **nlcom**,
	[R] **predict**, [R] **predictnl**, [R] **test**, [R] **testnl**, [R] **vce**
Related:	[TS] **arch**, [TS] **arima**, [TS] **var**, [TS] **var svar**, [TS] **varirf create**,
	[R] **reg3**, [R] **regress**, [R] **sureg**
Background:	[U] **16.5 Accessing coefficients and standard errors**,
	[U] **23 Estimation and post-estimation commands**,
	[TS] **var intro**

Title

> **varfcast** — Introduction to dynamic forecasts after var or svar

Syntax

varfcast *subcommand* ...

Description

The are three commands for producing, analyzing and managing dynamic forecasts of the endogenous variables in a VAR(p). `varfcast compute` computes the predictions, the estimated confidence intervals, and the standard errors of the predictions. `varfcast graph` graphs the predictions, the confidence intervals, and the observed values. `varfcast clear` drops the variables previously created by `varfcast compute`.

Command	Documentation	Purpose
varfcast compute	[TS] **varfcast compute**	obtains dynamic forecasts
varfcast graph	[TS] **varfcast graph**	graphs dynamic forecast obtained from varfcast compute
varfcast clear	[TS] **varfcast clear**	drop variables created by varfcast compute

Remarks

A common reason for fitting a VAR(p) is to obtain predictions. One-step-ahead forecasts are predictions of the values that the endogenous variables will take next period, conditional on their current and lagged values and the values of any exogenous variables in the model. This type of prediction is most easily obtained from `predict` after `var` or `svar`. One-step-ahead forecasts are a special case of h-step-ahead forecasts. An h-step-ahead forecast begins as a one-step-ahead forecast, but then it continues, using the forecasted values of the endogenous variables as data to calculate the 2-step-ahead forecasts and so on. Due to the recursive nature of h-step-ahead forecasts, they are known as dynamic forecasts. Stata offers `varfcast compute` for computing dynamic forecasts for a previously fitted VAR(p). After the forecasts have been computed, they can be analyzed and presented using `varfcast graph`. `varcast clear` drops the variables created by `varfcast compute`.

Also See

Complementary:	[TS] **var** [TS] **var svar**, [TS] **varbasic**, [TS] **varfcast clear**, [TS] **varfcast compute**, [TS] **varfcast graph**
Related:	[TS] **arima**, [TS] **vargranger**, [TS] **varirf**, [TS] **varlmar**, [TS] **varnorm**, [TS] **varsoc**, [TS] **varstable**, [TS] **varwle**
Background:	[U] **14.4.3 Time-series varlists**, [TS] **var intro**

Title

> **varfcast clear** — Drop variables containing previous forecasts from varfcast

Syntax

varfcast <u>cl</u>ear [, all]

Description

varfcast clear drops the variables containing the forecasts previously created by varfcast compute.

Options

all specifies that varfcast clear should drop the variables created in all prior calls to varfcast compute. By default, varfcast clear drops the variables created by the immediately preceding call to varfcast compute.

Remarks

There are times when one wants to drop the variables created by varfcast compute. varfcast clear makes this easy.

▷ Example

Below, we fit a VAR(2) model and use varfcast compute to obtain dynamic forecasts of the dependent variables.

```
. use http://www.stata-press.com/data/r8/lutkepohl
(Quarterly SA West German macro data, Bil DM, from Lutkepohl 1993 Table E.1)
. qui var dlinvestment dlincome dlconsumption
. varfcast compute, step(4) nose
```

Since we specified the nose option, varfcast compute will create three new variables: dlinvestment_f, dlincome_f, and dlconsumption_f. In the output below, we use varfcast clear to drop these three variables.

```
. varfcast clear
```

◁

▷ Example

In the previous example, since we made only one call to varfcast compute, there was only one set of variables created by varfcast compute. In this example, we illustrate that the all option will drop the variables from all previous calls to varfcast compute. In the output below we fit two VARs, with distinct endogenous variables, and use varfcast compute to obtain dynamic forecasts for each of the two sets of endogenous variables.

```
. qui var dlinvestment dlincome dlconsumption
. varfcast compute, step(4) nose
. qui var inv_us inc_us con_us
. varfcast compute, step(4) nose
```

If at this point we were to type

```
. varfcast clear
```

then Stata would only drop the variables inv_us_f, inc_us_f, and con_us_f. If instead we were to type

```
. varfcast clear, all
```

then in addition to dropping the variables inv_us_f, inc_us_f, and con_us_f, Stata would also drop the variables dlinvestment_f, dlincome_f, and dlconsumption_f created by the first call to varfcast compute.

◁

Methods and Formulas

varfcast clear is implemented as an ado-file.

Also See

Complementary:	[TS] **var**, [TS] **var svar**, [TS] **varbasic**, [TS] **varfcast compute**
Related:	[TS] **varirf drop**
Background:	[TS] **var intro**, [TS] **varfcast**

Title

varfcast compute — Compute dynamic forecasts of dependent variables after var or svar

Syntax

varfcast <u>c</u>ompute [, <u>s</u>tep(*#*) <u>d</u>ynamic(*time_constant*) bs bsp <u>bsc</u>entile

  <u>r</u>eps(*#*) <u>nodo</u>ts <u>s</u>aving(*filename* [, replace]) nose <u>l</u>evel(*#*)

  clear <u>est</u>imates(*estname*)]

varfcast compute can only be used with var or svar e() results; see [TS] **var** and [TS] **var svar**.

varfcast compute is for use with time-series data; see [TS] **tsset**. You must tsset your data before using varfcast compute.

Description

varfcast compute produces dynamic forecasts of the dependent variables in a previously fitted VAR model. varfcast compute will create new variables and, if necessary, extend the time frame of the dataset to contain the prediction horizon.

Options

step(*#*) specifies the number of periods to be forecast. The default is 1.

dynamic(*time_constant*) specifies the period to begin the dynamic forecasts. The default is the period after the last observation in the estimation sample. The dynamic() option accepts either a Stata date function that returns an integer or an integer that corresponds to a date using the current tsset format. dynamic() must specify a date in or beyond the estimation sample. After fitting a VAR(p), varfcast compute needs to have p lags of the endogenous variables available to begin the dynamic forecasts. For this reason, dynamic() must specify a date such that the p values of the endogenous variables immediately preceding the date specified in dynamic() are not available.

bs specifies that the confidence bounds for the forecasts are to be estimated by a simulation method based on bootstrapping the residuals. bs may not be combined with bsp.

bsp specifies that the confidence bounds for the forecasts are to be estimated via simulation in which the innovations are drawn from a multivariate normal distribution. bsp may not be combined with bs.

bscentile specifies that the lower and upper bounds in the simulation based estimates of the confidence intervals are to be estimated using centiles of the bootstrapped dataset. By default, the lower and upper bounds of the simulation-based estimates of the confidence intervals are estimated using the simulation estimates of the standard errors and the standard normal quantile determined by level(). If bscentile is specified, then either bs or bsp must also be specified.

reps(*#*) gives the number of repetitions used in the simulations. The default is 200. reps() is only valid with either bs or bsp.

nodots specifies that no dots are to be displayed while obtaining the simulation-based standard errors. By default, for each replication, a dot is displayed. If nodots is specified, then either bs or bsp must also be specified.

saving(*filename* [, replace]) specifies the name of the file to hold the Stata dataset that contains the bootstrap replications. The replace option is used to overwrite any file with this name. saving() may only be specified with either bs or bsp.

nose specifies that the asymptotic standard errors, and thus, the asymptotic confidence intervals, are not to be calculated. By default, the asymptotic standard errors and the asymptotic confidence intervals are calculated.

level(#) specifies the confidence level, in percent, for confidence intervals. The default is level(95) or as set by set level; see [U] **23.6 Specifying the width of confidence intervals**.

clear causes varfcast compute to drop the variables created in the previous call to varfcast compute.

estimates(*estname*) specifies that varfcast compute should use the VAR results obtained from var or var svar that are stored as *estname*. By default, varfcast compute uses the currently active e() results. See [R] **estimates** for more on saving and restoring previously obtained estimates.

Remarks

VARs are frequently used to construct forecasts. Recall that a VAR(p) with endogenous variables $\mathbf{y}_t$ and exogenous variables $\mathbf{x}_t$ can be written as

$$\mathbf{y}_t = \mathbf{v} + \mathbf{A}_1\mathbf{y}_{t-1} + \cdots + \mathbf{A}_p\mathbf{y}_{t-p} + \mathbf{B}\mathbf{x}_t + \mathbf{u}_t \tag{1}$$

where

$t = 1, \ldots, T$
$\mathbf{y}_t = (y_{1t}, \ldots, y_{Kt})'$ is a $K \times 1$ random vector
the $\mathbf{A}_i$ are fixed $(K \times K)$ matrices of parameters
$\mathbf{x}_t$ is an $(M \times 1)$ vector of exogenous variables
$\mathbf{B}$ is a $(K \times M)$ matrix of coefficients
$\mathbf{v}$ is a $(K \times 1)$ vector of fixed parameters
$\mathbf{u}_t$ is assumed to be white noise; that is,
$\quad E(\mathbf{u}_t) = \mathbf{0}_K$
$\quad E(\mathbf{u}_t\mathbf{u}_t') = \mathbf{\Sigma}$
$\quad E(\mathbf{u}_t\mathbf{u}_s') = \mathbf{0}_K$ for $t \neq s$

After obtaining the estimates $\widehat{\mathbf{v}}$, $\widehat{\mathbf{A}}_i$ and $\widehat{\mathbf{B}}$, via var or svar, an optimal h-step ahead forecast is given by

$$\widehat{\mathbf{y}}_t(h) = \widehat{\mathbf{v}} + \sum_{i=1}^{p} \widehat{\mathbf{A}}_i\widehat{\mathbf{y}}_t(h - i) + \widehat{\mathbf{B}}\mathbf{x}_{t+h} \tag{2}$$

where $\widehat{\mathbf{y}}_t(j) = \mathbf{y}_{t+j}$ for $j \leq 0$. The notation, $\widehat{\mathbf{y}}_t(h)$, in (2) highlights the fact that the predictions are a function of h, the number of steps. The forecast equation is also recursive; i.e., the prediction at step $h + 1$ depends on the prediction at step h. It is this recursive aspect that causes these forecasts to be known as dynamic forecasts. Also, since the values of any exogenous variables, $\mathbf{x}_t$, are taken as given, if there are exogenous variables in the VAR, the forecasts are conditional on the values of those variables.

Sometimes interest centers on predicting the values the dependent variables will take in the next period, given their current values, p lagged values, and the values of any exogenous variables in the model. These one-step ahead forecasts are also available from `predict` (see [TS] **var** for a discussion of this point).

If there are K endogenous variables in the previously specified VAR, then `varfcast compute` will generate $4K$ new variables, according to the following naming conventions:

> There are K new forecasted variables. The name of each forecasted variable is constructed by appending the suffix "_f" to the name of the original variable.

> There are K estimated lower bounds for the forecast interval. The name of each lower-bound variable is constructed by appending the suffix "_f_L" to the name of the original variable.

> There are K estimated upper bounds for the forecast interval. The name of each upper-bound variable is constructed by appending the suffix "_f_U" to the name of the original variable.

> There are K estimated standard errors of the forecast. The name of each standard error variable is constructed by appending the suffix "_f_se" to the name of the original variable.

If `varfcast compute` cannot calculate the asymptotic standard errors and neither of the bootstrap options is specified, then only the K new forecast variables are created. The asymptotic standard errors and their corresponding confidence intervals are not available when

1. the `nose` option is specified,

2. the forecasts do not begin the period after the estimation sample ends, and

3. there are exogenous variables in the model.

Specifying either `bs` or `bsp` will put the corresponding simulation-based standard errors and confidence intervals into the new *varname*_f_se, *varname*_f_U, and *varname*_f_L variables.

▷ Example

Here is an example illustrating how to obtain dynamic forecasts of the endogenous variables after fitting a VAR(2). Since we have already fit this particular VAR(2) in [TS] **var**, here we fit it `quietly`. Still, it is worth noting that we are using the time-series first-difference operator in specifying the variables. See [U] **14.4.3 Time-series varlists** for a discussion of time-series operators.

```
. use http://www.stata-press.com/data/r8/lutkepohl
(Quarterly SA West German macro data, Bil DM, from Lutkepohl 1993 Table E.1)
. quietly var d.linvestment d.lincome d.lconsumption if qtr <= q(1978q4), dfk
. varfcast compute, step(8)
```

Since we have not used the `dynamic()` option to specify otherwise, the predictions begin at the period immediately following the estimation sample. Also, since there are no exogenous variables, and we have not specified `nose`, `bs`, or `bsp`, `varfcast compute` calculates the asymptotic 95% upper and lower confidence bounds for the forecasts.

To illustrate what it is that `varfcast compute` does, we `describe`, the resulting data.

```
. describe
Contains data from http://www.stata-press.com/data/r8/lutkepohl.dta
  obs:            92                          Quarterly SA West German macro
                                              data, Bil DM, from Lutkepohl
                                              1993 Table E.1
  vars:           22                          22 Apr 2001 14:53
  size:       12,328 (98.8% of memory free)
```

variable name	storage type	display format	value label	variable label
investment	int	%8.0g		Investment
income	int	%8.0g		Income
consumption	int	%8.0g		Consumption
qtr	float	%tq		
linvestment	float	%9.0g		
dlinvestment	float	%9.0g		
lincome	float	%9.0g		
dlincome	float	%9.0g		
lconsumption	float	%9.0g		
dlconsumption	float	%9.0g		
D_linvestment_f	double	%10.0g		forecasted D.linvestment, dyn(77)
D_lincome_f	double	%10.0g		forecasted D.lincome, dyn(77)
D_lconsumptio~f	double	%10.0g		forecasted D.lconsumption, dyn(77)
D_linvestment~L	double	%10.0g		95% lower bound:D_linvestment_f
D_linvestment~U	double	%10.0g		95% upper bound:D_linvestment_f
D_linvestment~e	double	%10.0g		s.e. for D_linvestment_f
D_lincome_f_L	double	%10.0g		95% lower bound:D_lincome_f
D_lincome_f_U	double	%10.0g		95% upper bound:D_lincome_f
D_lincome_f_se	double	%10.0g		s.e. for D_lincome_f
D_lconsumptio~L	double	%10.0g		95% lower bound:D_lconsumption_f
D_lconsumptio~U	double	%10.0g		95% upper bound:D_lconsumption_f
D_lconsumptio~e	double	%10.0g		s.e. for D_lconsumption_f

```
Sorted by:  qtr
     Note:  dataset has changed since last saved
```

The output from describe indicates that varfcast compute has created twelve new variables. The new variables are constructed using the names of the original variables and a suffix that indicates what they are. The three forecasted variables bear the "_f" suffix. The three upper bounds of the confidence intervals are constructed using the "_f_U" suffix, while the three lower bounds use the suffix "_f_L". The three estimated standard errors are constructed by appending the suffix "_f_se". Also note that the "." that separates the time-series operator from the variable name has been replaced by "_". Finally, note that all the new variables have labels indicating what they are.

Once the new variables have been created, they are ready for presentation or further analysis. Here we use varfcast graph to create a matrix graph of the forecasted variables, their 95% confidence intervals, and the actual realizations.

(Continued on next page)

. varfcast graph d.linvestment d.lincome d.lconsumption, observed

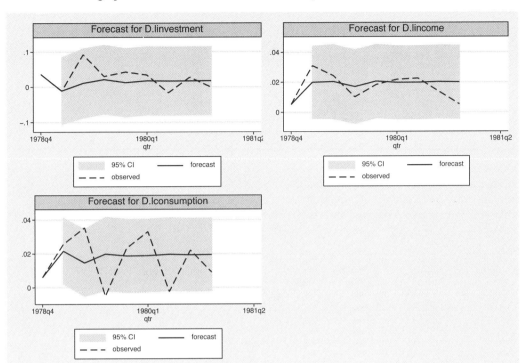

The graphs reveal that the model did a fair job of predicting the variables one period out, but that after one or two steps the accuracy deteriorates notably. In particular, our model has not captured the volatility in the change in consumption. Also note that after 3 or 4 periods, the forecasts taper off to the long-run conditional means of the variables. With no exogenous variables in the model, the fact that the variables are covariance stationary ensures that this must occur. How fast the forecasts settle down depends on the number of lags in the model and the degree of autocorrelation. If there are stochastic exogenous variables in the model, then the forecasts will not necessarily settle down to their long-run conditional means.

◁

▷ Example

As mentioned above, the asymptotic confidence bounds are not available if there are exogenous variables in the model or if the origin of the dynamic forecast, specified in dynamic(), is inside the estimation sample. The two simulation methods, however, are still available.

The following example illustrates how to obtain the same forecasts as above, but with the upper and lower confidence bounds estimated via a multivariate normal parametric simulation. After fitting a VAR(p) model, the two simulation algorithms use the estimated coefficients, the first p observations on the endogenous variables, the observations on any exogenous variables, and a sample of pseudo-innovations to obtain each simulation sample. Thus, the simulation-based standard errors are conditional on this information. In contrast, the asymptotic standard errors include a term that estimates the uncertainty arising from using estimated coefficients instead of the true coefficients. The details of the standard error computations are described in the *Methods and Formulas* section.

Let's begin by saving off the forecast variables for d.lconsumption and dropping the remaining forecast variables.

```
. rename D_lconsumption_f D_lconsumption_af
. label var D_lconsumption_af "forecast for D_lconsumption"
. rename D_lconsumption_f_se D_lconsumption_af_se
. label var D_lconsumption_af_se "s.e. of forecast for D_lconsumption"
. rename D_lconsumption_f_L D_lconsumption_af_L
. label var D_lconsumption_af_L "95% asymp LB D_lconsumption"
. rename D_lconsumption_f_U D_lconsumption_af_U
. label var D_lconsumption_af_U "95% asymp UB D_lconsumption"
. drop *_f*
```

Now, let's use `varfcast compute` to obtain the forecasts with simulation estimates of their standard errors, and the upper and lower limits of 95% confidence intervals computed using these standard errors and a normal approximation.

```
. set seed 123456
. varfcast compute, step(8) bsp reps(250) nodots
```

To compare the two estimated confidence intervals, first we graph the forecast and the two estimated lower bounds, and then we graph the forecast and the two estimated upper bounds.

```
. line    D_lconsumption_f                      ///
          D_lconsumption_af_L                    ///
          D_lconsumption_f_L                     ///
          qtr                                    ///
          if qtr>=q(1978q4) & qtr< q(1981q1),    ///
          ytitle("")                             ///
          legend(span size(small))              ///
          title("Sim vs Asymp estimated lower bounds for D.lconsumption")
```

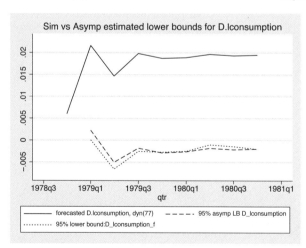

```
. line    D_lconsumption_f                      ///
          D_lconsumption_af_U                    ///
          D_lconsumption_f_U                     ///
          qtr                                    ///
          if qtr>=q(1978q4) & qtr< q(1981q1),    ///
          ytitle("")                             ///
          legend(span size(small))              ///
          title("Sim vs Asymp estimated upper bounds for D.lconsumption")
```

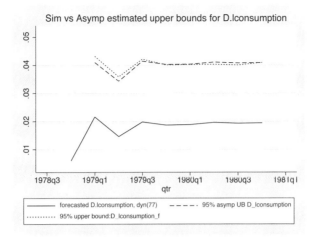

While the graphs reveal that the simulation-based confidence intervals are slightly wider than the asymptotic confidence intervals for d.lconsumption, the estimated intervals are not very different in this case. Since the asymptotic standard errors include a term for the uncertainty arising from using estimated coefficients while the simulation method uses the original estimates in all simulations, the fact that the simulation-based confidence intervals are wider indicates that the asymptotic estimate of the uncertainty due to the innovations is too low in this case.

◁

❑ Technical Note

varfcast compute calls tsappend to add any additional observations to the dataset that are needed to compute and store the forecasts (see [TS] **tsappend**). tsappend will also fill in any gaps in the data with missing values. This implies that varfcast compute will fill any gaps in your data with missing values, just as if you had used tsfill (see [TS] **tsset**). For example, suppose that you had data from 1960:q1 through 1982:q4, with a gap in the data at 1980:q1. (By a gap, we mean that there are no observations in your dataset for 1980:q1.) Further suppose that you were interested in using data up through 1978:q4 for estimation and then obtaining dynamic forecasts of the variables up through 1980:q4. Then, varfcast compute would fill in the gap at 1980:q1 with missing values.

❑

▷ Example

In an example in [TS] **var**, we fit a version of the VAR(2) model considered above subject to two constraints. Here we quietly repeat the estimation.

```
. drop *_f*
. constraint define 1 [D_lconsumption]LD.linvestment = 0
. constraint define 2 [D_lincome]L2D.lconsumption = 0
. quietly var d.linvestment d.lincome d.lconsumption  if qtr>=q(1961q2) &
> qtr <= q(1978q4), constraints(1 2)
. varfcast compute, step(8)
```

Below, we use graphs to compare our forecasts and the estimated standard errors of the forecasts.

```
. label var D_lconsumption_f "constrained forecast"

. label var D_lconsumption_af "unconstrained forecast"

. label var dlconsumption "actual D.lconsumption"

. line    D_lconsumption_f                              ///
          D_lconsumption_af                             ///
          dlconsumption                                 ///
          qtr                                           ///
          if qtr>=q(1978q4) & qtr< q(1981q1),           ///
          ytitle("")                                    ///
          title("Constrained vs unconstrained forecasts for D.lconsumption") ///
          legend(bexpand)
```

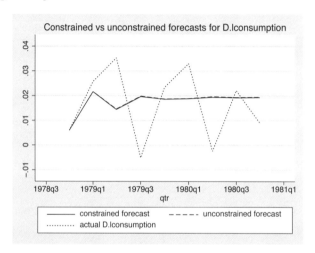

```
. label var D_lconsumption_f_se "constrained s.e."

. label var D_lconsumption_af_se "unconstrained s.e."

. line    D_lconsumption_f_se                           ///
          D_lconsumption_af_se                          ///
          qtr                                           ///
          if qtr>=q(1978q4) & qtr< q(1981q1),           ///
          legend(                                       ///
               span                                     ///
          )                                             ///
          ytitle("")                                    ///
          title("Constrained vs unconstrained s.e. of forecasts for
               D.lconsumption")
```

(Continued on next page)

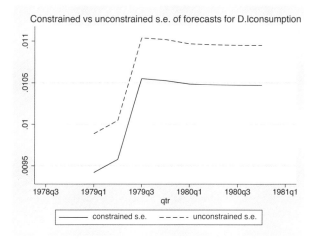

Constrained vs unconstrained s.e. of forecasts for D.lconsumption

◁

While the graphs above reveal that our forecasts are nearly indistinguishable, the standard error of the forecast has fallen substantially. Intuitively, the estimated asymptotic standard errors have fallen since the constraints reduce the number of parameters to estimate.

❏ Technical Note

The estimates of the asymptotic standard errors depend on the estimated VCE of the parameters. As discussed in [TS] **var**, the `dfk` option specifies that a small sample adjustment is made in estimating $\widehat{\Sigma}$. Since $\widehat{\Sigma}$ is used in estimating the VCE of the coefficients, specifying `dfk` will change the estimated asymptotic standard errors of the forecast.

❏

Methods and Formulas

`varfcast compute` is implemented as an ado-file.

Recall that a VAR(p) with endogenous variables $\mathbf{y}_t$ and exogenous variables $\mathbf{x}_t$ can be written as

$$\mathbf{y}_t = \mathbf{v} + \mathbf{A}_1 \mathbf{y}_{t-1} + \cdots + \mathbf{A}_p \mathbf{y}_{t-p} + \mathbf{B} \mathbf{x}_t + \mathbf{u}_t$$

where

$t = 1, \ldots, T$
$\mathbf{y}_t = (y_{1t}, \ldots, y_{Kt})'$ is a $K \times 1$ random vector,
the $\mathbf{A}_i$ are fixed $(K \times K)$ matrices of parameters,
$\mathbf{x}_t$ is an $(M \times 1)$ vector of exogenous variables,
$\mathbf{B}$ is a $(K \times M)$ matrix of coefficients,
$\mathbf{v}$ is a $(K \times 1)$ vector of fixed parameters, and
$\mathbf{u}_t$ is assumed to be white noise; that is,
$\quad E(\mathbf{u}_t) = \mathbf{0}_K$
$\quad E(\mathbf{u}_t \mathbf{u}_t') = \mathbf{\Sigma}$
$\quad E(\mathbf{u}_t \mathbf{u}_s') = \mathbf{0}_K$ for $t \neq s$

`varfcast compute` will dynamically predict the variables in the vector $\mathbf{y}_t$ conditional on p initial values of the endogenous variables and any exogenous $\mathbf{x}_t$. Adopting the notation from Lütkepohl (1993, 334) to fit the case at hand, the optimal h-step ahead forecast of $\mathbf{y}_{t+h}$ conditional on $\mathbf{x}_t$ is

$$\mathbf{y}_t(h) = \widehat{\mathbf{v}} + \widehat{\mathbf{A}}_1 \mathbf{y}_t(h-1) + \cdots + \widehat{\mathbf{A}}_p \mathbf{y}_t(h-p) + \widehat{\mathbf{B}}\mathbf{x}_t \tag{3}$$

If there are no exogenous variables, then (3) becomes

$$\mathbf{y}_t(h) = \widehat{\mathbf{v}} + \widehat{\mathbf{A}}_1 \mathbf{y}_t(h-1) + \cdots + \widehat{\mathbf{A}}_p \mathbf{y}_t(h-p)$$

When there are no exogenous variables, `varfcast compute` can compute the asymptotic confidence bounds.

As shown by Lütkepohl (1993, 177–178), the asymptotic estimator of the covariance matrix of the prediction error is given by

$$\widehat{\boldsymbol{\Sigma}}_{\widehat{y}}(h) = \widehat{\boldsymbol{\Sigma}}_y(h) + \frac{1}{T}\widehat{\boldsymbol{\Omega}}(h) \tag{4}$$

where

$$\widehat{\boldsymbol{\Sigma}}_y(h) = \sum_{i=0}^{h-1} \widehat{\boldsymbol{\Phi}}_i \widehat{\boldsymbol{\Sigma}} \widehat{\boldsymbol{\Phi}}'_i$$

$$\widehat{\boldsymbol{\Omega}}(h) = \frac{1}{T}\sum_{t=0}^{T} \left[\sum_{i=0}^{h-1} \mathbf{Z}'_t \left(\mathbf{B}'\right)^{h-1-i} \otimes \widehat{\boldsymbol{\Phi}}_i\right] \widehat{\boldsymbol{\Sigma}}_\beta \left[\sum_{i=0}^{h-1} \mathbf{Z}'_t \left(\mathbf{B}'\right)^{h-1-i} \otimes \widehat{\boldsymbol{\Phi}}_i\right]' \tag{5}$$

$$\widehat{\mathbf{B}} = \begin{bmatrix} 1 & \mathbf{0} & \mathbf{0} & \dots & \mathbf{0} & \mathbf{0} \\ \widehat{\mathbf{v}} & \widehat{\mathbf{A}}_1 & \widehat{\mathbf{A}}_2 & \dots & \widehat{\mathbf{A}}_{p-1} & \widehat{\mathbf{A}}_p \\ \mathbf{0} & \mathbf{I}_K & \mathbf{0} & \dots & \mathbf{0} & \mathbf{0} \\ \mathbf{0} & \mathbf{0} & \mathbf{I}_K & & \mathbf{0} & \mathbf{0} \\ \vdots & \vdots & & \ddots & & \vdots \\ \mathbf{0} & \mathbf{0} & \mathbf{0} & \dots & \mathbf{I}_K & \mathbf{0} \end{bmatrix}$$

$$\mathbf{Z}_t = (1, \mathbf{y}'_t, \dots, \mathbf{y}'_{t-p-1})'$$

$$\widehat{\boldsymbol{\Phi}}_0 = \mathbf{I}_K$$

$$\widehat{\boldsymbol{\Phi}}_i = \sum_{j=1}^{i} \widehat{\boldsymbol{\Phi}}_{i-j} \widehat{\mathbf{A}}_j \qquad i = 1, 2, \dots$$

$$\widehat{\mathbf{A}}_j = \mathbf{0} \quad \text{for } j > p$$

$\widehat{\boldsymbol{\Sigma}}$ is the estimate of the covariance matrix of the innovations, and $\widehat{\boldsymbol{\Sigma}}_\beta$ is the estimated VCE of the coefficients in the VAR(p). The formula in (5) is general enough to handle the case in which constraints are placed on the coefficients in the VAR(p).

Equation (4) is made up of two terms. $\widehat{\boldsymbol{\Sigma}}_y(h)$ is the estimated mean squared error MSE of the forecast. $\widehat{\boldsymbol{\Sigma}}_y(h)$ estimates the error in the forecast arising from the unseen innovations. $\frac{1}{T}\widehat{\boldsymbol{\Omega}}(h)$ estimates the error in the forecast that is due to the fact that we are using estimated coefficients instead of the true coefficients. As the sample size grows, uncertainty with respect to the coefficient estimates decreases, and $\frac{1}{T}\widehat{\boldsymbol{\Omega}}(h)$ goes to zero.

Assuming that $\mathbf{y}_t$ is normally distributed, the bounds for the asymptotic $(1 - \alpha)100\%$ interval around the forecast for the kth component of $\mathbf{y}_t$, h periods ahead are

$$\widehat{\mathbf{y}}_{k,t}(h) \pm z_{(\frac{\alpha}{2})}\widehat{\sigma}_k(h) \tag{5}$$

where $\widehat{\sigma}_k(h)$ is the kth diagonal element of $\widehat{\mathbf{\Sigma}}_{\widehat{y}}(h)$.

Specifying the bs option causes the standard errors to be computed via simulation using bootstrapped residuals. Both var and svar contain estimators for the coefficients of a VAR(p) that are conditional on the first p observations on the endogenous variables in the data. Similarly, these algorithms are conditional on the first p observations of the endogenous variables in the data. However, the simulation-based estimates of the standard errors are also conditional on the estimated coefficients. The asymptotic standard errors are not conditional on the coefficient estimates because the second term on the right-hand side of (4) accounts for the uncertainty arising from using estimated parameters.

For a simulation with R repetitions, this method uses the following algorithm:

1. Fit the model and save the estimated coefficients.

2. Use the estimated coefficients to calculate the residuals.

3. Repeat steps 3a to 3c R times.

    3a. Draw a simple random sample with replacement, of size $T + h$, from the residuals. When the tth observation is drawn, all K residuals are selected; this preserves any contemporaneous correlation among the residuals.

    3b. Use the sampled residuals, p initial values of the endogenous variables, any exogenous variables, and the estimated coefficients to construct a new sample dataset.

    3c. Save the simulated endogenous variables for the h forecast periods in the bootstrapped dataset.

4. For each endogenous variable and each of the forecast periods, the simulated standard error is the estimated standard error of the R simulated forecasts. By default, the upper and lower bounds of the $(1 - \alpha)100\%$ are estimated using the simulation-based estimates of the standard errors and the normality assumption as in (5). If the bscentile option is specified, then the sample centiles for the upper and lower bounds of the R simulated forecasts are used for the upper and lower bounds of the confidence intervals.

If the bsp option is specified, then a parametric simulation algorithm is used. Specifically, everything is as above with the exception that 3a is replaced by 3a(bsp):

3a(bsp). Draw $T + h$ observations from a multivariate normal distribution with covariance matrix $\widehat{\mathbf{\Sigma}}$.

The algorithm above assumes that h forecast periods come after the original sample of T observations. If the h forecast periods lie within the original sample, smaller simulated datasets are sufficient.

References

Hamilton, J. D. 1994. *Time Series Analysis*. Princeton: Princeton University Press.

Lütkepohl, H. 1993. *Introduction to Multiple Time Series Analysis*. 2d ed. New York: Springer.

Also See

Complementary:	[TS] **var**, [TS] **var svar**, [TS] **varbasic**, [TS] **varfcast graph**
Related:	[TS] **arima**, [TS] **vargranger**, [TS] **varirf**, [TS] **varlmar**, [TS] **varnorm**, [TS] **varsoc**, [TS] **varstable**, [TS] **varwle**
Background:	[U] **14.4.3 Time-series varlists**, [TS] **var intro**, [TS] **varfcast**

Title

varfcast graph — Graph forecasts of dependent variables after var or svar

Syntax

varfcast graph *varlist* [if *exp*] [in *range*] [, noci o̲bserved i̲ndividual

    i̲saving(*filenamestub* [, replace]) iname(*namestub* [, replace]) c̲ilines

    ciopts(*rarea_options* | *rline_options*) *connected_options*

    combine_options twoway_options]

Description

varfcast graph graphs dynamic forecasts of the endogenous variables from a VAR(*p*) that have already been obtained from varfcast compute. *varlist* specifies the names of the endogenous variables whose dynamic forecasts are to be graphed.

Options

noci specifies that the confidence intervals are to be suppressed.

observed specifies that observed values of the predicted variables are to be included in the graph.

individual specifies that the graphs of the forecasted variables are not to be combined into a matrix graph, but instead graphed individually.

isaving(*filenamestub* [, replace]) specifies that all the individual graphs are to be saved to disk under the names obtained by concatenating *filenamestub* with the names of the dependent variables in the model. If the replace option is specified, then any existing files with these names will be replaced. If the forecasted dependent variable contains time-series operators, then the "." will be replaced by "_" in the new filename.

iname(*namestub* [, replace]) specifies that all the individual graphs are to be saved in memory under the names obtained by concatenating *filenamestub* with the names of the dependent variables in the model. If the replace option is specified, then any existing graphs with these names will be replaced. If the forecasted dependent variable contains time-series operators, then the "." will be replaced by "_" in the new name.

cilines causes the confidence intervals to be displayed as lines. By default, the confidence intervals are displayed as shaded areas. cilines may not be combined with noci.

ciopts(*rarea_options* | *rline_options*) affect the rendition of the confidence bands for the forecast(s); see [G] **graph twoway rarea** or [G] **graph twoway rline**. By default, the confidence intervals are displayed as shaded areas so the options for [G] **graph twoway rarea** can be specified. When cilines is specified, the confidence intervals are displayed as lines and the [G] **graph twoway rline** options can be specified. ciopts() cannot be combined with noci.

connected_options affect the rendition of the plotted points corresponding to the forecast and the observed values. See [G] **graph twoway connected**.

combine_options are any of the options documented in [G] **graph combine**. These include options for titling the combined graph (see [G] *title_options*) and options for saving the graph to disk or memory (see [G] *saving_option* and [G] *name_option*, respectively). *combine_options* cannot be specified with the individual option.

twoway_options are any of the options documented in [G] *twoway_options* excluding by(). Note that the saving() and name() (from [G] *twoway_options*) options cannot be specified with the individual option.

Remarks

varfcast graph graphs dynamic forecasts of the endogenous variables from a VAR(*p*) that have been computed using varfcast compute. The *varlist* specifies the names of endogenous variables whose dynamic forecasts are to be graphed. By default, for each specified endogenous variable, varfcast graph makes a graph of the corresponding dynamic forecast and the confidence interval of the forecast. Then varfcast graph combines these graphs into a single matrix graph. The individual option specifies that varfcast graph should not combine the individual graphs into a matrix graph.

▷ Example

Let's begin by quietly fitting the VAR(2) model previously fit in [TS] **var**. Below, we fit the VAR(2) and use varfcast compute to obtain the estimated dynamic predictions and their confidence intervals.

```
. use http://www.stata-press.com/data/r8/lutkepohl
(Quarterly SA West German macro data, Bil DM, from Lutkepohl 1993 Table E.1)
. quietly var d.linvestment d.lincome d.lconsumption if qtr<=q(1978q4), dfk
. varfcast compute, step(8)
```

Since we used time-series operators when we fit the VAR(2), and since a "." cannot appear in the name of variable, the dynamic forecast of d.lincome will be saved as D_lincome, and similarly for the other two variables. (See [TS] **varfcast compute** for more on this point.)

Having obtained the predictions using the commands above, we use varfcast graph to graph them in the output below.

```
. varfcast graph d.linvestment d.lincome d.lconsumption, observed
```

In the call to varfcast graph, we specified the names of the endogenous variables as we specified them in the call to var. Note that we did not replace the "." with an "_" in the *varlist* specified to varfcast graph. Below is the matrix graph that varfcast graph produced.

(Continued on next page)

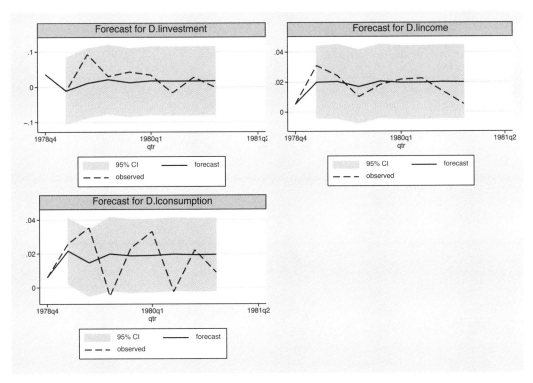

Since we specified three variables, there are three individual graphs in the matrix graph. In each graph, the title indicates which endogenous variable is being graphed. Similarly, the legends on each individual graph clarify which lines correspond to the predictions and which correspond to the confidence intervals. The `observed` option caused `varfcast graph` to include the actual values of the dynamically predicted variables.

The graphs reveal that all three of the predictions do rather well in the first period, but then the quality of the fit declines. It is interesting that the predictions for `D.lincome` are rather good, even after several periods, while `D.lconsumption` is much more volatile than the model predicts.

◁

(Continued on next page)

Saved Results

varfcast graph saves in r():

Macros

r(graph*i*)	endogenous variable for results in subgraph *i*
r(variables*i*)	variables included in subgraph *i*
r(tmin*i*)	first time-period included in subgraph *i*
r(tmax*i*)	last time-period included in subgraph *i*
r(fname*i*)	filename under which subgraph *i* is saved, if isaving() specified
r(name*i*)	name under which subgraph *i* is saved in memory, if iname() specified
r(individual)	individual, if specified
r(ciopts)	*ciopts*
r(twowaycon)	*twoway_connected_options*
r(comb_opts)	*graph_combine_options*
r(timevar)	name of time variable set by tsset
r(tsfmt)	format of time variable set by tsset

Methods and Formulas

varfcast clear is implemented as an ado-file.

Also See

Complementary:	[TS] **var**, [TS] **var svar**, [TS] **varbasic**, [TS] **varfcast compute**
Related:	[TS] **arima**, [TS] **vargranger**, [TS] **varirf**, [TS] **varlmar**,
	[TS] **varnorm**, [TS] **varsoc**, [TS] **varstable**, [TS] **varwle**
Background:	[U] **14.4.3 Time-series varlists**,
	[TS] **var intro**, [TS] **varfcast**

Title

> **vargranger** — Perform pairwise Granger causality tests after var or svar

Syntax

vargranger $\left[\,,\ \underline{\text{est}}\text{imates}(estname)\ \underline{\text{sepa}}\text{rator}(\#)\ \right]$

vargranger can only be used with var or svar e() results; see [TS] **var** and [TS] **var svar**.

Description

vargranger performs a set of Granger causality tests for each equation in a VAR(p), providing a convenient alternative to test.

Options

estimates(*estname*) specifies that vargranger is to use the previously obtained set of var or svar estimates saved as *estname*. By default, vargranger uses the currently active results. See [R] **estimates** for information on saving and restoring e() results.

separator(#) specifies how often separator lines should be drawn between rows. By default, separator lines appear every K lines, where K is the number of equations in the VAR(p) under analysis. Note that separator(0) implies that lines should not appear in the table.

Remarks

After fitting a VAR(p), one question of interest is whether one variable "Granger causes" another (Granger 1969). A variable x is said to Granger cause a variable y if, given the past values of y, past values of x are useful for predicting y. A common method for testing Granger causality is to regress y on its own lagged values and on lagged values of x, and test the null hypothesis that the estimated coefficients on the lagged values of x are jointly zero. Failure to reject the null hypothesis is equivalent to failing to reject the hypothesis that x does not Granger cause y.

For each equation, and each endogenous variable that is not the dependent variable in that equation, vargranger computes and reports Wald tests that the coefficients on all the lags of an endogenous variable are jointly zero. In other words, for each equation in a VAR(p), vargranger tests the hypotheses that each of the other endogenous variables does not Granger cause the dependent variable in that equation.

Since it may be interesting to investigate these types of hypotheses using the VAR(p) that underlies an SVAR, vargranger can also produce these tests using the e() results from an svar. When vargranger uses svar e() results, the hypotheses are with respect to the underlying var estimates.

See [TS] **var** and [TS] **var svar** for an introduction to fitting VARs and SVARs in Stata. See Lütkepohl (1993), Hamilton (1994), and Amisano and Gianinni (1997) for an introduction to Granger causality, and for information on VARs and SVARs in general.

▷ Example

Let's fit a VAR(2), and then run vargranger.

```
. use http://www.stata-press.com/data/r8/lutkepohl
(Quarterly SA West German macro data, Bil DM, from Lutkepohl 1993 Table E.1)
. var dlinvestment dlincome dlconsumption if qtr<=q(1978q4),
> lags(1/2) dfk small

Vector autoregression

Sample:  1960q4   1978q4
```

Equation	Obs	Parms	RMSE	R-sq	F	P
dlinvestment	73	7	.046148	0.1286	1.622818	0.1547
dlincome	73	7	.011719	0.1142	1.418048	0.2210
dlconsumpt~n	73	7	.009445	0.2513	3.691827	0.0032

```
Model lag order selection statistics
```

FPE	AIC	HQIC	SBIC	LL	Det(Sigma_ml)
2.183e-11	-16.035809	-15.773227	-15.37691	606.30704	1.226e-11

	Coef.	Std. Err.	t	P>|t|	[95% Conf. Interval]	
dlinvestment						
dlinvestment						
L1	-.3196318	.1254564	-2.55	0.013	-.5701135	-.0691501
L2	-.1605508	.1249066	-1.29	0.203	-.4099349	.0888333
dlincome						
L1	.1459851	.5456664	0.27	0.790	-.9434729	1.235443
L2	.1146009	.5345709	0.21	0.831	-.9527043	1.181906
dlconsumpt~n						
L1	.9612288	.6643086	1.45	0.153	-.3651061	2.287564
L2	.9344001	.6650949	1.40	0.165	-.3935047	2.262305
_cons	-.0167221	.0172264	-0.97	0.335	-.0511157	.0176714
dlincome						
dlinvestment						
L1	.0439309	.0318592	1.38	0.173	-.0196781	.1075399
L2	.0500302	.0317196	1.58	0.120	-.0133001	.1133605
dlincome						
L1	-.1527311	.1385702	-1.10	0.274	-.4293954	.1239332
L2	.0191634	.1357525	0.14	0.888	-.2518753	.290202
dlconsumpt~n						
L1	.2884992	.168699	1.71	0.092	-.0483193	.6253176
L2	-.0102	.1688987	-0.06	0.952	-.3474171	.3270171
_cons	.0157672	.0043746	3.60	0.001	.0070331	.0245013
dlconsumpt~n						
dlinvestment						
L1	-.002423	.0256763	-0.09	0.925	-.0536874	.0488414
L2	.0338806	.0255638	1.33	0.190	-.0171591	.0849203
dlincome						
L1	.2248134	.1116778	2.01	0.048	.0018415	.4477852
L2	.3549135	.1094069	3.24	0.002	.1364755	.5733515
dlconsumpt~n						
L1	-.2639695	.1359595	-1.94	0.056	-.5354213	.0074823
L2	-.0222264	.1361204	-0.16	0.871	-.2939996	.2495467
_cons	.0129258	.0035256	3.67	0.000	.0058867	.0199649

```
. vargranger
```

Granger causality Wald tests

Equation	Excluded	F	df	df_r	Prob > F
dlinvestment	dlincome	0.0485	2	66	0.9527
dlinvestment	dlconsumption	1.5004	2	66	0.2306
dlinvestment	ALL	1.5917	4	66	0.1869
dlincome	dlinvestment	1.7683	2	66	0.1786
dlincome	dlconsumption	1.7184	2	66	0.1873
dlincome	ALL	1.9466	4	66	0.1130
dlconsumption	dlinvestment	0.9715	2	66	0.3839
dlconsumption	dlincome	6.1465	2	66	0.0036
dlconsumption	ALL	3.7746	4	66	0.0080

Since the `estimates()` option was not specified, `vargranger` used the currently active e() results. Consider the results of the three tests for the first equation. The first is a Wald test that the coefficients on the two lags of `dlincome` that appear in the equation for `dlinvestment` are jointly zero. The null hypothesis that `dlincome` does not Granger cause `dlinvestment` cannot be rejected. Similarly, we cannot reject the null hypothesis that the coefficients on the two lags of `dlconsumption` in the equation for `dlinvestment` are jointly zero, so we cannot reject the hypothesis that `dlconsumption` does not Granger cause `dlinvestment`. The third test is with respect to the null hypothesis that the coefficients on the two lags of all the other endogenous variables are jointly zero. Since this cannot be rejected, we cannot reject the null hypothesis that `dlincome` and `dlconsumption`, jointly, do not Granger cause `dlinvestment`.

Since we failed to reject most of these null hypotheses, we might be interested in imposing some constraints on the coefficients. See [TS] **var** for more on fitting VAR(p) models with constraints on the coefficients.

◁

▷ Example

We could have used `test` to compute these Wald tests, but `vargranger` saves a significant amount of typing. Still, it is useful to see how to use `test` to obtain the results reported by `vargranger`.

```
. test [dlinvestment]L.dlincome [dlinvestment]L2.dlincome

 ( 1)  [dlinvestment]L.dlincome = 0
 ( 2)  [dlinvestment]L2.dlincome = 0

       F( 2,    66) =     0.05
            Prob > F =     0.9527

. test [dlinvestment]L.dlconsumption [dlinvestment]L2.dlconsumption, accumulate

 ( 1)  [dlinvestment]L.dlincome = 0
 ( 2)  [dlinvestment]L2.dlincome = 0
 ( 3)  [dlinvestment]L.dlconsumption = 0
 ( 4)  [dlinvestment]L2.dlconsumption = 0

       F( 4,    66) =     1.59
            Prob > F =     0.1869
```

```
. test [dlinvestment]L.dlinvestment [dlinvestment]L2.dlinvestment, accumulate

 ( 1)  [dlinvestment]L.dlincome = 0
 ( 2)  [dlinvestment]L2.dlincome = 0
 ( 3)  [dlinvestment]L.dlconsumption = 0
 ( 4)  [dlinvestment]L2.dlconsumption = 0
 ( 5)  [dlinvestment]L.dlinvestment = 0
 ( 6)  [dlinvestment]L2.dlinvestment = 0

       F(  6,    66) =     1.62
            Prob > F =    0.1547
```

The first two calls to `test` show how `vargranger` obtains its results. The first test reproduces the first test reported for the `dlinvestment` equation. The second test reproduces the `ALL` entry for the first equation. The third test reproduces the standard F statistic for the `dlinvestment` equation, reported in the header of the `var` output in the previous example. Note that the standard F statistic also includes the lags of the dependent variable, as well as any exogenous variable(s) in the equation. This illustrates that the test performed by `vargranger` of the null hypothesis that the coefficients on all the lags of all the other endogenous variables are jointly zero for a particular equation, i.e., the `All` test, is not the same as the standard F statistic for that equation.

◁

▷ Example

When `vargranger` is run on `svar` estimates, the null hypotheses are with respect to the underlying `var` estimates. We run `vargranger` after using `svar` to fit an SVAR that has the same underlying VAR as our model in the previous example.

```
. matrix A = (., 0 ,0 \ ., ., 0\ .,.,.)

. matrix B = I(3)

. quietly svar dlinvestment dlincome dlconsumption if qtr<=q(1978q4),
> lags(1/2) dfk small aeq(A) beq(B)

. vargranger
```

                          Granger causality Wald tests

Equation	Excluded	F	df	df_r	Prob > F
dlinvestment	dlincome	0.0485	2	66	0.9527
dlinvestment	dlconsumption	1.5004	2	66	0.2306
dlinvestment	ALL	1.5917	4	66	0.1869
dlincome	dlinvestment	1.7683	2	66	0.1786
dlincome	dlconsumption	1.7184	2	66	0.1873
dlincome	ALL	1.9466	4	66	0.1130
dlconsumption	dlinvestment	0.9715	2	66	0.3839
dlconsumption	dlincome	6.1465	2	66	0.0036
dlconsumption	ALL	3.7746	4	66	0.0080

As we expected, the `vargranger` results are identical to those in the first example.

◁

Saved Results

vargranger saves in r():

Matrices

r(gstats) χ^2, df, and p-values (if e(small)=="")

r(gstats) F, df, df_r, and p-values (if e(small)!="")

Methods and Formulas

vargranger is implemented as an ado-file.

vargranger uses test to obtain Wald statistics of the hypothesis that all coefficients on the lags of variable x are jointly zero in the equation for variable y. vargranger will use the e() results saved by var or svar to determine whether to calculate and report small-sample F statistics or large-sample χ^2 statistics.

References

Amisano, G. and C. Giannini. 1997. *Topics in Structural* VAR *Econometrics*. 2d ed. Heidelberg: Springer.

Granger, C. W. J. 1969. Investigating causal relations by econometric models and cross-spectral methods. *Econometrica* 37: 424–438.

Hamilton, J. D. 1994. *Time Series Analysis*. Princeton: Princeton University Press.

Lütkepohl, H. 1993. *Introduction to Multiple Time Series Analysis*. 2d ed. New York: Springer.

Also See

Complementary:	[TS] **var**, [TS] **var svar**, [TS] **varbasic**
Background:	[TS] **var intro**

Title

varirf — An introduction to the varirf commands

Syntax

varirf *subcommand* ...

varirf may only be used with var or svar e() results; see [TS] **var** and [TS] **var svar**.

Description

Stata contains a suite of varirf commands for obtaining, maintaining, and analyzing a variety of impulse–response functions, variance decompositions, and their standard errors. This entry provides an overview of these commands, and it defines some terms that are used throughout these entries.

Stata has three types varirf commands. There is one varirf command that creates a varirf file that contains all the results. The second type of command presents results. The third type of command manages the results files.

Creating the VARIRF results

create	[TS] **varirf create**	Obtain impulse–response functions and forecast-error variance decompositions

Analyzing results

graph	[TS] **varirf graph**	Graph impulse–response functions and FEVDs
cgraph	[TS] **varirf cgraph**	Make combined graphs of impulse–response functions and FEVDs
ograph	[TS] **varirf ograph**	Graph overlaid impulse–response functions and FEVDs
table	[TS] **varirf table**	Create tables of impulse–response functions and FEVDs
ctable	[TS] **varirf ctable**	Make combined tables of impulse–response functions and FEVDs

Managing files

add	[TS] **varirf add**	Add VARIRF results from one VARIRF file to another
describe	[TS] **varirf describe**	Describe VARIRF file
dir	[TS] **varirf dir**	List the VARIRF files in a directory
drop	[TS] **varirf drop**	Drop VARIRF results from the active VARIRF file
erase	[TS] **varirf erase**	Erase a VARIRF file
rename	[TS] **varirf rename**	Rename a VARIRF result in a VARIRF file
set	[TS] **varirf set**	Set active VARIRF file

Remarks

Remarks are presented under the headings

An outline of how to estimate and analyze IRFs *and* FEVDs *in Stata*
Scope and nomenclature
An extended example
Managing varirf files

An outline of how to estimate and analyze IRFs and FEVDs in Stata

Usually, three simple steps are sufficient to estimate and analyze impulse response functions, IRFs, and forecast-error variance decompositions, FEVDs, in Stata.

1. Fit a VAR or an SVAR using `var` or `svar`.

2. Use `varirf create` to estimate the IRFs, FEVDs and their standard errors.

3. Use one or more of the four analysis commands to present the results. The analysis commands are

   a. `varirf graph` makes it easy to combine graphs of several IRFs or FEVDs into one or more matrix graphs using a single command.

   b. `varirf cgraph` makes a series of graphs of IRFs or FEVDs.

   c. `varirf ograph` displays plots of `varirf` results on a single graph.

   d. `varirf table` makes it easy to combine sub-tables of several IRFs or FEVDs into one or more tables using a single command.

   e. `varirf ctable` makes a series of tables of IRFs or FEVDs.

In the course of performing more advanced analysis, sometimes it will be necessary to maintain the files that hold the IRF and FEVD estimates. To deal with these situations, there are seven commands, which are described in the section *Managing VARIRF files* below.

Scope and nomenclature

The table below lists the quantities that Stata's `varirf create` can estimate and the names assigned to these quantities.

Quantity	Name
impulse–response functions	irf
orthogonalized impulse–response functions	oirf
cumulative impulse–response functions	cirf
cumulative orthogonalized impulse–response functions	coirf
Cholesky forecast-error variance decomposition	fevd
structural impulse–response functions	sirf
structural forecast-error variance decomposition	sfevd
standard error of the impulse–response functions	stdirf
standard error of the orthogonalized impulse response functions	stdoirf
standard error of the cumulative impulse response functions	stdcirf
standard error of the cumulative orthogonalized impulse–response functions	stdcoirf
standard error of the Cholesky forecast-error variance decomposition	stdfevd
standard error of the structural impulse–response functions	stdsirf
standard error of the structural forecast-error variance decomposition	stdsfevd

Since this list is somewhat cumbersome, we use VARIRF estimates as shorthand for any collection of the above estimates. While this is a slight abuse of terminology, it sets the foundation for defining

"VARIRF results" and "VARIRF file". When we speak of "VARIRF results" we mean a named collection of the above estimates. For example, suppose that we have fitted a VAR and that we then use `varirf create` to estimate the above functions for s periods for two different Cholesky orderings. For each Cholesky ordering, we call the collection of estimated functions a set of "VARIRF results". To distinguish each set of VARIRF results, they must be named. The name of a set of VARIRF results is known as a VARIRF name. VARIRF results are stored in files, the particulars of which are discussed in [TS] **varirf create**. The files that hold VARIRF results are called VARIRF files. A VARIRF file can hold several sets of VARIRF results, although each set of VARIRF results within a given file must have its own unique name.

An extended example

Here we use an example to show how to use the VARIRF commands to obtain and present VARIRF results. In the process, we illustrate the nomenclature discussed above.

▷ Example

Let's begin by fitting a VAR(2) and using `varirf create` to obtain the VARIRF results for two different Cholesky orderings. Since we have already displayed the results from this particular VAR(2) in [TS] **var**, we quietly fit this VAR(2).

```
. use http://www.stata-press.com/data/r8/lutkepohl
(Quarterly SA West German macro data, Bil DM, from Lutkepohl 1993 Table E.1)
. qui var dlinvestment dlincome dlconsumption if qtr<=q(1978q4), lags(1/2) dfk
```

Now we use `varirf create` to obtain VARIRF results.

```
. varirf create order1, step(10) set(myirf1, new)
current varirf data file is myirf1.vrf
file myirf1.vrf saved
```

While the full syntax is described in [TS] **varirf create**, note that we have specified three pieces of information in this command. First, we said that we wanted to call this set of VARIRF results "order1". Second, we specified that we wanted the VARIRF results computed for 10 steps. Third, we used the `set()` option to open a new VARIRF file named "myirf1". Since we did not specify the `order()` option, the Cholesky ordering is the order in which the variables were specified in the call to `var`; i.e., in this case the order is `dlinvestment dlincome dlconsumption`.

Now suppose that we wanted to compare the `order1` VARIRF results with those obtained from the same VAR(2) parameter estimates but with the Cholesky ordering `dlincome dlinvestment dlconsumption`. Below, we obtain the VARIRF results for this Cholesky ordering and store them in the same VARIRF file with the VARIRF name `order2`.

```
. varirf create order2, step(10) order(dlincome dlinvestment dlconsumption)
file myirf1.vrf saved
```

`varirf graph` and `varirf cgraph` are commands for creating graphs of VARIRF results. As described in [TS] **varirf graph**, `varirf graph` is a powerful tool for creating many graphs with a single command. [TS] **varirf cgraph** describes `varirf cgraph` which allows users to create a series of customized graphs via a series of specific graph commands. Here we use `varirf graph` to create two graphs of the orthogonalized impulse–response functions from `dlincome` to `dlconsumption` from the `order1` and `order2` results.

. varirf graph oirf, irf(order1 order2) impulse(dlincome) response(dlconsumption)

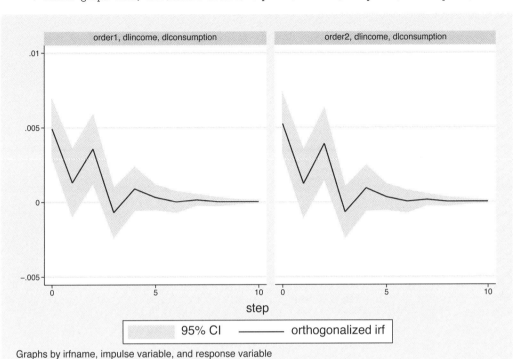

Graphs by irfname, impulse variable, and response variable

The horizontal titles inform us that the left-hand graph is of the oirf from the innovations in dlincome to dlconsumption from the order1 VARIRF results and that the graph on the right-hand side of the same function from the order2 VARIRF results. These graphs show that the basic forms of the two orthogonalized impulse–response functions are essentially the same. In both functions, an increase in the orthogonalized innovation to dlincome causes a short series of increases in dlconsumption that dies out after four or five periods.

However, we might also be interested in comparing their values in a table. varirf table (see [TS] **varirf table**) is analogous to the varirf graph command; it creates one large or many smaller tables using one combined command. varirf ctable (see [TS] **varirf ctable**) is analogous to the varirf cgraph command; it creates one large or many smaller tables from a series of specific table commands.

The output below illustrates how we can use varirf table to obtain a single table of the orthogonalized impulse–response functions from the order1 and order2 VARIRF results.

(Continued on next page)

```
. varirf table oirf, irf(order1 order2) impulse(dlincome)
> response(dlconsumption) noci std
```

Results from order1 order2

step	(1) oirf	(1) S.E.	(2) oirf	(2) S.E.
0	.004934	.000979	.005244	.001017
1	.001309	.001143	.001235	.001146
2	.003573	.001168	.00391	.001208
3	-.000692	.000837	-.000677	.000852
4	.000905	.000738	.00094	.000773
5	.000328	.000423	.000341	.000439
6	.000021	.000355	.000042	.000375
7	.000154	.000184	.000161	.000193
8	.000026	.00014	.000027	.000147
9	.000026	.000075	.00003	.000079

(1) irfname = order1, impulse = dlincome, and response = dlconsumption
(2) irfname = order2, impulse = dlincome, and response = dlconsumption

The keys at the top of each column are defined in the footer of the table. The keys tell us that in all four columns, the innovations to dlincome are the impulses and that dlconsumption is the response variable. They also inform us that the first two columns are from the order1 VARIRF results and that third and fourth column are from the order2 VARIRF results. The table reveals that the values as well as the standard errors of the two orthogonalized impulse–response functions are quite similar.

In our example, the orthogonalized impulse–response functions do not vary much when we change the Cholesky ordering from dlinvestment dlincome dlconsumption to dlincome dlinvestment dlconsumtion. It is important to keep in mind that this is frequently not the case. Stock and Watson (2001) provide a very nice example in which changing the Cholesky ordering produces very different orthogonalized impulse–response functions.

◁

Managing varirf files

Since different VARIRF results may be stored in distinct VARIRF files, there are several VARIRF commands to manage these results. While the details of these commands are provided in their separate manual entries, it worth noting that these commands answer three frequently asked questions.

The first question is, "On which VARIRF file is a command operating?" All the VARIRF commands operate on the active VARIRF file. varirf set is the command that makes a file the active VARIRF file. [TS] **varirf set**, provides details of this command. Another feature worthy of note is that most VARIRF commands provide a set() option. The set() option allows users to have varirf set make the specified file the active VARIRF file before proceeding with the command. Additional documentation about this option is provided in the manual entries for those VARIRF commands that offer this option.

The second question is, "What VARIRF results do I have and where are they?" The VARIRF commands varirf describe and varirf dir can answer this question. As documented in [TS] **varirf dir**, varirf dir provides a listing of the all the VARIRF files in a given directory. varirf describe provides summary or detailed information about the different VARIRF results that are in a specific VARIRF file (see [TS] **varirf describe** for details).

The third question is, "How can I copy, move or delete VARIRF results or files?" There are four commands for this type of problem.

1. `varirf add` is used to move one or more sets of VARIRF results from one VARIRF file to another. Optionally, some sets of VARIRF results may be renamed. See [TS] **varirf add** for details.

2. `varirf rename` performs the more limited task of renaming a set of VARIRF results within a given VARIRF file. See [TS] **varirf rename** for details.

3. `varirf drop` removes a list of VARIRF results from the active VARIRF file. See [TS] **varirf drop** for details.

4. `varirf erase` removes an entire VARIRF file. See [TS] **varirf erase** for details.

Also See

Complementary:	[TS] **var**, [TS] **var svar**, [TS] **varbasic**, [TS] **varirf add**, [TS] **varirf cgraph**, [TS] **varirf create**, [TS] **varirf ctable**, [TS] **varirf describe**, [TS] **varirf dir**, [TS] **varirf drop**, [TS] **varirf erase**, [TS] **varirf graph**, [TS] **varirf ograph**, [TS] **varirf rename**, [TS] **varirf set**, [TS] **varirf table**
Related:	[TS] **arima**
Background:	[U] **14.4.3 Time-series varlists**, [TS] **var intro**

Title

> **varirf add** — Add VARIRF results from one VARIRF file to another

Syntax

> varirf a̲dd *addlist*, using(*varirf_filename*)

where *addlist* is

> *addelement* [*addlist*]

and *addelement* is

> { _all | [*newname*=]*oldname* }

Description

> varirf add gets VARIRF results specified in *addlist* that are saved in the VARIRF file specified in using() and adds them to the VARIRF file made active by varirf set.

Options

> using(*varirf_filename*) specifies the VARIRF filename from which the results are to be extracted. using() is a required option.

Remarks

> If you have not read [TS] **varirf**, please do so.

> In general, varirf add is useful when you want to extract VARIRF results stored in one file and place them into the active VARIRF file.

> In analyzing VARIRF results, it is quite common to compare how the innovations to a given impulse variable affect a response variable under different underlying assumptions. For instance, one might want to see how the impulse–response function from D.lincome to D.lconsumption changes when different orderings are specified for the Cholesky decomposition. At times, these different VARIRF results are stored in distinct VARIRF files. This is the type of problem that varirf add is designed to solve.

> The *addlist* in the syntax diagram contains a list of the VARIRF results that are to be added from the using() VARIRF file into the active VARIRF file. Since users might want to change the name of the VARIRF results before adding them to the active VARIRF file, the *addlist* is made up of elements that allow for renaming the VARIRF results. In addition, since some users might want to add all the VARIRF results from the using() file into the active VARIRF file, _all is also a possible element in the *addlist*. Note that explicitly specifying an *oldname* in two different *addelement*s is not allowed. However, users may specify an *oldname* once and use _all to specify the remaining *varirf name*s.

▷ Example

Suppose that you have two sets of VARIRF results that were computed with different Cholesky orderings that are stored in distinct VARIRF files. Specifically, suppose that in the current directory, the file luta.vrf holds the VARIRF results named order_a, and the file lutb.vrf holds the VARIRF results named order_b. We can use varirf describe to get a description, which reminds us that these are the results that we want to compare. (See [TS] **varirf describe** for more on varirf describe.)

```
. use http://www.stata-press.com/data/r8/lutkepohl
(Quarterly SA West German macro data, Bil DM, from Lutkepohl 1993 Table E.1)
. qui var D.linvestment d.lincome d.lconsumption
. varirf create order_a, set(luta, replace)
(output omitted )
. varirf create order_b, order(D.linvestment D.lconsumption D.lincome)
> set(lutb, replace)
(output omitted )
. qui var d.lincome d.lconsumption, exog(D.linvestment)
. varirf create order_a, set(spec2, replace)
(output omitted )
. varirf create order_b, order(D.lconsumption D.lincome)
file spec2.vrf saved
. qui var d.lincome d.lconsumption, exog(D.linvestment) lags(1/4)
. varirf create l4order_a
file spec2.vrf saved
. varirf describe, using(luta) detail
order_a:
        model: var
        order: D.linvestment D.lincome D.lconsumption
         exog: none
     constant: constant
         lags: 1 2
         tmin: 3
         tmax: 91
      timevar: qtr
        tsfmt: %tq
       varcns: unconstrained
      svarcns: .
         step: 8
      stderror: asymptotic
         reps: .
      version: 1.0
. varirf describe, using(lutb) detail
order_b:
        model: var
        order: D.linvestment D.lconsumption D.lincome
         exog: none
     constant: constant
         lags: 1 2
         tmin: 3
         tmax: 91
      timevar: qtr
        tsfmt: %tq
       varcns: unconstrained
      svarcns: .
         step: 8
      stderror: asymptotic
         reps: .
      version: 1.0
```

To put the two sets of results into a single file, we could use `varirf add` to add `order_a` from `luta.vrf` to `lutb.vrf`, or to add `order_b` from `lutb.vrf` to `luta.vrf`. But, suppose that we wanted to extract these two sets of results and combine them into the new VARIRF file `compare.vrf`. Below, we illustrate how we could use `varirf set` to create the new file, make it the active VARIRF file, and then use `varirf add` to add `order_a` and `order_b` to this new VARIRF file. (See [TS] **varirf set** for more on `varirf set`.)

```
. varirf set compare, new
current varirf data file is compare.vrf

. varirf add order_a, using(luta)
file compare.vrf saved

. varirf add order_b, using(lutb)
file compare.vrf saved
```

We now illustrate one quick way to check to see that we were successful.

```
. varirf describe
The irfnames in compare.vrf are:
```

irfname	order
order_a	D.linvestment D.lincome D.lconsumption
order_b	D.linvestment D.lconsumption D.lincome

◁

Sometimes the situation is more complicated. Since VARIRF names must be unique in each VARIRF file, it is not possible to combine two distinct VARIRF results into a single VARIRF file under the same name. Hence, sometimes it is necessary to rename the VARIRF results when copying them from the `using()` VARIRF file into the active VARIRF file.

▷ Example

Consider a case in which `spec2.vrf` contains several different VARIRF results that were computed under assumptions different from the `order_a` in `luta.vrf` and the `order_b` in `lutb.vrf`, but unfortunately two of these new results re-use the names `order_a` and `order_b`.

```
. varirf describe, using(spec2) detail

order_a:
          model: var
          order: D.lincome D.lconsumption
           exog: D.linvestment
       constant: constant
           lags: 1 2
           tmin: 3
           tmax: 91
        timevar: qtr
          tsfmt: %tq
         varcns: unconstrained
         svarcns: .
           step: 8
       stderror: asymptotic
           reps: .
        version: 1.0
```

```
order_b:
          model: var
          order: D.lconsumption D.lincome
           exog: D.linvestment
       constant: constant
           lags: 1 2
           tmin: 3
           tmax: 91
        timevar: qtr
          tsfmt: %tq
         varcns: unconstrained
        svarcns: .
           step: 8
        stderror: asymptotic
           reps: .
        version: 1.0
l4order_a:
          model: var
          order: D.lincome D.lconsumption
           exog: D.linvestment
       constant: constant
           lags: 1 2 3 4
           tmin: 5
           tmax: 91
        timevar: qtr
          tsfmt: %tq
         varcns: unconstrained
        svarcns: .
           step: 8
        stderror: asymptotic
           reps: .
        version: 1.0
```

The following output shows how to use `varirf add` to add order_a to the active file under the name order_a2, to add order_b under the name order_b2, and not to add the VARIRF results named l4order_a.

```
. varirf set
current varirf data file is compare.vrf

. varirf add order_a2=order_a order_b2=order_b, using(spec2)
file compare.vrf saved

. varirf describe
The irfnames in compare.vrf are:
```

irfname	order
order_a	D.linvestment D.lincome D.lconsumption
order_b	D.linvestment D.lconsumption D.lincome
order_a2	D.lincome D.lconsumption
order_b2	D.lconsumption D.lincome

◁

▷ Example

Sometimes users want to move all the VARIRF results from the using() VARIRF file into the active VARIRF file, with the caveat that some of them are stored under new names. For this reason, it is possible to specify _all and to have additional elements in the list that rename some VARIRF results. This allows us to handle cases that are more complex than the previous examples. Imagine that there are several sets of VARIRF results in a VARIRF file called results.vrf. Suppose further that you want to add all of them into the active file, with the caveat that you want to rename irf_a and irf_b to irf_a2 and irf_b2, respectively. This can be accomplished via the command

. varirf add _all irf_a2 = irf_a irf_b2=irf_b, using(results)

◁

Methods and Formulas

varirf add is implemented as an ado-file.

Also See

Complementary:	[TS] **var**, [TS] **var svar**, [TS] **varbasic**, [TS] **varirf cgraph**, [TS] **varirf ctable**, [TS] **varirf describe**, [TS] **varirf dir**, [TS] **varirf drop**, [TS] **varirf erase**, [TS] **varirf graph**, [TS] **varirf ograph**, [TS] **varirf rename**, [TS] **varirf set**, [TS] **varirf table**
Related:	[TS] **arima**, [TS] **varirf create**
Background:	[U] **14.4.3 Time-series varlists** [TS] **var intro**, [TS] **varirf**

Title

> **varirf cgraph** — Make combined graphs of impulse–response functions and FEVDs

Syntax

> varirf cgraph (*spec*$_1$) $\big[$ (*spec*$_2$) $\big]$ ... $\big[$ (*spec*$_N$) $\big]$ $\big[$, individual
>
> set(*setcmd*) *common_options twoway_options combine_options* $\big]$

where (*spec*$_k$) is

> (*irfname impulse response statlist* $\big[$, *common_options twoway_options* $\big]$)

irfname is the name of a set VARIRF results in the active VARIRF file, *impulse* is an impulse variable in *irfname*, *response* is a response variable in *irfname*, and *statlist* is a list of the names of the statistics from the table below.

Statistic	Name
impulse–response functions	irf
orthogonalized impulse–response functions	oirf
cumulative impulse–response functions	cirf
cumulative orthogonalized impulse–response functions	coirf
Cholesky forecast-error variance decomposition	fevd
structural impulse–response functions	sirf
structural forecast-error variance decomposition	sfevd

and *common_options* are

> lstep(*#*) ustep(*#*) noci level(*#*) cilines
>
> ciopts1(*rarea_options* | *rline_options*) ciopts2(*rarea_options* | *rline_options*)
>
> plot1(*line_options*) plot2(*line_options*) plot3(*line_options*) plot4(*line_options*)

varirf cgraph operates on the active VARIRF file, see [TS] **varirf set**.

Three limits are placed on the *statlist*:

1. No statistic may appear more than once in *statlist*.

2. If confidence intervals are included, only two statistics may be included on the graph.

3. If confidence intervals are suppressed, only four statistics may be included on the graph.

Description

varirf cgraph makes a series of graphs. Each block within a pair of matching parentheses, i.e., each (*spec*$_i$), defines a specific graph. By default, varirf cgraph combines these graphs into one matrix of graphs. Specifying individual causes varirf cgraph to create a separate graph for each specific graph.

Options

individual specifies that each graph is to be placed in its own graph window. By default, varirf cgraph combines all the individual graphs into a matrix graph.

set(*setcmd*) causes varirf cgraph to make the file specified in set() the active VARIRF file. set() accepts a *setcmd*, where *setcmd* is a combination of *filename* and *options* that can be understood by varirf set. Using set() implies that varirf cgraph will operate on the file specified in set() and that the VARIRF file specified in set() file will remain the active VARIRF file when varirf cgraph is done.

common_options may be specified within a plot specification, globally, or in both places. When specified in a plot specification, the *common_options* only affect the specification in which they are used. When supplied globally, the *common_options* apply to all plot specifications. When supplied in both places, options in the plot specification take precedence.

lstep(#) specifies the first step, or period, to be plotted. By default, the first step is zero.

ustep(#) specifies the maximum step, or period, to be included. By default, the maximum step is the largest step available.

noci specifies that the confidence intervals for each statistic in *statlist* are to be suppressed. By default, each statistic is graphed with a confidence interval. noci cannot be combined with level().

level(#) specifies the confidence level, in percent, for confidence intervals. The default is level(95) or as set by set level; see [U] **23.6 Specifying the width of confidence intervals**. The value set by the overall level() can be overridden by level() inside a (*spec_i*). level() cannot be combined with noci.

cilines specifies that the confidence intervals are to be displayed as lines. By default, confidence intervals are displayed as shaded areas. cilines cannot be specified with noci.

ciopts1(*rarea_options* | *rline_options*) affect the rendition of the confidence bands for the first statistic in *statlist*; see [G] **graph twoway rarea** or [G] **graph twoway rline**. By default, the confidence intervals are displayed as shaded areas so that the options for [G] **graph twoway rarea** can be specified. When cilines is specified, the confidence intervals are displayed as lines and the [G] **graph twoway rline** options can be specified. ciopts1() may not be combined with noci.

ciopts2(*rarea_options* | *rline_options*) affect the rendition of the confidence bands for the second statistic in *statlist*; see [G] **graph twoway rarea** or [G] **graph twoway rline**. By default, the confidence intervals are displayed as shaded areas so that the options for [G] **graph twoway rarea** can be specified. ciopts2() cannot be combined with noci. ciopts2() can only be specified if there are two statistics requested in *statlist*.

plot1(*line_options*) ... plot4(*line_options*) affect the rendition of the plotted statistics in *statlist* according to the order specified (i.e. options in plot1() affect the rendition of the first statistic in *statlist*, ..., options in plot4() affect the rendition of the fourth statistic in *statlist*. For an explanation of these options see [G] **graph twoway line**.

twoway_options are any of the options documented in [G] *twoway_options* excluding by(). These include options for titling the graph (see [G] *title_options*) and options for saving the graph to disk (see [G] *saving_option*).

combine_options affect the appearance of the combined graph; see [G] **graph combine**. *combine_options* cannot be specified with individual.

Remarks

If you have not read [TS] **varirf**, please do so.

`varirf cgraph` makes it easy to create a series of customized graphs from VARIRF results. The information enclosed within each set of parentheses, i.e., each ($spec_i$), is a specific graph command. Each specific graph command has four parts: (1) impulse–response identifier, (2) list of statistics to be graphed, (3) confidence interval options, and (4) graph options.

Let's briefly discuss each of the four parts. The first part is made up of the combination of *irfname*, *impulse*, and *response*. Together, these three elements uniquely identify a set of impulse–response function estimates or a set of variance decomposition estimates. The second part is a list of statistics that are to be put on the same graph, optionally, along with their confidence intervals. Putting statistics with very different scales in this list is not recommended. For instance, it is sometimes the case that the `sirf` and the `oirf` are on similar scales while the `irf` is on a different scale. In this case, while putting the `sirf` and the `oirf` on the same graph would look fine, combining either of these two with the `irf` would produce an uninformative graph. Also, two limits are imposed for legibility. When confidence intervals are included, only two statistics may be included on a single graph. When the confidence intervals are suppressed, only four statistics may be included on a single graph. The options `noci` and `ilevel()` make up the third part—confidence interval options. The fourth part is made up of the *scatter_options* for the graph. These options are discussed in [G] **graph twoway scatter**.

Each specific graph will display the requested statistics that correspond to the specified combination of *irfname*, *impulse* variable, and *response* variable. By default, all the individual graphs are combined into a single matrix graph. Also, by default, all the steps, i.e., time periods, available for a given set of VARIRF results are graphed; use the `lstep()` or `ustep()` options to impose a common maximum for all the specific graphs.

▷ Example

Suppose that we have VARIRF results generated from two different SVAR models. We are interested in determining whether the shapes of the structural impulse–response functions and the structural forecast error variance decompositions are similar over the two models. We are also interested in knowing whether or not the structural impulse–response functions and the structural forecast-error variance decompositions differ significantly from their Cholesky counterparts. Below, we use `varirf cgraph` to create a matrix of 4 graphs to help us answer these questions.

```
. use http://www.stata-press.com/data/r8/lutkepohl
(Quarterly SA West German macro data, Bil DM, from Lutkepohl 1993 Table E.1)
. mat a = (., 0, 0\0,.,0\.,.,.)
. mat b = I(3)
. svar dlinvestment dlincome dlconsumption, aeq(a) beq(b)
 (output omitted)
. varirf create modela, set(results3, replace) step(8)
current varirf data file is results3.vrf
file results3.vrf saved
. svar dlincome dlinvestment dlconsumption, aeq(a) beq(b)
 (output omitted)
. varirf create modelb, step(8)
file results3.vrf saved
. varirf cgraph (modela dlincome dlconsumption oirf sirf)
> (modelb dlincome dlconsumption oirf sirf)
> (modela dlincome dlconsumption fevd sfevd, lstep(1))
> (modelb dlincome dlconsumption fevd sfevd, lstep(1)),
>  title("Results from modela and modelb")
```

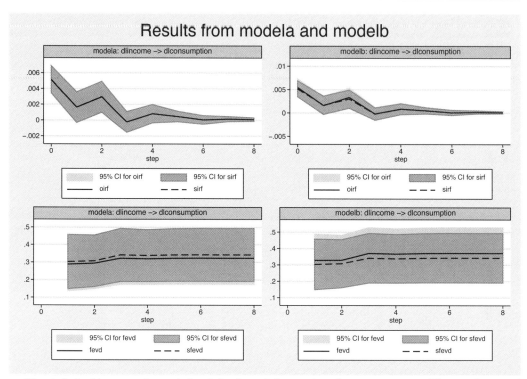

The default title on each graph is of the form *irfname*: *impulse –> response*. The graphs reveal that the shapes of the functions are very similar over the two models. The graphs illuminate only one minor difference over the two models. In `modela`, the estimated structural forecast-error variance is slightly larger than the Cholesky-based estimates, while in `modelb`, the Cholesky-based estimates are slightly larger than the structural estimates. However, in each model, the structural estimates are close to the center of the wide confidence interval for the Cholesky estimates and vice versa.

Now, let's focus on the results from `modela`. Suppose that we were interested in examining how `dlconsumption` responded to impulses in its own structural innovations, structural innovations to `dlincome`, and structural innovations to `dlinvestment`. Below, we illustrate how we could obtain three individual graphs, each containing one of the three impulse–response functions of interest.

(Continued on next page)

```
. varirf cgraph (modela dlinvestment dlconsumption sirf)
> (modela dlincome dlconsumption sirf)
> (modela dlconsumption dlconsumption sirf),
>  individual
```

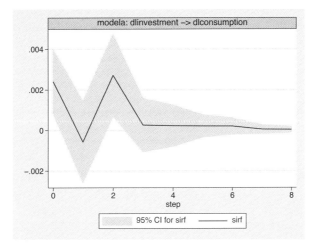

The first of the three graphs indicates that under the identification assumptions of modela, a positive shock to dlinvestment first increases, then decreases, and then increases dlconsumption. This effect quickly dies out.

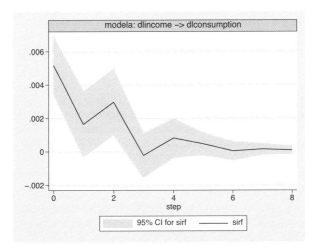

Let's now consider the graph of the structural impulse–response function from the innovations in dlincome on dlconsumption. In contrast to the first graph, the second graph indicates that a positive shock to dlincome causes an increase in dlconsumption that dies out after 4 or 5 periods.

(Continued on next page)

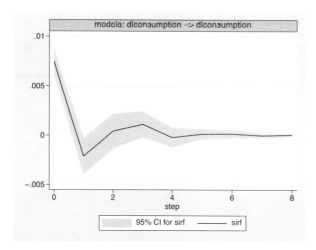

The third graph is of the structural impulse–response function from the innovations to dlcon-sumption to dlconsumption. We see that the identification restrictions used in modela imply that a positive shock to dlconsumption causes dlconsumption to oscillate for about 5 periods, after which the effect has dissipated.

◁

Saved Results

varirf cgraph saves in r():

Scalars
    r(k)   number of specific graph commands

Macros
    r(individual)   individual, if individual specified
    r(save)   saving filename and replace option
                        for combined graph, if specified
    r(name)   name and replace option
                        for combined graph, if specified
    r(title)   title of combined graph,
                        if individual not specified
    r(save*j*)   saving filename and replace option
                        for *j*th graph command, if specified
    r(name*j*)   name and replace option
                        for *j*th graph command, if specified
    r(title*j*)   title for *j*th graph
    r(ci*j*)   level applied to *j*th confidence interval or noci
    r(response*j*)   response in *j*th command
    r(impulse*j*)   impulse in *j*th command
    r(irfname*j*)   irfname in *j*th command
    r(stats*j*)   stats specified in *j*th command

Methods and Formulas

`varirf cgraph` is implemented as an ado-file.

Also See

Complementary:	[TS] **varirf add**, [TS] **varirf create**, [TS] **varirf describe**, [TS] **varirf dir**, [TS] **varirf drop**, [TS] **varirf erase**, [TS] **varirf rename**, [TS] **varirf set**
Related:	[TS] **arima**, [TS] **var**, [TS] **var svar**, [TS] **varirf ctable** [TS] **varirf graph**, [TS] **varirf ograph**, [TS] **varirf table**,
Background:	[U] **14.4.3 Time-series varlists** [TS] **varirf**

Title

> **varirf create** — Obtain impulse–response functions and FEVDs

Syntax

> varirf <u>cr</u>eate *irfname* $\left[\ ,\ \underline{\text{o}}\text{rder}(\textit{varlist})\ \underline{\text{st}}\text{ep}(\#)\ \text{replace set}(\textit{setcmd})\right.$
>
>   bs bsp <u>bsa</u>ving(*filename* $\left[\ ,\ \text{replace}\ \right]$) <u>r</u>eps(*integer*) nose <u>nod</u>ots
>
>   <u>est</u>imates(*estname*) <u>l</u>evel(#) $\left.\right]$

varirf create may only be used with var or svar e() results; see [TS] **var** and [TS] **var svar**.

varirf create is for use with time-series data; see [TS] **tsset**. You must tsset your data before using varirf create.

irfname is any valid Stata variable name, yet not to exceed 15 characters in length.

Description

varirf create estimates a series of impulse–response functions, forecast-error variance decompositions, and their standard errors from var or svar e() results. These estimates, known as VARIRF results, are stored in a VARIRF file under the name *irfname*.

Options

order(*varlist*) specifies the Cholesky ordering of the endogenous variables to be used in estimating the orthogonalized impulse response functions. By default, the order in which the variables were specified to var is used. order() may not be specified with svar e() results.

step(#) specifies the forecast horizon for the estimation. The default is 8 periods.

replace specifies that if results are stored in the VARIRF file under the name *irfname*, they should be replaced.

set(*setcmd*) specifies a varirf set command that will change the active VARIRF file. set() accepts a *setcmd*, a combination of *filename* and *options* that can be understood by [TS] **varirf set**. By default, varirf create works on the active VARIRF file. The file made active by the set() option will remain active after varirf create finishes.

bs requests that the standard errors of the estimated functions be obtained by bootstrapping the residuals. bs may not be specified if there are gaps in the data.

bsp requests that the standard errors of the estimated functions be obtained via a multivariate normal parametric bootstrap. bsp may not be specified if there are gaps in the data.

bsaving(*filename*$\left[\ ,\ \text{replace}\right]$) specifies the name of the file to hold the dataset of bootstrapped replications. This option may only be used in combination with bs or bsp.

reps(#) specifies the number of replications for the bootstrap. This option may only be specified in conjunction with bs or bsp. By default, bs and bsp use 200 replications. # must be an integer greater than or equal to 50.

nose requests that the asymptotic standard errors not be calculated. This option is implied by bs and bsp.

nodots specifies that dots are not to displayed. By default, varirf create displays a "." for each bootstrap replication.

estimates(*estname*) specifies that varirf create should compute VARIRF results from var or svar e() results stored as *estname*. By default, varirf create uses the currently active e() results. See [R] **estimates** for more on saving and restoring previously obtained estimates.

level(*#*) specifies the confidence level, in percent, for confidence intervals. The default is level(95) or as set by set level; see [U] **23.6 Specifying the width of confidence intervals**.

Remarks

If you have not read [TS] **varirf**, please do so.

The *Methods and Formulas* section discusses impulse–response functions and forecast-error variance decompositions. [TS] **varirf** contains a general discussion of how to estimate IRFs and FEVDs in Stata. In addition, [TS] **varirf** has a pair of examples that illustrate how to use varirf create and interpret the results. In this section, we provide a different example that illustrates how to use varirf create to obtain bootstrapped standard errors.

▷ Example

In this example, we will compare the bootstrapped and the asymptotic standard errors for a specific forecast-error variance decomposition. Let's begin by fitting a VAR(2) model to the Lütkepohl data and using varirf create to estimate the impulse-response functions, forecast-error variance decompositions, and their asymptotic standard errors.

```
. use http://www.stata-press.com/data/r8/lutkepohl
(Quarterly SA West German macro data, Bil DM, from Lutkepohl 1993 Table E.1)
. quietly var dlinvestment dlincome dlconsumption if qtr>=q(1961q2) &
> qtr <= q(1978q4), lags(1/2)
. varirf create asymp, step(8) set(results1, replace)
current varirf data file is results1.vrf
file results1.vrf saved
```

Below, we use varirf create to estimate the impulse response functions, the forecast-error variance decompositions, and bootstrap estimates of their standard errors.

```
. set seed 123456
. varirf create bs, step(8) bs reps(250) nodots
file results1.vrf saved
```

Now, we can use varirf ctable to compare our results.

(Continued on next page)

```
. varirf ctable (asymp dlincome dlconsumption fevd)
> (bs dlincome dlconsumption fevd), noci stderror
```

step	(1) fevd	(1) S.E.	(2) fevd	(2) S.E.
0	0	0	0	0
1	.282135	.087373	.282135	.106687
2	.278777	.083782	.278777	.101413
3	.33855	.090006	.33855	.102614
4	.339942	.089207	.339942	.100797
5	.342813	.090494	.342813	.100687
6	.343119	.090517	.343119	.100806
7	.343079	.090499	.343079	.100771

(1) irfname = asymp, impulse = dlincome, and response = dlconsumption
(2) irfname = bs, impulse = dlincome, and response = dlconsumption

Since our results only differ as to how standard errors were computed, our point estimates are the same. The bootstrap estimates of the standard errors are larger than the asymptotic estimates. This indicates that the sample size of 71 is not large enough for the distribution of the estimator of the forecast-error variance decomposition to be well approximated by the asymptotic distribution. In this case, we would expect our bootstrapped confidence interval to be more reliable than the confidence interval based on the asymptotic standard error.

◁

□ Technical Note

The details of the bootstrap algorithms are given in *Methods and Formulas*. One important point about these algorithms is that they are conditional on the first p observations, where p is the order of the fitted VAR(p). (In the case of an SVAR model, p is the order of the VAR(p) that underlies the SVAR.) The bootstrapped estimates are conditional on the first p observations, just like the estimators of the coefficients in VAR(p) models are conditional on the first p observations. In the case of the bootstrapped standard errors (the `bs` option), the p initial observations are used in conjunction with resampling the residuals to generate the bootstrap samples that are used for estimation. In the case of the more parametric bootstrap (the `bsp` option), the p initial observations are used in conjunction with draws from a multivariate normal distribution with variance–covariance matrix $\widehat{\Sigma}$, to generate the bootstrap samples that are used for estimation.

□

□ Technical Note

`varirf` uses $\widehat{\Sigma}$, the estimated variance matrix of the disturbances, in computing the asymptotic standard errors of all the functions. The point estimates of the orthogonalized impulse–response functions, the structural impulse–response functions, and all the variance decompositions also depend on $\widehat{\Sigma}$. As discussed in [TS] **var**, var uses the ML estimator of this matrix by default, but the `dfk` option uses an estimator that includes a small sample correction. Specifying `dfk` in the call to `var` or `svar` will change the estimate of $\widehat{\Sigma}$ and all the VARIRF results that depend upon it.

□

Estimates of some of the functions are not always available. When either the point estimate or standard error are not available, they are filled in with missing values. Below is a list of general restrictions.

1. The structural impulse–response functions, the structural forecast-error decompositions and both of their standard errors are only available when `varirf create` is applied to `svar e()` results. This is due to the fact that `varirf create` needs information about the estimated structural decomposition to compute the estimates.

2. As discussed in *Methods and Formulas*, the asymptotic standard errors of the structural impulse–response functions and the structural forecast-error decompositions are not available when `varirf create` is applied to `svar e()` results in which long-run constraints were applied. Bootstrap standard errors are available.

3. If `nose` is specified and neither of the bootstrap options are specified, all the standard errors will be set to missing.

Technical aspects of VARIRF files

The details provided here are not necessary reading for those using any of the VARIRF commands. This section is included for programmers and the very curious.

`varirf create` estimates a series of impulse–response functions and their standard errors. Although these estimates are stored in a VARIRF file, almost all users will never need to look at the contents of this file. The VARIRF commands are designed to provide the wherewithal to estimate, analyze, present, and manage VARIRF results. For the sake of completeness, we document some of the structure that is placed on these files. However, we would like to issue the following warning. If this structure is altered by users, some of the VARIRF commands may fail to work.

VARIRF files are simply Stata datasets that are stored with a `.vrf` extension rather than the default `.dta` extension. Furthermore, VARIRF files have a nested panel structure. The variable `irfname` contains the *irfname* specified in the call to `varirf create` that computed the results. The `impulse` variable holds the name of the endogenous variable whose innovations are the impulse. The `response` variable holds the name of the endogenous variable that is responding to the innovations in the `impulse` variable. Since there are K equations in the model, for each *irfname*, there are K^2 combinations of `impulse` and `response`. A combination of `irfname`, `impulse` and `response` uniquely define a set of impulse–response function estimates. The `step` variable records the periods for which these estimates were computed.

Below is a catalog of the quantities that `varirf create` estimates and the variable names under which they are stored in the VARIRF file.

(Continued on next page)

Quantity	Name
impulse–response functions	`irf`
orthogonalized impulse–response functions	`oirf`
cumulative impulse–response functions	`cirf`
cumulative orthogonalized impulse–response functions	`coirf`
Cholesky forecast-error decomposition	`fevd`
structural impulse–response functions	`sirf`
structural forecast-error decomposition	`sfevd`
standard error of the impulse–response functions	`stdirf`
standard error of the orthogonalized impulse–response functions	`stdoirfi`
standard error of the cumulative impulse–response functions	`stdcirf`
standard error of the cumulative orthogonalized impulse–response functions	`stdcoirf`
standard error of the Cholesky forecast-error decomposition	`stdfevd`
standard error of the structural impulse–response functions	`stdsirf`
standard error of the structural forecast-error decomposition	`stdsfevd`

In addition to the variables, information about the VARIRF results is stored in data characteristics. (See [P] **char** for more information on data characteristics.) To access this information, use `varirf describe`. `varirf describe` displays this information and returns this information in `r()`. Since this information is actually stored as characteristics, we document it here. The data characteristic `irfnames` contains a list of all the *irfnames* in the VARIRF file. The data characteristic `version` contains the version number of the VARIRF file. Also, for each *irfname*, there are a series of data characteristics. The table below provides the details of the characteristics that are stored for each *irfname*.

Name	Contents
`_dta[`*irfname*`_model]`	`var`, `short-run svar` or `long-run svar` depending on the `e()` results specified to `varirf create`
`_dta[`*irfname*`_order]`	Cholesky order used in VARIRF estimates
`_dta[`*irfname*`_exog]`	exogenous variables in VAR
`_dta[`*irfname*`_constant]`	`constant` or `noconstant` depending on whether `noconstant` was specified in VAR or SVAR
`_dta[`*irfname*`_lags]`	lags in model
`_dta[`*irfname*`_tmin]`	minimum value of `timevar` in estimation sample
`_dta[`*irfname*`_tmax]`	maximum value of `timevar` in estimation sample
`_dta[`*irfname*`_timevar]`	name of `tsset` timevar
`_dta[`*irfname*`_tsfmt]`	format of `timevar`
`_dta[`*irfname*`_varcns]`	`constrained` or colon-separated list of constraints placed on VAR coefficients
`_dta[`*irfname*`_svarcns]`	`constrained` or colon-separated list of constraints placed on VAR coefficients
`_dta[`*irfname*`_step]`	maximum step in VARIRF estimates *irfname* VARIRF results
`_dta[`*irfname*`_stderror]`	`asymptotic`, `bs`, `bsp` or `none` depending on type of standard errors requested
`_dta[`*irfname*`_reps]`	number of bootstrap replications performed
`_dta[`*irfname*`_version]`	version of VARIRF file that originally held

This information can be easily accessed by using `varirf describe`, so there is no need to open the VARIRF file and list the characteristics to view this information.

Methods and Formulas

varirf create is implemented as an ado-file.

A pth order vector autoregressive model VAR(p) with exogenous variables is given by

$$\mathbf{y}_t = \mathbf{v} + \mathbf{A}_1\mathbf{y}_{t-1} + \cdots + \mathbf{A}_p\mathbf{y}_{t-p} + \mathbf{B}\mathbf{x}_t + \mathbf{u}_t$$

where

$\mathbf{y}_t = (y_{1t}, \ldots, y_{Kt})'$ is a $(K \times 1)$ random vector,
the $\mathbf{A}_i$ are fixed $(K \times K)$ matrices of parameters,
$\mathbf{x}_t$ is an $(M \times 1)$ vector of exogenous variables,
$\mathbf{B}$ is a $(K \times M)$ matrix of coefficients,
$\mathbf{v}$ is a $(K \times 1)$ vector of fixed parameters, and
$\mathbf{u}_t$ is assumed to be white noise; that is,
$E(\mathbf{u}_t) = \mathbf{0}$
$E(\mathbf{u}_t\mathbf{u}_t') = \mathbf{\Sigma}$
$E(\mathbf{u}_t\mathbf{u}_s') = \mathbf{0}$ for $t \neq s$

As discussed in [TS] **varstable**, if the VAR(p) is stable, then it can be rewritten in moving-average form. Any exogenous variables are assumed to be covariance stationary. Since the functions of interest in this section only depend on the exogenous variables through the effect that they have on the estimated $\mathbf{A}_i$, we can simplify the notation by dropping them from the analysis. All the formulas given below still apply, with the caveat that the $\mathbf{A}_i$ are estimated jointly with $\mathbf{B}$ on the exogenous variables. Below, we discuss conditions under which the impulse–response functions and forecast-error variance decompositions have a causal interpretation. This interpretation also requires that any exogenous variables be strictly exogenous. Estimation only requires that the exogenous variables be predetermined; i.e., that $E[\mathbf{x}_{jt}u_{it}] = 0$ for all i, j and t. In contrast, strict exogeneity requires that $E[\mathbf{x}_{js}u_{it}] = 0$ for all i, j, s and t.

An introduction to impulse–response functions

Impulse–response functions describe how the innovations to one variable affect another variable after a given number of periods.

For an example of how impulse–response functions are interpreted, see Stock and Watson (2001), who use impulse–response functions to investigate the effect of surprise shocks to the Federal Funds rate on inflation and unemployment. In another interesting example, Christiano, et al. (1999) also use impulse–response functions to investigate how shocks to monetary policy affect other macroeconomic variables. Consider a VAR(p) without exogenous variables:

$$\mathbf{y}_t = \mathbf{v} + \mathbf{A}_1\mathbf{y}_{t-1} + \cdots + \mathbf{A}_p\mathbf{y}_{t-p} + \mathbf{u}_t \tag{1}$$

The VAR(p) represents the variables in $\mathbf{y}_t$ as functions of its own lags and serially uncorrelated innovations $\mathbf{u}_t$. All the information about contemporaneous correlations among the K variables in $\mathbf{y}_t$ is contained in $\mathbf{\Sigma}$. In fact, as discussed in [TS] **var svar**, a VAR(p) can be viewed as the reduced form of some dynamic simultaneous-equation model.

The most direct way to see how the innovations affect the variables in $\mathbf{y}_t$ after, say, i periods, is to re-write the model in its moving average form,

$$\mathbf{y}_t = \boldsymbol{\mu} + \sum_{i=0}^{\infty} \boldsymbol{\Phi}_i \mathbf{u}_{t-i} \qquad (2)$$

where $\boldsymbol{\mu}$ is the $K \times 1$ time-invariant mean of $\mathbf{y}_t$ and

$$\boldsymbol{\Phi}_i = \begin{cases} \mathbf{I}_K & \text{if } i = 0 \\ \sum_{j=1}^{i} \boldsymbol{\Phi}_{i-j} \mathbf{A}_j & \text{if } i = 1, 2, \ldots \end{cases}$$

We can only rewrite a VAR(p) in the moving-average form if it is stable. While the issue of stability is discussed in greater detail in [TS] **varstable**, essentially, a VAR(p) is stable if the variables are covariance stationary and none of the autocorrelations are too high.

The $\boldsymbol{\Phi}_i$ are the simple impulse–response functions. The j, k element of $\boldsymbol{\Phi}_i$ gives the effect of a one-time unit increase in the kth element of $\mathbf{u}_t$ on the jth element of $\mathbf{y}_t$ after i periods, holding everything else constant. Unfortunately, these effects have no causal interpretation. While Hamilton (1994, Section 11.4) provides a very nice discussion of why this is true, we summarize the intuition here. A causal interpretation would require us to be able to answer the question, "How does an innovation to variable k, holding everything else constant, affect variable j after i periods?" Since the $\mathbf{u}_t$ are contemporaneously correlated, the assumption of "holding everything else constant" is not valid. Contemporaneous correlation among the $\mathbf{u}_t$ implies that a shock to one variable is likely to be accompanied by shocks to some of the other variables, so it does not make sense to proceed with the thought experiment of shocking one variable and holding everything else constant. For this reason, (2) cannot yield this type of causal interpretation.

This shortcoming may be overcome by rewriting (2) in terms of mutually uncorrelated innovations. Suppose that we had a matrix $\mathbf{P}$ such that $\boldsymbol{\Sigma} = \mathbf{P}\mathbf{P}'$. If we had such a $\mathbf{P}$, then $\mathbf{P}^{-1}\boldsymbol{\Sigma}\mathbf{P}'^{-1} = \mathbf{I}_K$. Since

$$E\{\mathbf{P}^{-1}\mathbf{u}_t(\mathbf{P}^{-1}\mathbf{u}_t)'\} = \mathbf{P}^{-1}E\{(\mathbf{u}_t\mathbf{u}_t')\mathbf{P}'^{-1}\} = \mathbf{P}^{-1}\boldsymbol{\Sigma}\mathbf{P}'^{-1} = \mathbf{I}_K$$

this implies that we could use $\mathbf{P}^{-1}$ to orthogonalize the $\mathbf{u}_t$.

Thus, we could rewrite (2) as

$$\mathbf{y}_t = \boldsymbol{\mu} + \sum_{i=0}^{\infty} \boldsymbol{\Phi}_i \mathbf{P}\mathbf{P}^{-1} \mathbf{u}_{t-i}$$

$$= \boldsymbol{\mu} + \sum_{i=0}^{\infty} \boldsymbol{\Theta}_i \mathbf{P}^{-1} \mathbf{u}_{t-i}$$

$$= \boldsymbol{\mu} + \sum_{i=0}^{\infty} \boldsymbol{\Theta}_i \mathbf{w}_{t-i}$$

where $\boldsymbol{\Theta}_i = \boldsymbol{\Phi}_i \mathbf{P}$ and $\mathbf{w}_t = \mathbf{P}^{-1}\mathbf{u}_t$. If we had such a $\mathbf{P}$, then the $\mathbf{w}_k$ would be mutually orthogonal, and no information would be lost in the "holding everything else constant" assumption, implying that the $\boldsymbol{\Theta}_i$ would have the causal interpretation that we seek.

Choosing a $\mathbf{P}$ is very similar to placing identification restrictions on a system of dynamic simultaneous equations. The simple impulse–response functions do not identify the causal relationships that we wish to analyze. Thus, we seek at least as many identification restrictions as necessary to identify the causal impulse–response functions.

So, where do we get such a $\mathbf{P}$? Sims (1980) popularized the method of choosing $\mathbf{P}$ to be the Cholesky decomposition of $\widehat{\mathbf{\Sigma}}$. The impulse–response functions based on this choice of $\mathbf{P}$ are known as the *orthogonalized impulse–response functions*. Choosing $\mathbf{P}$ to be the Cholesky decomposition of $\widehat{\mathbf{\Sigma}}$ is equivalent to imposing a recursive structure for the corresponding dynamic structural-equation model. The ordering of the recursive structure is the same as the ordering imposed in the Cholesky decomposition. Since this choice is arbitrary, some researchers will look at the orthogonalized impulse–response functions with different orderings assumed in the Cholesky decomposition. The order() option varirf create facilitates this type of analysis.

The SVAR approach integrates the need to identify the causal impulse–response functions into the model specification and estimation process. Sufficient identification restrictions can be obtained by placing either short-run or long-run restrictions on the model. It is easiest if we begin with the short-run restrictions. The VAR(p) in (1) can be rewritten as

$$\mathbf{y}_t - \mathbf{v} - \mathbf{A}_1 \mathbf{y}_{t-1} - \cdots - \mathbf{A}_p \mathbf{y}_{t-p} = \mathbf{u}_t \tag{3}$$

A short-run SVAR model can be written as

$$\mathbf{A}(\mathbf{y}_t - \mathbf{v} - \mathbf{A}_1 \mathbf{y}_{t-1} - \cdots - \mathbf{A}_p \mathbf{y}_{t-p}) = \mathbf{A}\mathbf{u}_t = \mathbf{B}\mathbf{e}_t \tag{4}$$

where $\mathbf{A}$ and $\mathbf{B}$ are $K \times K$ nonsingular matrices of parameters to be estimated and $\mathbf{e}_t$ is a $K \times 1$ vector of disturbances with $\mathbf{e}_t \sim N(\mathbf{0}, \mathbf{I}_K)$, and $E[\mathbf{e}_t \mathbf{e}_s'] = \mathbf{0}_K$ for all $s \neq t$. Sufficient constraints must be placed on $\mathbf{A}$ and $\mathbf{B}$ so that $\mathbf{P}$ is identified. One way to see the connection is to draw out the implications of the latter equality in (4). From (4) it can be shown that

$$\mathbf{\Sigma} = \mathbf{A}^{-1}\mathbf{B}(\mathbf{A}^{-1}\mathbf{B})'$$

As discussed in [TS] **var svar**, the estimates $\widehat{\mathbf{A}}$ and $\widehat{\mathbf{B}}$ are obtained by maximizing the concentrated log-likelihood function based on the $\widehat{\mathbf{\Sigma}}$ obtained from the underlying VAR(p). The short-run SVAR approach chooses $\mathbf{P} = \widehat{\mathbf{A}}^{-1}\widehat{\mathbf{B}}$ to identify the causal impulse–response functions. The long-run SVAR approach works similarly, with $\mathbf{P} = \widehat{\mathbf{C}} = \widehat{\mathbf{A}}^{-1}\widehat{\mathbf{B}}$, where $\widehat{\mathbf{A}}^{-1}$ is the matrix of estimated long-run or accumulated effects of the reduced form VAR shocks.

There is one important difference between long-run and short-run SVAR models. As discussed by Amisano and Gianinni (1997, Chapter 6), in the short-run model, the constraints are applied directly to the parameters in $\mathbf{A}$ and $\mathbf{B}$. Then, $\mathbf{A}$ and $\mathbf{B}$ interact with the estimated parameters of the underlying VAR(p). In contrast, in a long-run model, the constraints are placed on functions of the estimated VAR(p) parameters. While estimation and inference of the parameters in $\mathbf{C}$ is straightforward, obtaining the asymptotic standard errors of the structural impulse–response functions requires untenable assumptions. For this reason, varirf create does not estimate the asymptotic standard errors of the structural impulse–response functions generated by long-run SVAR models. However, bootstrapped standard errors are still available.

Impulse–response function formulas

The previous discussion implies that there are three different choices of $\mathbf{P}$ that can be used to obtain distinct $\mathbf{\Theta}_i$. $\mathbf{P}$ is chosen to be the Cholesky decomposition of $\mathbf{\Sigma}$ for the orthogonalized impulse–response functions. For the structural impulse–response functions, $\mathbf{P} = \mathbf{A}^{-1}\mathbf{B}$ for short-run models and $\mathbf{P} = \mathbf{C}$ for long-run models. We shall distinguish between the three by defining $\mathbf{\Theta}_i^o$ to be the orthogonalized impulse response functions, $\mathbf{\Theta}_i^{sr}$ to be the short-run structural impulse–response functions, and $\mathbf{\Theta}_i^{lr}$ to be the long-run structural impulse–response functions . Similarly, we define $\mathbf{P}_c$ to be the Cholesky decomposition of $\mathbf{\Sigma}$, $\mathbf{P}_{sr} = \mathbf{A}^{-1}\mathbf{B}$ to be the short-run structural decomposition, and $\mathbf{P}_{lr} = \mathbf{C}$ to be the long-run structural decomposition.

Given estimates of the $\widehat{\mathbf{A}}_i$ and $\widehat{\mathbf{\Sigma}}$ from var or svar, the estimates of the simple impulse–response functions and the orthogonalized impulse–response functions are, respectively,

$$\widehat{\mathbf{\Phi}}_i = \sum_{j=1}^{i} \widehat{\mathbf{\Phi}}_{i-j}\widehat{\mathbf{A}}_j$$

$$\widehat{\mathbf{\Theta}}_i^o = \widehat{\mathbf{\Phi}}_i\widehat{\mathbf{P}}_c$$

where $\widehat{\mathbf{A}}_j = \mathbf{0}_K$ for $j > p$ and $\widehat{\mathbf{P}}_c$ is the Cholesky decomposition of $\widehat{\mathbf{\Sigma}}$

Given the estimates $\widehat{\mathbf{A}}$ and $\widehat{\mathbf{B}}$, or $\widehat{\mathbf{C}}$, from svar, the estimates of the structural impulse–response functions are either

$$\widehat{\mathbf{\Theta}}_i^{sr} = \widehat{\mathbf{\Phi}}_i\widehat{\mathbf{P}}_{sr}$$

or

$$\widehat{\mathbf{\Theta}}_i^{lr} = \widehat{\mathbf{\Phi}}_i\widehat{\mathbf{P}}_{lr}$$

where

$$\widehat{\mathbf{P}}_{sr} = \widehat{\mathbf{A}}^{-1}\widehat{\mathbf{B}}$$

$$\widehat{\mathbf{P}}_{lr} = \widehat{\mathbf{C}}$$

It is important to note that the estimated structural impulse–response functions stored in a VARIRF file may be from either a short-run or a long-run model. Since a given set of svar e() results must have been generated from either short-run or long-run restrictions, it is not possible to generate both short-run and long-run structural impulse–response functions from the same svar e() results. Since they cannot both exist at the same time, it would be wasteful to store the short-run and long-run structural impulse–response variables as separate variables. For this reason, all estimated structural impulse–response functions are stored in variable sirf. As discussed in [TS] **varirf describe**, one can easily determine whether the structural impulse–response functions were generated from a short-run or a long-run SVAR model using varirf describe.

Following Lütkepohl (1993, Section 3.7), estimates of the cumulative impulse–response functions and the cumulative orthogonalized impulse response functions (COIRFs) at period n are, respectively,

$$\widehat{\mathbf{\Psi}}_n = \sum_{i=0}^{n} \widehat{\mathbf{\Phi}}_i$$

$$\widehat{\mathbf{\Xi}}_n = \sum_{i=0}^{n} \widehat{\mathbf{\Theta}}_i$$

The asymptotic standard errors of the different impulse–response functions are obtained by applications of the delta method. See Lütkepohl (1993, Section 3.7) and Amisano and Giannini (1997, Chapter 4) for the derivations. See Serfling (1980, Section 3.3) for a discussion of the delta method. While the basic idea is straightforward, the algebra is rather complicated. Some additional complications arise from the fact that the coefficients in the VAR(p), or the VAR(p) underlying SVAR, may be subject to linear constraints.

The various estimators of the different impulse–response functions are presented above in matrix form. Below, we present the formulas for their VCEs. In order to discuss the VCE of an estimator in matrix form, the estimator must first be vectorized. Hence, we make extensive use of the vec() operator, where vec($\mathbf{X}$) is the vector obtained by stacking the columns of $\mathbf{X}$.

Lütkepohl (1993, Section 3.7) derives the asymptotic VCEs of vec($\mathbf{\Phi}_i$), vec($\mathbf{\Theta}_i^o$), vec($\widehat{\mathbf{\Psi}}_n$), and vec($\widehat{\mathbf{\Xi}}_n$). Since vec($\mathbf{\Phi}_i$) is $K^2 \times 1$, the asymptotic VCE of vec($\mathbf{\Phi}_i$) is $K^2 \times K^2$, and it is given by

$$\mathbf{G}_i \widehat{\mathbf{\Sigma}}_{\widehat{\alpha}} \mathbf{G}_i'$$

where

$$\mathbf{G}_i = \sum_{m=0}^{i-1} \mathbf{J}(\widehat{\mathbf{M}}')^{(i-1-m)} \otimes \widehat{\mathbf{\Phi}}_m \qquad \mathbf{G}_i \text{ is } K^2 \times K^2 p$$

$$\mathbf{J} = (\mathbf{I}_K, \mathbf{0}_K, \ldots, \mathbf{0}_K) \qquad \mathbf{J} \text{ is } K \times Kp$$

$$\widehat{\mathbf{M}} = \begin{bmatrix} \widehat{\mathbf{A}}_1 & \widehat{\mathbf{A}}_2 & \ldots & \widehat{\mathbf{A}}_{p-1} & \widehat{\mathbf{A}}_p \\ \mathbf{I}_K & \mathbf{0}_K & \ldots & \mathbf{0}_K & \mathbf{0}_K \\ \mathbf{0}_K & \mathbf{I}_K & & \mathbf{0}_K & \mathbf{0}_K \\ \vdots & & \ddots & \vdots & \vdots \\ \mathbf{0}_K & \mathbf{0}_K & \ldots & \mathbf{I}_K & \mathbf{0}_K \end{bmatrix} \qquad \widehat{\mathbf{M}} \text{ is } Kp \times Kp$$

The $\widehat{\mathbf{A}}_i$ are the estimates of the coefficients on the lagged variables in the VAR(p), and $\widehat{\mathbf{\Sigma}}_{\widehat{\alpha}}$ is the VCE matrix of $\widehat{\boldsymbol{\alpha}} = \mathrm{vec}(\widehat{\mathbf{A}}_1, \ldots, \widehat{\mathbf{A}}_p)$. $\widehat{\mathbf{\Sigma}}_{\widehat{\alpha}}$ is a $K^2 p \times K^2 p$ matrix whose elements come from the VCE of the VAR(p) coefficient estimator. As such, this VCE is the VCE of the constrained estimator if there are any constraints placed on the VAR(p) coefficients.

The $K^2 \times K^2$ asymptotic VCE matrix for the vec($\widehat{\mathbf{\Psi}}_n$) after n periods is given by

$$\mathbf{F}_n \widehat{\mathbf{\Sigma}}_{\widehat{\alpha}} \mathbf{F}_n'$$

where

$$\mathbf{F}_n = \sum_{i=1}^{n} \mathbf{G}_i$$

The $K^2 \times K^2$ asymptotic VCE matrix of the vectorized orthogonalized impulse–response functions at horizon i, vec($\mathbf{\Theta}_i^o$), is

$$\mathbf{C}_i \widehat{\mathbf{\Sigma}}_{\widehat{\alpha}} \mathbf{C}_i' + \overline{\mathbf{C}}_i \widehat{\mathbf{\Sigma}}_{\widehat{\sigma}} \overline{\mathbf{C}}_i'$$

where

$$\mathbf{C}_0 = \mathbf{0} \qquad\qquad \mathbf{C}_0 \text{ is } K^2 \times K^2 p$$

$$\mathbf{C}_i = (\widehat{\mathbf{P}}'_c \otimes \mathbf{I}_K)\mathbf{G}_i, \quad i = 1, 2, \dots \qquad \mathbf{C}_i \text{ is } K^2 \times K^2 p$$

$$\overline{\mathbf{C}}_i = (\mathbf{I}_K \otimes \boldsymbol{\Phi}_i)\mathbf{H}, \quad i = 0, 1, \dots \qquad \overline{\mathbf{C}}_i \text{ is } K^2 \times K^2$$

$$\mathbf{H} = \mathbf{L}'_K \left\{ \mathbf{L}_K \mathbf{N}_K (\widehat{\mathbf{P}}_c \otimes \mathbf{I}_K) \mathbf{L}'_K \right\}^{-1} \qquad \mathbf{H} \text{ is } K^2 \times K\frac{(K+1)}{2}$$

$\mathbf{L}_K$ solves: $\qquad \mathrm{vech}(\mathbf{F}) = \mathbf{L}_K \, \mathrm{vec}(\mathbf{F}) \qquad\qquad \mathbf{L}_K \text{ is } K\frac{(K+1)}{2} \times K^2$

$$\text{for } \mathbf{F} \; K \times K \text{ and symmetric}$$

$\mathbf{K}_K$ solves: $\quad \mathbf{K}_K \mathrm{vec}(\mathbf{G}) = \mathrm{vec}(\mathbf{G}') \text{ for any } K \times K \text{ matrix } \mathbf{G} \qquad \mathbf{K}_K \text{ is } K^2 \times K^2$

$$\mathbf{N}_K = \tfrac{1}{2}\left(\mathbf{I}_{K^2} + \mathbf{K}_K\right) \qquad\qquad \mathbf{N}_K \text{ is } K^2 \times K^2$$

$$\widehat{\boldsymbol{\Sigma}}_{\widehat{\sigma}} = 2\mathbf{D}_K^+ (\widehat{\boldsymbol{\Sigma}} \otimes \widehat{\boldsymbol{\Sigma}})\mathbf{D}_K^+ \qquad\qquad \widehat{\boldsymbol{\Sigma}}_{\widehat{\sigma}} \text{ is } K\frac{(K+1)}{2} \times K\frac{(K+1)}{2}$$

$$\mathbf{D}_K^+ = (\mathbf{D}'_K \mathbf{D}_K)^{-1} \mathbf{D}'_K \qquad\qquad \mathbf{D}_K^+ \text{ is } K\frac{(K+1)}{2} \times K^2$$

$\mathbf{D}_K$ solves: $\quad \mathbf{D}_K \mathrm{vech}(\mathbf{F}) = \mathrm{vec}(\mathbf{F}) \quad \text{for } \mathbf{F} \; K \times K \text{ and symmetric} \qquad \mathbf{D}_K \text{ is } K^2 \times K\frac{(K+1)}{2}$

$$\mathrm{vech}(\mathbf{X}) = \begin{bmatrix} x_{11} \\ x_{21} \\ \vdots \\ x_{K1} \\ x_{22} \\ \vdots \\ x_{K2} \\ \vdots \\ x_{KK} \end{bmatrix} \qquad \text{for } \mathbf{X} \; K \times K \qquad \mathrm{vech}(\mathbf{X}) \text{ is } K\frac{(K+1)}{2} \times 1$$

Note that $\widehat{\boldsymbol{\Sigma}}_{\widehat{\sigma}}$ is the VCE of $\mathrm{vech}(\widehat{\boldsymbol{\Sigma}})$. Also note that more details about $\mathbf{L}_K$, $\mathbf{K}_K$, $\mathbf{D}_K$ and vech() are available in Lütkepohl (1993, Appendix A.12). Finally, as Lütkepohl (1993, 101) discusses, $\mathbf{D}_K^+$ is the Moore–Penrose inverse of $\mathbf{D}_K$.

As discussed in Amisano and Giannini (1997, Chapter 6), the asymptotic standard errors of the structural impulse–response functions are available for short-run SVAR models, but not for long-run SVAR models. Following Amisano and Giannini (1997, Chapter 5), the asymptotic $K^2 \times K^2$ VCE of the (short-run) structural impulse–response functions after i periods, when a maximum of h periods are estimated is the i, i block of

$$\widehat{\boldsymbol{\Sigma}}(h)_{ij} = \widetilde{\mathbf{G}}_i \widehat{\boldsymbol{\Sigma}}_{\widehat{\alpha}} \widetilde{\mathbf{G}}'_j + \left\{ \mathbf{I}_K \otimes (\mathbf{J}\widehat{\mathbf{M}}^i \mathbf{J}') \right\} \boldsymbol{\Sigma}(0) \left\{ \mathbf{I}_K \otimes (\mathbf{J}\widehat{\mathbf{M}}^j \mathbf{J}') \right\}'$$

where

$$\widetilde{\mathbf{G}}_0 = \mathbf{0}_K \qquad\qquad \mathbf{G}_0 \text{ is } K^2 \times K^2 p$$

$$\widetilde{\mathbf{G}}_i = \sum_{k=0}^{i-1} \left[\left\{ \widehat{\mathbf{P}}_{sr}' \mathbf{J} (\widehat{\mathbf{M}}')^{i-1-k} \otimes \left(\mathbf{J} \widehat{\mathbf{M}}^k \mathbf{J}' \right) \right\} \right] \qquad \mathbf{G}_i \text{ is } K^2 \times K^2 p$$

$$\widehat{\boldsymbol{\Sigma}}(0) = \mathbf{Q}_2 \widehat{\boldsymbol{\Sigma}}_W \mathbf{Q}_2' \qquad\qquad \widehat{\boldsymbol{\Sigma}}(0) \text{ is } K^2 \times K^2$$

$$\widehat{\boldsymbol{\Sigma}}_W = \mathbf{Q}_1 \widehat{\boldsymbol{\Sigma}}_{AB} \mathbf{Q}_1' \qquad\qquad \widehat{\boldsymbol{\Sigma}}_W \text{ is } K^2 \times K^2$$

$$\mathbf{Q}_2 = \widehat{\mathbf{P}}_{sr}' \otimes \widehat{\mathbf{P}}_{sr} \qquad\qquad \mathbf{Q}_2 \text{ is } K^2 \times K^2$$

$$\mathbf{Q}_1 = \left[(\mathbf{I}_K \otimes \widehat{\mathbf{B}}^{-1}), (-\widehat{\mathbf{P}}_{sr}'^{-1} \otimes \mathbf{B}^{-1}) \right] \qquad \mathbf{Q}_1 \text{ is } K^2 \times 2K^2$$

and $\widehat{\boldsymbol{\Sigma}}_{AB}$ is the $2K^2 \times 2K^2$ VCE of the estimator of $\text{vec}(\mathbf{A}, \mathbf{B})$.

An introduction to forecast-error variance decompositions

Another measure of the effect of the innovations in variable k on variable j is the forecast-error variance decomposition (FEVD). This method, which is also known as *innovation accounting*, measures the fraction of the error in forecasting variable j after h periods that is attributable to the orthogonalized innovations in variable k. Since the derivation of the FEVD requires orthogonalizing the $\mathbf{u}_t$ innovations, the FEVD is always predicated upon a choice of $\mathbf{P}$.

Lütkepohl (1993, Section 2.2.2) shows that the h-step forecast error can be written as

$$\mathbf{y}_{t+h} - \widehat{\mathbf{y}}_t(h) = \sum_{i=0}^{h-1} \boldsymbol{\Phi}_i \mathbf{u}_{t+h-i} \qquad (6)$$

where y_{t+h} is the value observed at time $t + h$ and $\widehat{y}_t(h)$ is the h-step ahead predicted value for y_{t+h} that was made at time t.

Since the $\mathbf{u}_t$ are contemporaneously correlated, their distinct contributions to the forecast error cannot be ascertained. However, if we choose a $\mathbf{P}$ such that $\boldsymbol{\Sigma} = \mathbf{P}\mathbf{P}'$, then, as above, we can orthogonalize the $\mathbf{u}_t$ into $\mathbf{w}_t = \mathbf{P}^{-1}\mathbf{u}_t$. We could then ascertain the relative contribution of the distinct elements of $\mathbf{w}_t$. On this basis, we can rewrite (6) as

$$\mathbf{y}_{t+h} - \widehat{\mathbf{y}}_t(h) = \sum_{i=0}^{h-1} \boldsymbol{\Phi}_i \mathbf{P}\mathbf{P}^{-1} \mathbf{u}_{t+h-i}$$

$$= \sum_{i=0}^{h-1} \boldsymbol{\Theta}_i \mathbf{w}_{t+h-i} \qquad (7)$$

Since the forecast errors can be written in terms of the orthogonalized errors, it follows that the forecast-error variance can be written in terms of the orthogonalized error variances. Forecast error variance decompositions measure the fraction of the total forecast-error variance that is attributable to each of the orthogonalized shocks.

❑ Technical Note

The details in this note are not critical to the discussion that follows. A forecast-error variance decomposition is derived for a given $\mathbf{P}$. Following Lütkepohl (1993, Section 2.3.3), letting $\theta_{mn,i}$ be the m,nth element of $\mathbf{\Theta}_i$, we can express the h-step forecast error of the jth component of $\mathbf{y}_t$ as

$$
\mathbf{y}_{j,t+h} - \widehat{\mathbf{y}}_j(h) = \sum_{i=0}^{h-1} \theta_{j1,1} \mathbf{w}_{1,t+h-i} + \cdots + \theta_{jK,i} \mathbf{w}_{K,t+h-i}
$$

$$
= \sum_{k=1}^{K} \theta_{jk,0} \mathbf{w}_{k,t+h} + \cdots + \theta_{jk,h-1} \mathbf{w}_{k,t+1}
$$

The $\mathbf{w}_t$, which were constructed using $\mathbf{P}$, are mutually orthogonal with unit variance. This allows us to easily compute the mean squared error, MSE, of the forecast of variable j at horizon h in terms of the contributions of the components of $\mathbf{w}_t$. Specifically,

$$
E[\{y_{j,t+h} - y_{j,t}(h)\}^2] = \sum_{k=1}^{K} (\theta_{jk,0}^2 + \cdots + \theta_{jk,h-1}^2)
$$

The kth term in the sum above is interpreted as the contribution of the orthogonalized innovations in variable k to the h-step forecast error of variable j. Note that the kth element in the sum above can be rewritten as

$$
(\theta_{jk,0}^2 + \cdots + \theta_{jk,h-1}^2) = \sum_{i=0}^{h-1} \left(\mathbf{e}_j' \mathbf{\Theta}_k \mathbf{e}_k \right)^2
$$

where $\mathbf{e}_i$ is the ith column of $\mathbf{I}_K$. Normalizing by the forecast error for variable j at horizon h yields

$$
\omega_{jk,h} = \frac{\sum_{i=0}^{h-1} \left(\mathbf{e}_j' \mathbf{\Theta}_k \mathbf{e}_k \right)^2}{MSE[y_{j,t}(h)]}
$$

where $MSE[y_{j,t}(h)] = \sum_{i=0}^{h-1} \sum_{k=1}^{K} \theta_{jk,i}^2$.

❑

Since the FEVD depends on the choice of $\mathbf{P}$, there are different forecast-error variance decompositions associated with distinct $\mathbf{P}$. varirf create can estimate the FEVD for a VAR(p) or an SVAR. For a VAR(p), $\mathbf{P}$ is chosen to be the Cholesky decomposition of $\widehat{\mathbf{\Sigma}}$. For an SVAR, $\mathbf{P}$ is chosen to be the estimated structural decomposition, $\mathbf{P} = \widehat{\mathbf{A}}^{-1}\widehat{\mathbf{B}}$ for short-run models and $\mathbf{P} = \widehat{\mathbf{C}}$ for long-run SVAR models. Due to the same complications that arose with the structural impulse–response functions, the asymptotic standard errors of the structural FEVD are not available after long-run SVAR models, but bootstrapped standard errors are still available.

Forecast error variance decomposition formulas

This section provides the details of how `varirf create` estimates the Cholesky forecast-error variance decompositions, the structural forecast-error variance decompositions, and their standard errors. Beginning with the Cholesky-based forecast-error decompositions, the fraction of the h-step ahead forecast-error variance of variable j that is attributable to the Cholesky orthogonalized innovations in variable k can be estimated as

$$\widehat{\omega}_{jk,h} = \frac{\sum_{i=0}^{h-1}(\mathbf{e}_j'\widehat{\boldsymbol{\Theta}}_i\mathbf{e}_k)^2}{\widehat{\mathrm{MSE}}_j(h)}$$

where $\mathrm{MSE}_j(h)$ is the jth diagonal element of

$$\sum_{i=0}^{h-1}\widehat{\boldsymbol{\Phi}}_i\widehat{\boldsymbol{\Sigma}}\widehat{\boldsymbol{\Phi}}_i'$$

(See Lütkepohl (1993, 97) for a discussion of this result.) $\widehat{\omega}_{jk,h}$ and $\mathrm{MSE}_j(h)$ are scalars. The square of the standard error of $\widehat{\omega}_{jk,h}$ is

$$\mathbf{d}_{jk,h}\widehat{\boldsymbol{\Sigma}}_\alpha\mathbf{d}_{jk,h}' + \overline{\mathbf{d}}_{jk,h}\widehat{\boldsymbol{\Sigma}}_\sigma\overline{\mathbf{d}}_{jk,h}$$

where

$$\mathbf{d}_{jk,h} = \frac{2}{\mathrm{MSE}_j(h)^2}\sum_{i=0}^{h-1}\left\{\mathrm{MSE}_j(h)(\mathbf{c}_j'\widehat{\boldsymbol{\Phi}}_i\widehat{\mathbf{P}}_c\mathbf{e}_k)(\mathbf{e}_k'\widehat{\mathbf{P}}_c'\otimes\mathbf{e}_j')\mathbf{G}_i\right.$$
$$\left.-(\mathbf{e}_j'\boldsymbol{\Phi}_i\widehat{\mathbf{P}}_c\mathbf{e}_k)^2\sum_{m=0}^{h-1}(\mathbf{e}_j'\widehat{\boldsymbol{\Phi}}_m\widehat{\boldsymbol{\Sigma}}\otimes\mathbf{e}_j')\mathbf{G}_m\right\} \qquad \mathbf{d}_{jk,h} \text{ is } 1\times K^2p$$

$$\overline{\mathbf{d}}_{jk,h} = \sum_{i=0}^{h-1}\left\{\mathrm{MSE}_j(h)(\mathbf{e}_j'\widehat{\boldsymbol{\Phi}}_i\mathbf{P}_c\mathbf{e}_k)(\mathbf{e}_k'\otimes\mathbf{e}_j'\widehat{\boldsymbol{\Phi}}_i)\mathbf{H}\right.$$
$$\left.-(\mathbf{e}_j'\widehat{\boldsymbol{\Phi}}_i\widehat{\mathbf{P}}_c\mathbf{e}_k)^2\sum_{m=0}^{h-1}(\mathbf{e}_j'\widehat{\boldsymbol{\Phi}}_m\otimes\mathbf{e}_j\widehat{\boldsymbol{\Phi}}_m)\mathbf{D}_K\right\}\frac{1}{\mathrm{MSE}_j(h)^2} \qquad \overline{\mathbf{d}}_{jk,h} \text{ is } 1\times K\frac{(K+1)}{2}$$

$$\mathbf{G}_0 = \mathbf{0} \qquad\qquad \mathbf{G}_0 \text{ is } K^2\times K^2p$$

and $\mathbf{D}_K$ is the $K^2 \times K\frac{(K+1)}{2}$ duplication matrix defined previously.

For the structural forecast-error decompositions, we follow Amisano and Giannini (1997, Section 5.2). They define the matrix of structural forecast-error decompositions at horizon s, when a maximum of h periods are estimated, as

(Continued on next page)

$$\widehat{\mathbf{W}}_s = \widehat{\mathbf{F}}_s^{-1}\widehat{\widetilde{\mathbf{M}}}_s \qquad \text{for } s = 1,\ldots,h+1$$

$$\widehat{\mathbf{F}}_s = \left(\sum_{i=0}^{s-1} \widehat{\mathbf{\Theta}}_i^{sr}\widehat{\mathbf{\Theta}}_i^{sr\prime}\right) \odot \mathbf{I}_K$$

$$\widehat{\widetilde{\mathbf{M}}}_s = \sum_{i=0}^{s-1} \widehat{\mathbf{\Theta}}_i^{sr} \odot \widehat{\mathbf{\Theta}}_i^{sr}$$

where $\odot$ is the Hadamard (or element-by-element) product.

The $K^2 \times K^2$ asymptotic VCE of $\text{vec}(\widehat{\mathbf{W}}_s)$ of is given by

$$\widetilde{\mathbf{Z}}_s \mathbf{\Sigma}(h)\widetilde{\mathbf{Z}}_s'$$

where $\widehat{\mathbf{\Sigma}}(h)$ is as derived previously, and

$$\widetilde{\mathbf{Z}}_s = \left[\frac{\partial\text{vec}(\widehat{\mathbf{W}}_s)}{\partial\text{vec}(\widehat{\mathbf{\Theta}}_0^{sr})}, \frac{\partial\text{vec}(\widehat{\mathbf{W}}_s)}{\partial\text{vec}(\widehat{\mathbf{\Theta}}_1^{sr})}, \ldots, \frac{\partial\text{vec}(\widehat{\mathbf{W}}_s)}{\partial\text{vec}(\widehat{\mathbf{\Theta}}_h^{sr})}\right]$$

$$\frac{\partial\text{vec}(\widehat{\mathbf{W}}_s)}{\partial\text{vec}(\widehat{\mathbf{\Theta}}_j^{sr})} = 2\left\{(\mathbf{I}_K \otimes \widehat{\mathbf{F}}_s^{-1})\widetilde{\mathbf{D}}(\widehat{\mathbf{\Theta}}_j^{sr}) - (\widehat{\mathbf{W}}_s' \otimes \widehat{\mathbf{F}}_s^{-1})\widetilde{\mathbf{D}}(\mathbf{I}_K)\mathbf{N}_K(\widehat{\mathbf{\Theta}}_j^{sr} \otimes I_K)\right\}$$

If $\mathbf{X}$ is an $n \times n$ matrix, then $\widetilde{\mathbf{D}}(\mathbf{X})$ is the $n^2 \times n^2$ matrix with $\text{vec}(\mathbf{X})$ on the diagonal and zeros in all the off-diagonal elements, and $\mathbf{N}_K$ is as defined previously.

Algorithms for bootstrapping the standard errors

`varirf create` offers two bootstrap algorithms for estimating the standard errors of the various impulse–response functions and forecast-error variance decompositions. Both `var` and `svar` contain estimators for the coefficients in a VAR(p) that are conditional on the first p observations. The two bootstrap algorithms are also conditional on the first p observations.

Specifying the `bs` option causes the standard errors to be calculated by bootstrapping the residuals. For a bootstrap with R repetitions, this method uses the following algorithm:

1. Fit the model and save off the estimated parameters.

2. Use the estimated coefficients to calculate the residuals.

3. Repeat steps 3a to 3c R times.

    3a. Draw a simple random sample of size T with replacement from the residuals. The random samples are drawn over the $K \times 1$ vectors of residuals. When the tth vector is drawn, all K residuals are selected. This preserves the contemporaneous correlations among the residuals.

    3b. Use the p initial observations, the sampled residuals, and the estimated coefficients to construct a new sample dataset.

    3c. Fit the model and calculate the different impulse response functions and forecast-error variance decompositions.

    3d. Save these estimates off as observation r in the bootstrapped dataset.

4. For each impulse–response function and forecast-error variance decomposition, the estimated standard deviation from the R bootstrapped estimates is the estimated standard error of that impulse–response function or forecast-error variance decomposition.

Specifying the `bsp` option causes the standard errors to be estimated by a multivariate normal parametric bootstrap. The algorithm for the multivariate normal parametric bootstrap is identical to the one above, with the exception that 3a is replaced by 3a(bsp):

3a(bsp). Draw T pseudo-variates from a multivariate normal distribution with covariance matrix $\widehat{\Sigma}$.

References

Amisano, G. and C. Giannini. 1997. *Topics in Structural* VAR *Econometrics.* 2d ed. Heidelberg: Springer.

Christiano, L. J., M. Eichenbaum, and C. L. Evans. 1999. Monetary Policy Shocks: What have we learned and to what end? In *Handbook of Macroeconomics*, vol. 1, eds. J. B. Taylor and M. Woodford. New York: Elsevier Science.

Hamilton, J. D. 1994. *Time Series Analysis.* Princeton: Princeton University Press.

Lütkepohl, H. 1993. *Introduction to Multiple Time Series Analysis.* 2d ed. New York: Springer.

Serfling, R. J. 1980. *Approximation Theorems of Mathematical Statistics.* New York: John Wiley & Sons.

Sims, C. A. 1980. Macroeconomics and reality. *Econometrica* 48: 1–48.

Stock, J. H. and M. W. Watson. 2001. Vector autoregressions. *Journal of Economic Perspectives* 15: 101–115.

Also See

Complementary:	[TS] **var**, [TS] **var svar**, [TS] **varbasic**, [TS] **varirf add**, [TS] **varirf cgraph**, [TS] **varirf ctable**, [TS] **varirf describe**, [TS] **varirf dir**, [TS] **varirf drop**, [TS] **varirf erase**, [TS] **varirf graph**, [TS] **varirf ograph**, [TS] **varirf rename**, [TS] **varirf set**, [TS] **varirf table**
Related:	[TS] **arima**
Background:	[U] **14.4.3 Time-series varlists**, [TS] **var intro**, [TS] **varirf**

Title

> **varirf ctable** — Make combined tables of impulse–response functions and FEVDs

Syntax

> varirf $\underline{\text{ct}}$able $(spec_1)$ $\big[$ $(spec_2)$ $\big]$ ... $\big[$ $(spec_N)$ $\big]$ $\big[$, $\underline{\text{in}}$dividual $\underline{\text{st}}$ep($\#$)
>
> set($setcmd$) $\underline{\text{ti}}$tle("$text$") $common\_options$ $\big]$

where $(spec_k)$ is

> ($irfname$ $impulse$ $response$ $statlist$ $\big[$, $\underline{\text{it}}$itle("$text$") $common\_options$ $\big]$)

$irfname$ is the name of a set of VARIRF results in the active VARIRF file, $impulse$ is an impulse in $irfname$, $response$ is a response in $irfname$, and $statlist$ is a list of the names of the statistics defined in the table below.

Statistic	Name
impulse–response functions	irf
orthogonalized impulse–response functions	oirf
cumulative impulse–response functions	cirf
cumulative orthogonalized impulse–response functions	coirf
Cholesky forecast-error variance decomposition	fevd
structural impulse–response functions	sirf
structural forecast-error variance decomposition	sfevd

and $common\_options$ are

> $\big[$noci $|$ $\underline{\text{l}}$evel($\#$)$\big]$ $\underline{\text{st}}$derror

varirf ctable operates on the active VARIRF file; see [TS] **varirf set**.

Description

varirf ctable makes table(s) of VARIRF results. Each block within a pair of matching parentheses, i.e., each $(spec_k)$, specifies the information for a specific table. By default, varirf ctable combines these tables into one table. Specifying individual causes varirf ctable to create a separate table for each block.

Options

individual specifies that each block, or $(spec_k)$, is to be placed in its own table. By default, varirf ctable combines all the blocks into one table.

step($\#$) specifies a common maximum step, or period, for all tables. By default, each table is constructed using all the steps available in the corresponding set of VARIRF results.

272

set(*setcmd*) specifies that varirf ctable should make the file specified in set() the active VARIRF file. set() accepts a *setcmd*, where *setcmd* is a combination of *filename* and *options* that can be understood by varirf set. Using set() implies that varirf ctable will operate on the file specified in set() and that the newly set() file will remain the active VARIRF file after varirf ctable is done.

title("*text*") specifies a title for the table, or the set of tables.

common_options may be specified within a table specification, globally, or both. When specified in a table specification, the *common_options* only affect the specification in which they are used. When supplied globally, the *common_options* affect all table specifications. When specified in both places, options in the table specification take precedence.

noci specifies that for all tables, the confidence intervals for each statistic in *statlist* are to be suppressed. By default, a confidence interval for each statistic is reported. noci may not be combined with level().

level(#) specifies the default confidence level, in percent. The default is level(95) or as set by set level; see [U] **23.6 Specifying the width of confidence intervals**. level() may only be specified for a specific table when the individual option is specified. level() may not be combined with noci.

stderror specifies that a standard error for each statistic is to be included.

ititle("*text*") specifies an individual subtitle for a specific table. ititle() may only be specified when the individual option is specified.

Remarks

If you have not read [TS] **varirf**, please do so.

varirf ctable makes it easy to create a series of customized tables from VARIRF results. The information enclosed within each set of parentheses is a specific table command. Each specific table command has three parts: (1) set identifier, (2) list of statistics to be included in the table, and (3) confidence interval and standard error options.

The first part is made up of the combination of *irfname*, *impulse*, and *response*. Together, these three elements uniquely identify a set of impulse–response function estimates or a set of variance decomposition estimates. The second part is a list of statistics that are to be included in the same table along with their confidence intervals or standard errors. Note that *statlist* may not contain the same statistic more than once. The options noci, level(), and stderror make up the third part.

Each specific table will display the requested statistics that correspond to the specified combination of *irfname*, *impulse* variable, and *response* variable over the estimation horizon. By default, all the individual tables are combined into a single table. Also, by default, all the steps, or periods, available are included in the table. Use the step() option to impose a common maximum for all tables.

▷ Example

Here we consider an example that was also discussed in [TS] **varirf table**. In [TS] **varirf table**, we create the same tables using the varirf table command. Comparing the example below with the one in [TS] **varirf table** illustrates that varirf table issues a combination of the specific table commands used in varirf ctable. It is easier to display lots of information using varirf table, but varirf ctable gives the user more control over how the tables are formed.

While graphs quickly allow one to examine the shapes of the estimated functions, tables are sometimes more useful when comparing the actual values. Suppose that we wanted to see if the values of the orthogonalized impulse–response functions or the forecast-error variance decompositions changed significantly between two different Cholesky orderings. Further suppose that we were most interested in whether the effect of the innovations to dlincome on dlconsumption, or its forecast-error variance, changed over two Cholesky orderings denoted by ordera and orderb. Below, we illustrate how we could use varirf ctable to create a table that would allow us to quickly compare the values of the estimates of these functions.

```
. use http://www.stata-press.com/data/r8/lutkepohl
(Quarterly SA West German macro data, Bil DM, from Lutkepohl 1993 Table E.1)
. var dlinvestment dlincome dlconsumption
 (output omitted )
. varirf create ordera, set(results4, replace) step(8)
(current varirf data file is results4.vrf
file results4.vrf saved
. varirf create orderb, order(dlincome dlinvestment dlconsumption) step(8)
file results4.vrf saved
. varirf ctable (ordera dlincome dlconsumption oirf fevd)
> (orderb dlincome dlconsumption oirf fevd),
> title("Ordera versus orderb") noci std
```

Ordera versus orderb

step	(1) oirf	(1) S.E.	(1) fevd	(1) S.E.
0	.005123	.000878	0	0
1	.001635	.000984	.288494	.077483
2	.002948	.000993	.294288	.073722
3	-.000221	.000662	.322454	.075562
4	.000811	.000586	.319227	.074063
5	.000462	.000333	.322579	.075019
6	.000044	.000275	.323552	.075371
7	.000151	.000162	.323383	.075314

step	(2) oirf	(2) S.E.	(2) fevd	(2) S.E.
0	.005461	.000925	0	0
1	.001578	.000988	.327807	.08159
2	.003307	.001042	.328795	.077519
3	-.00019	.000676	.370775	.080604
4	.000846	.000617	.366896	.079019
5	.000491	.000349	.370399	.079941
6	.000069	.000292	.371487	.080323
7	.000158	.000172	.371315	.080287

```
(1) irfname = ordera, impulse = dlincome, and response = dlconsumption
(2) irfname = orderb, impulse = dlincome, and response = dlconsumption
```

The output is composed of a single table. Since the table did not fit horizontally, it automatically wrapped. At the bottom of the table is a list of the keys that appear at the top of each column. The results in the table above indicate that the orthogonalized impulse–response functions do not change by much. Since the estimated forecast-error variances do change, we might want to produce two tables that contain the estimated forecast-error variance decompositions and their 95% confidence intervals. Below, we illustrate how to use varirf ctable to obtain these tables for this specific example.

```
. varirf ctable (ordera dlincome dlconsumption fevd)
> (orderb dlincome dlconsumption fevd),
> individual
Table 1
```

step	(1) fevd	(1) Lower	(1) Upper
0	0	0	0
1	.288494	.13663	.440357
2	.294288	.149797	.43878
3	.322454	.174356	.470552
4	.319227	.174066	.464389
5	.322579	.175544	.469613
6	.323552	.175826	.471277
7	.323383	.17577	.470995

```
95% lower and upper bounds reported
(1) irfname = ordera, impulse = dlincome, and response = dlconsumption
Table 2
```

step	(2) fevd	(2) Lower	(2) Upper
0	0	0	0
1	.327807	.167893	.487721
2	.328795	.17686	.48073
3	.370775	.212794	.528757
4	.366896	.212022	.52177
5	.370399	.213718	.52708
6	.371487	.214058	.528917
7	.371315	.213956	.528674

```
95% lower and upper bounds reported
(2) irfname = orderb, impulse = dlincome, and response = dlconsumption
```

Since we specified the individual options, the output contains two tables, one for each specific table command. At the bottom of each table, there is a list of the keys used in that table and a note indicating the level of the confidence intervals that we requested. The results from Table 1 and Table 2 indicate that both of the estimated functions are well within the confidence interval of the other, so we conclude that the functions are not significantly different.

◁

(Continued on next page)

Saved Results

varirf ctable saves in r():

Scalars

r(ncols)	number of columns in all tables
r(k_umax)	number of distinct keys
r(k)	number of specific table commands

Macros

r(key$i$)	ith key
r(tnotes)	list of keys applied to each column

Methods and Formulas

varirf ctable is implemented as an ado-file.

Also See

Complementary:	[TS] **var**, [TS] **var svar**, [TS] **varbasic**, [TS] **varirf add**, [TS] **varirf create**, [TS] **varirf describe**, [TS] **varirf dir**, [TS] **varirf drop**, [TS] **varirf erase**, [TS] **varirf rename**, [TS] **varirf set**
Related:	[TS] **arima**, [TS] **varirf cgraph**, [TS] **varirf graph**, [TS] **varirf ograph**, [TS] **varirf table**
Background:	[U] **14.4.3 Time-series varlists**, [TS] **var intro**, [TS] **varirf**

Title

> **varirf describe** — Describe a VARIRF file

Syntax

> varirf <u>d</u>escribe [*varirf_resultslist*] [, set(*setcmd*) using(*varirf_filename*) <u>d</u>etail
>
> <u>v</u>ariables irfwording]

Description

> varirf describe describes the VARIRF results in a VARIRF file.

Options

> set(*setcmd*) specifies a varirf set command that will change the active VARIRF file. By default, varirf describe works on the VARIRF file made active by [TS] **varirf set**.

> using(*varirf_filename*) specifies that varirf describe is to describe the VARIRF file specified in using() without resetting the active VARIRF file.

> detail specifies that varirf describe should display information about each set of VARIRF results in the VARIRF file. By default, varirf describe displays information about the VARIRF results in the VARIRF file and the order used in computing the orthogonalized impulse response functions (OIRFs). When a *varirf_resultslist* is specified, detail is implied, and detailed information about each VARIRF name in the list is displayed.

> variables specifies that the output from describe should be displayed.

> irfwording is a programmer's option that changes the header of the default display.

Remarks

> If you have not read [TS] **varirf**, please do so.

> varirf describe describes the VARIRF results in a VARIRF file. By default, varirf describe displays the *irfnames* in the active VARIRF file and the order used in computing the orthogonalized impulse response functions (OIRFs). By specifying the using() option, one can display information about a VARIRF file that is not the active VARIRF file without making the specified file the active VARIRF file. Alternatively, specifying the set() option changes the active VARIRF file, and varirf describe displays information about the results in this newly activated VARIRF file.

> Besides choices about the VARIRF file, varirf describe also offers choices about how much information to provide. By default, the *varirf_resultslist* is empty and the detail option is not specified. Together, these imply that only the irfnames in the VARIRF file and the order used in computing their OIRFs will be displayed. Specifying the detail option without specifying a *varirf_resultslist* causes detailed information about each set of the VARIRF results in the VARIRF file to be displayed. If a *varirf_resultslist* is given, then the detail option is implied and information about each of the specified sets of VARIRF results will be displayed.

Let's consider an example that illustrates how `varirf describe` behaves by default and how to interpret the detailed information it provides. Below, we check that there is an active VARIRF file, and then use `varirf describe` to display the `irfnames` in that file.

```
. use http://www.stata-press.com/data/r8/lutkepohl
(Quarterly SA West German macro data, Bil DM, from Lutkepohl 1993 Table E.1)
. var dlincome dlconsumption, exog(l(0/1).dlinvestment)
  (output omitted)
. varirf create order1, step(5) set(varirfdes, replace)
current varirf data file is varirfdes.vrf
file varirfdes.vrf saved
. varirf create order2, step(5) order(dlconsumption dlincome)
file varirfdes.vrf saved
. varirf
current varirf data file is varirfdes.vrf
. varirf describe
The irfnames in varirfdes.vrf are:
```

irfname	order
order1	dlincome dlconsumption
order2	dlconsumption dlincome

`varirf describe` has informed us that there are two sets of VARIRF results in the VARIRF file varirfdes.vrf. The VARIRF results are stored under the VARIRF names `order1` and `order2`. In `order1`, the order `dlincome dlconsumption` was used in computing the OIRFs. order2. In `order2`, the order `dlconsumption dlincome` was used in computing the OIRFs.

Now, let's get some detailed information about the VARIRF results in the active VARIRF file.

```
. varirf describe, detail
order1:
         model: var
         order: dlincome dlconsumption
          exog: dlinvestment L.dlinvestment
      constant: constant
          lags: 1 2
          tmin: 3
          tmax: 91
       timevar: qtr
         tsfmt: %tq
        varcns: unconstrained
       svarcns: .
          step: 5
      stderror: asymptotic
          reps: .
       version: 1.0
```

(Continued on next page)

```
order2:
           model: var
           order: dlconsumption dlincome
            exog: dlinvestment L.dlinvestment
        constant: constant
            lags: 1 2
            tmin: 3
            tmax: 91
         timevar: qtr
          tsfmt: %tq
          varcns: unconstrained
         svarcns: .
            step: 5
         stderror: asymptotic
            reps: .
         version: 1.0
```

With the `detail` option, `varirf describe` displays information about each set of VARIRF results stored in the VARIRF file. For each set of VARIRF results, details about the VARIRF results and the `e()` results used to generate the VARIRF results are displayed. The following table describes the information displayed for each VARIRF result.

Name	Contents
model	var, short-run svar, or long-run svar depending on e() results specified to varirf create
order	Cholesky order assumed in VARIRF estimates
exog	exogenous variables in VAR or underlying VAR
constant	constant or noconstant, depending on whether noconstant was specified in VAR or SVAR
lags	lags in model
tmin	minimum value of timevar in estimation sample
tmax	maximum value of timevar in estimation sample
timevar	name of tsset timevar
tsfmt	format of timevar in estimation sample
varcns	unconstrained or colon-separated list of constraints placed on VAR coefficients
svarcns	'.' or colon-separated list of constraints placed on SVAR coefficients
step	maximum step in VARIRF estimates
stderror	asymptotic, bs, bsp, or none, depending on type of standard errors specified to varirf create
reps	'.' or number of bootstrap replications performed
version	version of VARIRF file that originally held *irfname* VARIRF results

When the `variables` option is specified, `varirf describe` uses `describe` to display information about the file as a whole. In particular, `describe` provides information about the number of observations and the variables in the VARIRF file. (See [R] **describe** for more information about `describe`.)

(Continued on next page)

Saved Results

varirf describe saves in r():

Scalars

r(N)	number of observations in VARIRF file
r(k)	number of variables in VARIRF file
r(width)	width of dataset in VARIRF file
r(N_max)	maximum number of observations
r(k_max)	maximum number of variables
r(widthmax)	maximum width of dataset
r(changed)	data–have–changed–since–last–saved flag

Macros

r(version)	version of VARIRF results file
r(irfnames)	names of VARIRF results in VARIRF file
r(*irfname*_model)	var, short-run svar or long-run svar
r(*irfname*_order)	Cholesky order assumed in VARIRF estimates
r(*irfname*_exog)	exogenous variables in VAR or underlying VAR
r(*irfname*_constant)	constant or noconstant
r(*irfname*_lags)	lags in model
r(*irfname*_tmin)	minimum value of timevar in estimation sample
r(*irfname*_tmax)	maximum value of timevar in estimation sample
r(*irfname*_timevar)	name of tsset timevar
r(*irfname*_tsfmt)	format of timevar in estimation sample
r(*irfname*_varcns)	unconstrained or colon-separated list of constraints placed on VAR coefficients
r(*irfname*_svarcns)	'.' or colon-separated list of constraints placed on SVAR coefficients
r(*irfname*_step)	maximum step in VARIRF estimates
r(*irfname*_stderror)	asymptotic, bs, bsp, or none, depending on type of standard errors specified to varirf create
r(*irfname*_reps)	'.' or number of bootstrap replications performed
r(*irfname*_version)	version of VARIRF file that originally held *irfname* VARIRF results

Methods and Formulas

varirf describe is implemented as an ado-file.

Also See

Complementary:	[TS] **var**, [TS] **var svar**, [TS] **varbasic**, [TS] **varirf add**, [TS] **varirf cgraph**, [TS] **varirf create**, [TS] **varirf ctable**, [TS] **varirf describe**, [TS] **varirf dir**, [TS] **varirf drop**, [TS] **varirf erase**, [TS] **varirf graph**, [TS] **varirf ograph**, [TS] **varirf rename**, [TS] **varirf set**, [TS] **varirf table**,
Related:	[TS] **arima**
Background:	[U] **14.4.3 Time-series varlists**, [TS] **var intro**, [TS] **varirf**

Title

varirf dir — List the VARIRF files in a directory

Syntax

varirf <u>dir</u> [*base_directory*]

Description

varirf dir lists the VARIRF files in a directory. By default, it lists the VARIRF files in the current directory. If a *base_directory* is specified, then it lists the VARIRF files in the specified *base_directory*. The optional *base_directory* may or may not end in a directory separator. (The directory separator is the character that separates levels in a directory tree; e.g., "\" in Windows.) However, the *base_directory* should not contain wild cards; e.g., "myfiles*" will not produce what you would expect from wild card substitution.

❑ Technical Note

Actually, varirf dir provides a listing of all the files in the *base_directory* that end in the extension .vrf. While it is not a requirement that VARIRF files end in the .vrf extension, it is a good idea to use this default extension.

❑

Methods and Formulas

varirf dir is implemented as an ado-file.

Also See

Complementary: [TS] **var**, [TS] **var svar**, [TS] **varbasic**, [TS] **varirf add**, [TS] **varirf cgraph**, [TS] **varirf create**, [TS] **varirf ctable**, [TS] **varirf describe**, [TS] **varirf drop**, [TS] **varirf erase**, [TS] **varirf graph**, [TS] **varirf ograph**, [TS] **varirf rename**, [TS] **varirf set**, [TS] **varirf table**

Related: [TS] **arima**

Background: [U] **14.4.3 Time-series varlists**,
[TS] **var intro**, [TS] **varirf**

Title

> **varirf drop** — Drop VARIRF results from the active VARIRF file

Syntax

varirf drop *varirf_resultslist* [, set(*setcmd*)]

Description

varirf drop removes a list of VARIRF results from the active VARIRF file set by varirf set.

Options

set(*setcmd*) specifies that varirf drop should make the file specified in set() the active VARIRF file. Using set() implies that varirf drop will operate on the file specified in set(), and that the set() file will remain the active VARIRF file after varirf drop is done.

Remarks

If you have not read [TS] **varirf**, please do so.

varirf drop will drop a set of VARIRF results from the active VARIRF file. It is designed to handle circumstances in which one wants to remove a set of VARIRF results without replacing them with a new set.

▷ Example

Suppose that we want to remove the VARIRF results lags2 from the VARIRF file results2.vrf, without immediately replacing them with a new set of VARIRF results. In the output below, we begin by using varirf set to make results2.vrf the active VARIRF file. Then, to ensure that the lags2 results are the ones that we wanted to drop, we use varirf describe to describe the contents of results2.vrf.

(Continued on next page)

```
. use http://www.stata-press.com/data/r8/lutkepohl
(Quarterly SA West German macro data, Bil DM, from Lutkepohl 1993 Table E.1)
. var dlincome dlconsumption
  (output omitted )
. varirf create lags2, set(results2, replace)
current varirf data file is results2.vrf
file results2.vrf saved
. varirf set results2
current varirf data file is results2.vrf
. varirf describe, detail
lags2:
         model: var
         order: dlincome dlconsumption
          exog: none
      constant: constant
          lags: 1 2
          tmin: 3
          tmax: 91
       timevar: qtr
         tsfmt: %tq
        varcns: unconstrained
       svarcns: .
          step: 8
      stderror: asymptotic
          reps: .
       version: 1.0
```

Having reassured ourselves that these are the results that we want to drop, we remove them.

```
. varirf drop lags2
(lags2 dropped)
(note: dataset contains 0 observations)
file results2.vrf saved
```

◁

Saved Results

`varirf describe` saves in `r()`:

               Macros
                     `r(dropped)`   list of dropped VARIRF results

Methods and Formulas

`varirf drop` is implemented as an ado-file.

Also See

Complementary:	[TS] **var**, [TS] **var svar**, [TS] **varbasic**, [TS] **varirf add**, [TS] **varirf cgraph**, [TS] **varirf create**, [TS] **varirf ctable**, [TS] **varirf describe**, [TS] **varirf dir**, [TS] **varirf erase**, [TS] **varirf graph**, [TS] **varirf ograph**, [TS] **varirf rename**, [TS] **varirf set**, [TS] **varirf table**
Related:	[TS] **arima**
Background:	[U] **14.4.3 Time-series varlists**, [TS] **var intro**, [TS] **varirf**

Title

> **varirf erase** — Erase a VARIRF file

Syntax

varirf erase *varirf_filename*

Description

varirf erase removes a VARIRF file.

Remarks

If you have not read [TS] **varirf**, please do so.

varirf erase deletes a VARIRF file.

▷ Example

Suppose that the file `lut_old.vrf` constraints VARIRF results from an old version of a dataset. We are now using a new version of the dataset, and we want to remove the old results. We could use `varirf erase` to remove the VARIRF file `lut_old.vrf`.

```
. varirf erase lut_old
lut_old.vrf removed
```

◁

Methods and Formulas

varirf erase is implemented as an ado-file.

Also See

Complementary:	[TS] **var**, [TS] **var svar**, [TS] **varbasic**, [TS] **varirf add**, [TS] **varirf cgraph**, [TS] **varirf create**, [TS] **varirf ctable**, [TS] **varirf describe**, [TS] **varirf dir**, [TS] **varirf drop**, [TS] **varirf graph**, [TS] **varirf ograph**, [TS] **varirf rename**, [TS] **varirf set**, [TS] **varirf table**
Related:	[TS] **arima**
Background:	[U] **14.4.3 Time-series varlists**, [TS] **var intro**, [TS] **varirf**

Title

> **varirf graph** — Graph impulse–response functions and FEVDs

Syntax

> varirf graph *statlist* $\big[$, <u>irf</u>(*resultslist*) <u>i</u>mpulse(*impulselist*) <u>r</u>esponse(*responselist*)
>
> noci <u>l</u>evel(*#*) <u>l</u>step(*#*) <u>ust</u>ep(*#*) plot1(*line_options*) ... plot4(*line_options*)
>
> <u>ci</u>lines ciopts1(*rarea_options* | *rline_options*) ciopts2(*rarea_options* | *rline_options*)
>
> <u>by</u>opts(*by_option*) <u>in</u>dividual <u>isa</u>ving(*filenamestub* $\big[$, replace$\big]$)
>
> iname(*namestub* $\big[$, replace$\big]$) set(*setcmd*) *twoway_options* $\big]$

where *statlist* is a list of the names of the statistics from the table below.

Statistic	Name
impulse–response functions	irf
orthogonalized impulse–response functions	oirf
cumulative impulse–response functions	cirf
cumulative orthogonalized impulse–response functions	coirf
Cholesky forecast-error variance decomposition	fevd
structural impulse–response functions	sirf
structural forecast-error variance decomposition	sfevd

Three limits are placed on the *statlist*.

1. No statistic may appear more than once in *statlist*.

2. If confidence intervals are included (the default), only two statistics may be included in the graph.

3. If confidence intervals are suppressed (option noci), only four statistics may be included in the graph.

resultslist is a list of VARIRF results stored in the active VARIRF file, and *impulselist* and *responselist* are lists of impulses and responses, and these must be common to all the specified VARIRF results.

Description

varirf graph graphs impulse–response functions and forecast-error variance decompositions over time.

All of the statistics from the specified *statlist* are plotted on a single subgraph and a matrix of these subgraphs is created, with a subgraph for each impulse variable versus response variable combination. If multiple VARIRF results are specified, the matrix has subgraphs for each impulse–response variable combination for each result set. Specify the individual option to draw separate graphs for each combination, rather than a matrix of subgraphs.

The graphs are produced for the active VARIRF file. Use the set() option to set or change the active file.

Options

irf(*resultslist*) specifies the VARIRF results for which the subgraphs are to be constructed. By default, irf() is equal to all the VARIRF results in the active VARIRF file.

impulse(*impulselist*) and response(*responselist*) specify the list of impulse and response variables for which subgraphs are to be drawn. A separate subgraph is drawn for each impulse variable/response variable combination. If impulse() or response() is not specified, then graphs are drawn for all combinations of impulse variables and response variables.

noci specifies that the confidence intervals for each statistic are to be suppressed. By default, a confidence interval is graphed for each statistic. noci may not be specified with level().

level(*#*) specifies the confidence level in percent, for confidence intervals. The default is level(95) or as set by set level; see [U] **23.6 Specifying the width of confidence intervals**. Note that level() defines the confidence level for all graphs. See [TS] **varirf cgraph** for a graph command that allows the confidence levels to vary over the graphs. level() may not be combined with noci.

lstep(*#*) specifies the first step, or period, to be included in the graphs. By default, the first step is zero.

ustep(*#*) specifies the maximum step, or period, to be included in the graphs. By default, all the available steps are used. # must be greater than or equal to 1.

plot1(*line_options*) ... plot4(*line_options*) affect the rendition of the plotted statistics in *statlist* according to the order specified (i.e. options in plot1() affect the rendition of the first statistic in *statlist*, ..., options in plot4() affect the rendtion of the fourth statistic in *statlist*. For an explanation of these options see [G] **graph twoway line**.

cilines causes the confidence intervals to be displayed as lines. By default, the confidence intervals are displayed as shaded areas. cilines may not be combined with noci.

ciopts1(*rarea_options* | *rline_options*) and ciopts2(*rarea_options* | *rline_options*) affect the rendition of the confidence bands for the first (ciopts1() and ciopts2()) statistics in *statlist*; see [G] **graph twoway rarea** or [G] **graph twoway rline**. By default, the confidence intervals are displayed as shaded areas so the options for [G] **graph twoway rarea** can be specified. When cilines is specified, the confidence intervals are displayed as lines and the [G] **graph twoway rline** options may be specified. ciopts1() and ciopts2() may not be combined with noci.

byopts(*by_option*) are documented in [G] **by_option**. These options affect the appearance of the combined graph. byopts() may not be specified with individual.

individual specifies that each graph is to be displayed individually. By default, varirf graph combines all the subgraphs into a single matrix graph. individual may not be combined with byopts().

isaving(*filenamestub* [, replace]) specifies that the *i*th individual graph should be saved to disk in the current working directory under the name *filenamestubi*.gph. *filenamestub* must be a valid Stata name of 32 characters or less. isaving() may only be specified with the individual option.

iname(*namestub* [, replace]) specifies that the *i*th individual graph should be saved in memory under the name *namestubi*.gph. *namestub* must be a valid Stata name of 24 characters or less. iname() may only be specified with the individual option.

set(*setcmd*) specifies that varirf graph should make the file specified in set() the active VARIRF file. Using set() implies that varirf graph will operate on the file specified in set(), and that the file will remain the active VARIRF file after varirf graph has completed.

twoway_options are any of the options documented in [G] *twoway_options*, excluding by(). These include options for titling the graph (see [G] *title_options*) and options for saving the graph to disk (see [G] *saving_option*). Note that the saving() and name() options may not be combined with the individual option.

Remarks

If you have not read [TS] **varirf**, please do so.

For each specified set of VARIRF results, varirf graph graphs the statistics in *statlist* over steps for all combinations of the elements specified in *impulselist* and *responselist*. By default, varirf graph combines all the graphs into one matrix graph. When the individual option is specified, each graph is displayed individually.

Since all the statistics in *statlist* appear on the same graph, putting statistics with very different scales in this list is not recommended. For instance, it is sometimes the case that the sirf and the oirf are on similar scales, while the irf is on a different scale. In this case, while putting the sirf and the oirf on the same graph would look fine, combining either of these two with the irf would produce an uninformative graph. Also, two limits are imposed for legibility. When confidence intervals are included, only two statistics may be included on a single graph. When the confidence intervals are suppressed, only four statistics may be included on a single graph.

▷ Example

Here we consider an example that is also discussed in [TS] **varirf cgraph**. In [TS] **varirf cgraph**, we created similar graphs using varirf cgraph. It is easier to display lots of information using varirf graph, but varirf cgraph gives the user more control over how the graphs are formed.

Suppose that we have VARIRF results generated from two different SVAR models. We are interested in whether the shapes of the structural impulse–response functions and the structural forecast-error variance decompositions are similar over the two models. We are also interested in knowing whether the structural impulse–response functions and the structural forecast-error variance decompositions differ significantly from their Cholesky counterparts. Let's consider these two questions separately.

Below, we use varirf graph to create a combined graph from graphs of the oirf and sirf from each model, in order for us to answer the first question.

```
. use http://www.stata-press.com/data/r8/lutkepohl
(Quarterly SA West German macro data, Bil DM, from Lutkepohl 1993 Table E.1)

. mat a = (., 0, 0\0,.,0\.,.,.)

. mat b = I(3)

. svar dlinvestment dlincome dlconsumption, aeq(a) beq(b)

  (output omitted )

. varirf create modela, set(results3, replace) step(8)
current varirf data file is results3.vrf
file results3.vrf saved

. svar dlincome dlinvestment dlconsumption, aeq(a) beq(b)

  (output omitted )

. varirf create modelb, step(8)
file results3.vrf saved
```

. varirf graph oirf sirf, impulse(dlincome) response(dlconsumption)

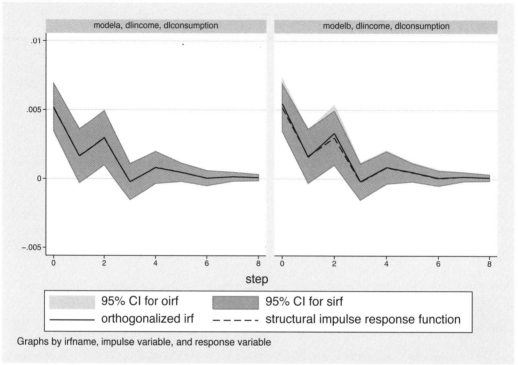

Figure 1.

The default title on each subgraph is of the form *irfname, impulse, response*. The legend at the bottom identifies what is plotted on the graphs. The graph reveals that the oirf and sirf estimates are essentially the same for both models and that the shapes of the functions are very similar for the two models.

Below, we use varirf graph to create a graph of fevd and sfevd from each model to help us answer the second question.

(*Continued on next page*)

```
. varirf graph fevd sfevd, impulse(dlincome) response(dlconsumption) lstep(1)
> legend(cols(1))
```

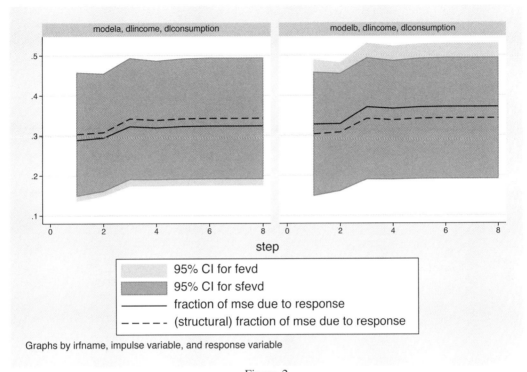

Figure 2.

This combined graph reveals that the shapes of these functions are also very similar for the two models. However, the graph illuminates one minor difference between the models. In modela, the estimated structural forecast-error variance is slightly larger than the Cholesky-based estimates, while in modelb the Cholesky-based estimates are slightly larger than the structural estimates. However, for both models, the structural estimates are close to the center of the wide confidence intervals for the two estimates.

Now, let's focus in on the results from modela. Suppose that we were interested in examining how dlconsumption responded to impulses in its own structural innovations, structural innovations to dlincome, and structural innovations to dlinvestment. Below, we illustrate how we can use varirf graph to obtain a matrix graph of the impulse–response functions of interest.

```
. varirf graph sirf, irf(modela) response(dlconsumption)
```

The upper-left graph of Figure 2 shows the structural impulse–response function of an innovation in dlconsumption on dlconsumption. It indicates that the identification restrictions used in modela imply that a positive shock to dlconsumption causes an increase in dlconsumption, followed by a decrease, followed by an increase, and so on, until the effect dies out after about 5 periods.

(Continued on next page)

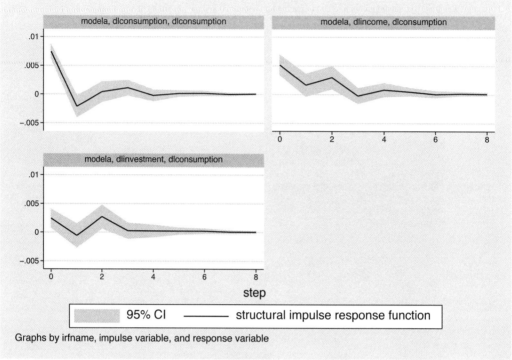

Figure 3.

The upper-right graph shows the structural impulse–response function of an innovation in `dlincome` on `dlconsumption`, and indicates that a positive shock to `dlincome` causes an increase in `dlconsumption`, which dies out after 4 or 5 periods.

◁

In the syntax diagram it was noted that any of the elements in *impulselist* or *responselist* must be in all of the specified VARIRF results. Since `impulse()` and `response()` default to all the impulses and responses in a given set of VARIRF results, it is possible to use these defaults to easily obtain graphs when the names of the impulses or responses vary over the sets of VARIRF results of interest. See [TS] **varirf table** for an example that illustrates this point.

(*Continued on next page*)

Saved Results

varirf graph saves in r():

Scalars
r(k)	number of graphs

Macros
r(stats)	*statlist*
r(irfname)	*resultslist*
r(impulse)	*impulselist*
r(response)	*responselist*
r(plot*i*)	contents of option plot*i*()
r(ci)	level applied to confidence intervals or noci
r(ciopts*i*)	contents of option ciopts*i*()
r(byopts)	contents of option byopts()
r(saving)	the supplied saving() option
r(name)	the supplied name() option
r(individual)	individual or blank
r(isaving)	contents of option saving()
r(iname)	contents of option name()
r(subtitle*j*)	subtitle for individual graph *j*

Methods and Formulas

varirf graph is implemented as an ado-file.

Also See

Complementary:	[TS] **var**, [TS] **var svar**, [TS] **varbasic**, [TS] **varirf add**, [TS] **varirf create**, [TS] **varirf describe**, [TS] **varirf dir**, [TS] **varirf drop**, [TS] **varirf erase**, [TS] **varirf rename**, [TS] **varirf set**
Related:	[TS] **arima**, [TS] **varirf cgraph**, [TS] **varirf ctable**, [TS] **varirf ograph**, [TS] **varirf table**
Background:	[U] **14.4.3 Time-series varlists**, [TS] **var intro**, [TS] **varirf**

Title

> **varirf ograph** — Graph overlaid impulse–response functions and FEVDs

Syntax

varirf o̲graph (*spec₁*) [(*spec₂*)]... [(*spec₁₅*)] [, set(*setcmd*) *ograph_options*

   twoway_options]

where (*spec*) is

   (*irfname impulse response stat* [, ciopts(*rarea_options*) *ograph_options line_options*])

irfname is a varirf named result (or "." which means the first named result in the active varirf file), *impulse* is an impulse in *irfname*, *response* is a response in *irfname*, *stat* is one of the names of the statistics from the table below:

Statistic	Name
impulse–response functions	irf
orthogonalized impulse–response functions	oirf
cumulative impulse–response functions	cirf
cumulative orthogonalized impulse–response functions	coirf
Cholesky forecast-error variance decomposition	fevd
structural impulse–response functions	sirf
structural forecast-error variance decomposition	sfevd

and *ograph_options* are

l̲step(#) u̲step(#) l̲evel(#) ci cil̲ines

Description

varirf ograph displays plots of varirf results on a single graph.

Options

set(*setcmd*) causes varirf ograph to set a file as the active VARIRF file. set() accepts a *setcmd*, a combination of *filename* and *options* that can be understood by [TS] **varirf set**. Using set() implies that varirf ograph will operate on the specified file and this file will remain the active VARIRF file after varirf ograph is finished.

(Continued on next page)

292

ograph_options may be supplied within a plot specification (like the `ciopts()` option) or globally (like the `set()` option) or in both places. When supplied in a plot specification, *ograph_options* only affect the specification in which they are used. When supplied globally, *ograph_options* affect all plot specifications. When supplied in both places, options in the plot specification take precedence.

`lstep(#)` specifies the first step, or period, to be included. By default, the first step is zero.

`ustep(#)` specifies the maximum step, or period, to be included. By default, the maximum step is the largest step available.

`level(#)` specifies the confidence level, in percent, for confidence bands; see [U] **23.6 Specifying the width of confidence intervals**.

`ci` adds confidence bands to the graph. The default is not to add confidence bands. The `noci` option may be used within a plot specification to suppress its confidence bands when the `ci` option is supplied globally.

`cilines` indicates that the confidence bands are to be plotted as lines instead of areas. This option implies the `ci` option.

`ciopts(`*rarea_options*`)` affect the rendition of the confidence bands; see [G] **graph twoway rarea**.

line_options affect the rendition of the plotted line(s); see [G] **graph twoway line**.

twoway_options are any of the options documented in [G] ***twoway_options***, excluding `by()`. These include options for titling the graph (see [G] ***title_options***) and options for saving the graph to disk (see [G] ***saving_option***).

Remarks

If you have not read [TS] **varirf**, please do so.

▷ Example

The first example in [TS] **varirf** uses `varirf graph` to graphically compare the `oirf` for impulse `dlincome` and response `dlinvestment` for two different Cholesky orderings. In the following, we use `varirf ograph` to place both plots on the same graph.

```
. use http://www.stata-press.com/data/r8/lutkepohl
(Quarterly SA West German macro data, Bil DM, from Lutkepohl 1993 Table E.1)
. var dlinvestment dlincome dlconsumption if qtr <= q(1978q4), lags(1/2) dfk
  (output omitted)
. varirf create order1, step(10) set(myirf1, new)
current varirf data file is myirf1.vrf
file myirf1.vrf saved
. varirf create order2, step(10) order(dlincome dlinvestment dlconsumption)
file myirf1.vrf saved
. varirf ograph    (order1 dlincome dlconsumption oirf)
>                  (order2 dlincome dlconsumption oirf),
>                  legend(textwidth(70))
```

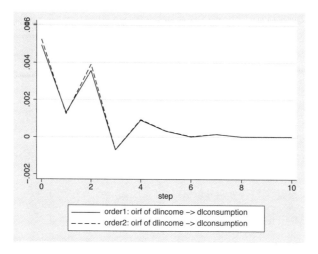

◁

❏ Technical Note

Graph options provide a way to change the rendition of each plot. The following graph contains the plots of the forecast-error variance decompositions (`fevd`) for impulse `dlincome` and each response using the results from the first collection of results in the active VARIRF file (using the "`.`" shortcut). In the second plot we supply the `clpat(dash)` option (an abbreviation for `clpattern(dash)`) in order to get a dashed pattern for the line. In the third plot we supply the `m(o) clpat(dashdot) recast(connected)` options in order to get small circles connected by a line with a dash-dot pattern; the `cilines` option causes the confidence bands to be plotted using lines instead of areas. We use the `title()` option to add a descriptive title to the graph and supply the `ci` option globally to add confidence bands to all the plots.

```
. varirf ograph   (. dlincome dlincome fevd)
>                 (. dlincome dlconsumption fevd, clpat(dash))
>                 (. dlincome dlinvestment fevd,
>                       cilines m(o) clpat(dashdot) recast(connected))
>                 , ci
>                 title("Comparison of forecast-error variance decomposition")
```

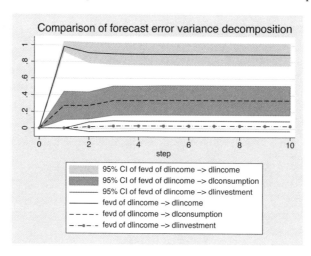

The clpattern() option is described in [G] *connect_options*, msymbol() is described in [G] *marker_options*, title() is described in [G] *title_options*, and recast() is described in [G] *advanced_options*.

❏

Saved Results

varirf ograph saves in r():

Scalars
r(plots)	number of plot specifications
r(ciplots)	number of plotted confidence bands

Macros
r(irfname$k$)	irfname from (*spec$_k$*)
r(impulse$k$)	impulse from (*spec$_k$*)
r(response$k$)	response from (*spec$_k$*)
r(stat$k$)	stat from (*spec$_k$*)
r(ci$k$)	level from (*spec$_k$*) or noci

Methods and Formulas

varirf ograph is implemented as an ado-file.

Also See

Complementary:	[TS] **var**, [TS] **var svar**, [TS] **varbasic**, [TS] **varirf add**, [TS] **varirf create**, [TS] **varirf describe**, [TS] **varirf dir**, [TS] **varirf drop**, [TS] **varirf erase**, [TS] **varirf rename**, [TS] **varirf set**
Related:	[TS] **arima**, [TS] **varirf cgraph**, [TS] **varirf ctable**, [TS] **varirf graph**, [TS] **varirf table**
Background:	[U] **14.4.3 Time-series varlists** [TS] **var intro**, [TS] **varirf**

Title

varirf rename — Rename a VARIRF result in a VARIRF file

Syntax

varirf <u>ren</u>ame *oldname newname* $\big[$, set(*setcmd*) $\big]$

Description

varirf rename changes the name of a set of VARIRF results from *oldname* to *newname* in the active VARIRF file. The set() option can be used to change the active VARIRF file.

Options

set(*setcmd*) specifies that varirf rename should make the file specified in set() the active VARIRF file. Using set() implies that the varirf rename will operate on the file specified in set(), and that the file will remain the active VARIRF file after varirf rename has completed.

Remarks

If you have not read [TS] **varirf**, please do so.

▷ Example

The best way to explain how varirf rename works is by example. Suppose that we have two VARIRF files called results1.vrf and results2.vrf. Each of these files contains a set of VARIRF results named order1. These VARIRF results were obtained from different models, and we are interested in comparing the two sets of results. Before comparing the results, we want to put them into the same file. But, before we can use varirf add to put them into the same file, we must rename at least one set of results. (See [TS] **varirf add** for more on this command.) Below, we illustrate how we can use varirf describe to verify that results1.vrf and results2.vrf contain VARIRF results obtained from distinct models. (See [TS] **varirf describe**.)

(Continued on next page)

296

```
. use http://www.stata-press.com/data/r8/lutkepohl
(Quarterly SA West German macro data, Bil DM, from Lutkepohl 1993 Table E.1)
. qui var dlinvestment dlincome dlconsumption
. varirf create order1, set(results1, replace)
current varirf data file is results1.vrf
file results1.vrf saved
. qui var dlincome dlconsumption, exog(dlinvestment)
. varirf create order1, set(results2, replace)
current varirf data file is results2.vrf
file results2.vrf saved
. varirf describe, using(results1) detail
order1:
        model: var
        order: dlinvestment dlincome dlconsumption
         exog: none
     constant: constant
         lags: 1 2
         tmin: 3
         tmax: 91
      timevar: qtr
        tsfmt: %tq
       varcns: unconstrained
      svarcns: .
         step: 8
     stderror: asymptotic
         reps: .
      version: 1.0
. varirf describe, using(results2) detail
order1:
        model: var
        order: dlincome dlconsumption
         exog: dlinvestment
     constant: constant
         lags: 1 2
         tmin: 3
         tmax: 91
      timevar: qtr
        tsfmt: %tq
       varcns: unconstrained
      svarcns: .
         step: 8
     stderror: asymptotic
         reps: .
      version: 1.0
```

The above indicates that the only difference in the two models used to generate these two sets of VARIRF results is that dlinvestment was included as an endogenous variable in the model used to create the VARIRF results in results1.vrf, while dlinvestment was an exogenous variable in the model used in generating the results in results2.vrf.

Suppose that we decide to use results2.vrf as our new base file. We can use varirf rename with the set() option to make results2.vrf the active VARIRF file and rename the VARIRF results from order1 to order1_r2.

```
. varirf rename order1 order1_r2, set(results2)
current varirf data file is results2.vrf
(36 real changes made)
order1 renamed to order1_r2
```

Now, let's use varirf describe to check that varirf rename did what we wanted.

```
. varirf describe, detail
order1_r2:
           model: var
           order: dlincome dlconsumption
            exog: dlinvestment
        constant: constant
            lags: 1 2
            tmin: 3
            tmax: 91
         timevar: qtr
           tsfmt: %tq
          varcns: unconstrained
         svarcns: .
            step: 8
        stderror: asymptotic
            reps: .
         version: 1.0
```

◁

Saved Results

varirf rename saves in r():

Macros
     r(irfnames)   irfnames after rename
     r(oldnew)   *oldname newname*

Methods and Formulas

varirf rename is implemented as an ado-file.

Also See

Complementary:	[TS] **var**, [TS] **var svar**, [TS] **varbasic**, [TS] **varirf add**, [TS] **varirf cgraph**, [TS] **varirf create**, [TS] **varirf ctable**, [TS] **varirf describe**, [TS] **varirf dir**, [TS] **varirf drop**, [TS] **varirf erase**, [TS] **varirf graph**, [TS] **varirf ograph**, [TS] **varirf set**, [TS] **varirf table**
Related:	[TS] **arima**
Background:	[U] **14.4.3 Time-series varlists**, [TS] **var intro**, [TS] **varirf**

Title

varirf set — Set active VARIRF file

Syntax

varirf set [*varirf_filename*] [, [<u>n</u>ew | replace] clear]

Description

varirf set makes the *varirf_filename* the active VARIRF file for all the varirf commands. If the *varirf_filename* does not end with a .vrf extension then varirf set will append the extension.

Options

<u>n</u>ew causes varirf set to check that the *varirf_filename* is a new filename and to create an empty VARIRF file to hold VARIRF results. new may not be specified with replace.

replace causes varirf set to replace any existing file with an empty VARIRF file to hold VARIRF results. replace may not be specified with new.

clear clears the active *varirf_filename* set. clear may not be specified with either a *varirf_filename* nor with any other option. clear does not verify that a *varirf_filename* has been previously set.

Remarks

If you have not read [TS] **varirf**, please do so.

All the varirf commands work on the active VARIRF file. varirf set makes a file the active VARIRF file.

By default, varirf set looks for an existing file. If the file exists, then varirf set makes it the active VARIRF file. If the file does not exist, then varirf set creates a file with the specified name. When the replace option is specified, then any existing file of the same name is replaced. The new option is used to check that the VARIRF file is new. When the new option is specified, varirf set will exit with a syntax error if a file of that name already exists.

▷ Example

Let's consider an example that illustrates how varirf set works. To begin with, let's use varirf set to see which file is the active VARIRF file.

```
. varirf set
no varirf file set
```

The output above indicates that we have not yet made any file the active VARIRF file. Let's continue by making a new file called results1.vrf the active VARIRF file.

```
. varirf set results1
file results1.vrf saved
current varirf data file is results1.vrf
```

If we wanted to make an existing VARIRF file the active VARIRF file, we could proceed as below.

```
. use http://www.stata-press.com/data/r8/lutkepohl
(Quarterly SA West German macro data, Bil DM, from Lutkepohl 1993 Table E.1)

. qui var dlincome dlconsumption, exog(l.dlinvestment)

. varirf create order1, set(results2, replace)
current varirf data file is results2.vrf
file results2.vrf saved

. varirf set results2
current varirf data file is results2.vrf
```

The output indicates that `varirf set` made the existing file `results2.vrf` the active VARIRF file.

The output from `varirf set` informed us that `results2.vrf` is now the active VARIRF file, but it did not tell us what is in `results2.vrf`. Below, we use `varirf describe` (see [TS] **varirf describe**) to find out what we have stored in `results2.vrf`.

```
. varirf describe, detail
order1:
          model: var
          order: dlincome dlconsumption
           exog: L.dlinvestment
       constant: constant
           lags: 1 2
           tmin: 3
           tmax: 91
        timevar: qtr
          tsfmt: %tq
         varcns: unconstrained
        svarcns: .
           step: 8
        stderror: asymptotic
           reps: .
        version: 1.0
```

The output from `varirf describe` provides us with some details about the one set of VARIRF results in `results2.vrf` which is stored as `order1`. The output from `varirf describe` indicates that the `order1` results were obtained from a VAR model. Suppose that we actually wanted these results from an SVAR model. In this case, we might be interested in replacing the file `results2.vrf` with another file. Below, we illustrate how we could use `varirf set` to replace `results2.vrf` with a new VARIRF file.

```
. varirf set results2, replace
file results2.vrf saved
current varirf data file is results2.vrf
```

To highlight the fact that we replaced the existing `results2.vrf` with a new file, we use `varirf describe`.

```
. varirf describe
no irfnames in  results2.vrf
```

The output from `varirf describe` indicates that `results2.vrf` no longer contains any VARIRF results.

◁

Saved Results

varirf set saves in r():

Macros
    r(vrffile)   name of active VARIRF file, if there an active VARIRF

Methods and Formulas

varirf set is implemented as an ado-file.

Also See

Complementary: [TS] **var**, [TS] **var svar**, [TS] **varbasic**, [TS] **varirf add**, [TS] **varirf cgraph**,
                    [TS] **varirf create**, [TS] **varirf ctable**, [TS] **varirf describe**, [TS] **varirf dir**,
                    [TS] **varirf drop**, [TS] **varirf erase**, [TS] **varirf graph**, [TS] **varirf ograph**,
                    [TS] **varirf rename**, [TS] **varirf set**, [TS] **varirf table**

Related: [TS] **arima**

Background: [U] **14.4.3 Time-series varlists**,
                    [TS] **var intro**, [TS] **varirf**

Title

> **varirf table** — Create tables of impulse–response functions and FEVDs

Syntax

> varirf <u>t</u>able *statlist* $\left[\,,\ \underline{\text{irf}}(\textit{varirf\_resultslist})\ \underline{\text{impulse}}(\textit{varirf\_impulselist})\right.$
>
> <u>res</u>ponse(*varirf_responselist*) $\left[\ \text{noci}\ \mid\ \underline{\text{level}}(\#)\right]$ <u>std</u>error <u>nostr</u>uctural
>
> <u>ste</u>p(*#*) <u>tit</u>le("*text*") <u>ind</u>ividual set(*setcmd*)

where *statlist* is a list of the names of the statistics from the table below, *varirf_resultslist* is a list of VARIRF results stored in the active VARIRF file, *varirf_impulselist* and *varirf_responselist* are lists of impulse variables and response variables that are in all the specified VARIRF results. By default, *statlist* is equal to all the names in the table below.

Statistic	Name
impulse–response functions	irf
orthogonalized impulse–response functions	oirf
cumulative impulse–response functions	cirf
cumulative orthogonalized impulse–response functions	coirf
Cholesky forecast-error variance decomposition	fevd
structural impulse–response functions	sirf
structural forecast-error variance decomposition	sfevd

Description

varirf table makes one or more tables from the specified VARIRF results from a common command. For each set of VARIRF results specified in irf(*varirf_resultslist*), each statistic in *statlist* is tabled for each combination of the impulse variables specified in *varirf_impulselist* and the response variables specified in *varirf_responselist*.

Options

irf(*varirf_resultslist*) specifies the VARIRF results from which the tables are to be constructed. By default, irf() is equal to all the VARIRF results in the active VARIRF file.

impulse(*varirf_impulselist*) specifies the list of VARIRF impulse variables for which the statistics are to be computed for each response. By default, for each set of VARIRF results, impulse() is equal to all the impulse variables for that set of VARIRF results. Note that since the impulses are actually the innovations to an endogenous variable, an "impulse variable" is the name of the endogenous variable.

response(*varirf_responselist*) specifies the list of VARIRF response variables for which the statistics are to tabled for each impulse variable. By default, for each set of VARIRF results, response() is equal to all the response variables for that set.

noci specifies that the confidence intervals for each statistic are to be suppressed. By default, a confidence interval is reported for each statistic. noci may not be specified with level().

level(#) specifies the default confidence level, in percent, for confidence intervals. The default is
  level(95) or as set by set level; see [U] **23.6 Specifying the width of confidence intervals**.
  Note that level() defines the confidence level for all tables. See [TS] **varirf ctable** for a table
  command that allows the confidence intervals to vary. level() may not be combined with noci.

stderror specifies that, for all tables, a standard error for each statistic is to be included in the
  table(s).

nostructural specifies that the default *statlist* should not include sirf nor sfevd. When nostructural is specified, the default *statlist* is irf oirf cirf coirf fevd.

step(#) specifies a common maximum step horizon for all tables. By default, each table is constructed
  using all the steps available in the corresponding set of VARIRF results.

title("*text*") specifies a title for the overall table. title() may not be combined with individual.

individual specifies that each set of VARIRF results is to be placed in its own table, with its own
  title and footer containing keys. By default, varirf table places all the VARIRF results in a
  single table with one title and one footer containing keys. individual may not be combined with
  title().

set(*setcmd*) specifies that varirf table should make the file specified in set() the active VARIRF
  file. Using set() implies that varirf table will operate on the file specified in set() and that
  the set() file will remain the active VARIRF file after varirf table has completed.

Remarks

If you have not read [TS] **varirf**, please do so.

varirf table makes it easy to put the requested statistics from one or more sets of VARIRF results
into one or more tables via one command. By default, varirf table creates one table in which all
the statistics in *statlist* are displayed for each specified set of VARIRF results, for all combinations
of the elements specified in *varirf_impulselist* and *varirf_responselist*. When the individual option
is specified, varirf table creates a table for each specified set of VARIRF results in which all the
statistics in *statlist* are displayed for all combinations of the elements specified in *varirf_impulselist*
and *varirf_responselist*.

▷ Example

Here we consider an example that was also discussed in [TS] **varirf ctable**. In [TS] **varirf ctable**,
we create the same tables using the varirf ctable command. Comparing the example below with
the one [TS] **varirf ctable** illustrates that varirf table issues a combination of the specific table
commands used in varirf ctable. It is easier to display lots of information using varirf table,
but varirf ctable gives the user more control over the how the tables are formed.

While graphs quickly allow one to examine the shapes of the estimated functions, tables are
sometimes more useful when comparing the actual values. Suppose that we wanted to see if the
values of the orthogonalized impulse–response functions or the forecast-error variance decompositions
changed significantly between two different Cholesky orderings. Further suppose that we were most
interested in whether the effect of the innovations to dlincome on dlconsumption, or its forecast-
error variance, changed over two Cholesky orderings denoted by ordera and orderb. Below, we
illustrate how we could use varirf table to quickly compare the values of the estimates of these
functions produced from the ordera and orderb results.

```
. use http://www.stata-press.com/data/r8/lutkepohl
(Quarterly SA West German macro data, Bil DM, from Lutkepohl 1993 Table E.1)
. var dlinvestment dlincome dlconsumption
  (output omitted )
. varirf create ordera, set(results4, replace) step(8)
current varirf data file is results4.vrf
file results4.vrf saved
. varirf create orderb, order(dlincome dlinvestment dlconsumption) step(8)
file results4.vrf saved
. varirf table oirf fevd, impulse(dlincome) response(dlconsumption)
> noci std title("Ordera versus orderb")
```

Ordera versus orderb

step	(1) oirf	(1) S.E.	(1) fevd	(1) S.E.
0	.005123	.000878	0	0
1	.001635	.000984	.288494	.077483
2	.002948	.000993	.294288	.073722
3	−.000221	.000662	.322454	.075562
4	.000811	.000586	.319227	.074063
5	.000462	.000333	.322579	.075019
6	.000044	.000275	.323552	.075371
7	.000151	.000162	.323383	.075314

step	(2) oirf	(2) S.E.	(2) fevd	(2) S.E.
0	.005461	.000925	0	0
1	.001578	.000988	.327807	.08159
2	.003307	.001042	.328795	.077519
3	−.00019	.000676	.370775	.080604
4	.000846	.000617	.366896	.079019
5	.000491	.000349	.370399	.079941
6	.000069	.000292	.371487	.080323
7	.000158	.000172	.371315	.080287

```
(1) irfname = ordera, impulse = dlincome, and response = dlconsumption
(2) irfname = orderb, impulse = dlincome, and response = dlconsumption
```

The output is composed of a single table. Since the table did not fit horizontally, it was wrapped automatically. At the bottom of the table is a definition of the keys that appear at the top of each column. The results in the table above indicate that the orthogonalized impulse–response functions do not change by much. Since the estimated forecast-error variances do change significantly, we might want to produce two tables that contain the estimated forecast-error variance decompositions and their 95% confidence intervals.

(Continued on next page)

. varirf table fevd, impulse(dlincome) response(dlconsumption) individual

Results from ordera

step	(1) fevd	(1) Lower	(1) Upper
0	0	0	0
1	.288494	.13663	.440357
2	.294288	.149797	.43878
3	.322454	.174356	.470552
4	.319227	.174066	.464389
5	.322579	.175544	.469613
6	.323552	.175826	.471277
7	.323383	.17577	.470995

95% lower and upper bounds reported
(1) irfname = ordera, impulse = dlincome, and response = dlconsumption

Results from orderb

step	(1) fevd	(1) Lower	(1) Upper
0	0	0	0
1	.327807	.167893	.487721
2	.328795	.17686	.48073
3	.370775	.212794	.528757
4	.366896	.212022	.52177
5	.370399	.213718	.52708
6	.371487	.214058	.528917
7	.371315	.213956	.528674

95% lower and upper bounds reported
(1) irfname = orderb, impulse = dlincome, and response = dlconsumption

Since we specified the individual option, the output contains two tables, one for each set of VARIRF results. Examining the results in the tables indicates that both of the estimated functions are well within the confidence interval of the other, so we conclude that the functions are not significantly different.

◁

▷ Example

In the syntax diagram above, it was noted that each of the elements in *varirf_impulselist* or *varirf_responselist* must be in all of the specified VARIRF results. This example highlights this fact, but it also explains how to use the default behavior of varirf table to easily obtain a combination of tables when the names of the impulse variables and response variables vary over the VARIRF results stored in the active VARIRF file.

Suppose that, after further analysis, we added another set of VARIRF results to our VARIRF file. Below, we use varirf describe to show that the new VARIRF results, stored as spec2, have different names for the impulse variables and response variables.

```
. varirf create spec2, step(8)
file results4.vrf saved

. varirf describe, detail
ordera:
            model: var
            order: dlinvestment dlincome dlconsumption
             exog: none
         constant: constant
             lags: 1 2
             tmin: 3
             tmax: 91
          timevar: qtr
            tsfmt: %tq
           varcns: unconstrained
          svarcns: .
             step: 8
          stderror: asymptotic
             reps: .
          version: 1.0
orderb:
            model: var
            order: dlincome dlinvestment dlconsumption
             exog: none
         constant: constant
             lags: 1 2
             tmin: 3
             tmax: 91
          timevar: qtr
            tsfmt: %tq
           varcns: unconstrained
          svarcns: .
             step: 8
          stderror: asymptotic
             reps: .
          version: 1.0
spec2:
            model: var
            order: D.lincome D.lconsumption
             exog: LD.linvestment
         constant: constant
             lags: 1 2
             tmin: 3
             tmax: 91
          timevar: qtr
            tsfmt: %tq
           varcns: unconstrained
          svarcns: .
             step: 8
          stderror: asymptotic
             reps: .
          version: 1.0
```

In particular, note that in the `spec2` results,

```
order: D.lincome D.lconsumption
```

Since the `order` contains the set of all possible impulses and responses, we can see that the names of the impulse variables and response variables are different under `spec2` than under the `ordera` and `orderb`. However, since we know that `dlincome` and `D.lincome` are actually the same variable, we might be interested in comparing the VARIRF results from, say, `ordera` and `spec2`. Since any impulse names given in VARIRF *impulselist* must be in all the specified sets of VARIRF results, we could not specify either D.lincome nor dlincome in the same list. However, by default, if *varirf_impulselist* is not specified, then varirf table will display results for all the impulses in that set of VARIRF results. Similarly, if *varirf_responselist* is not specified, then varirf table will display results for all the responses in that set of VARIRF results. This implies that one can still use varirf table to create combination tables. The command below illustrates how we could use varirf table to create a combination of tables for this example.

```
. varirf table oirf, irf(ordera spec2)
```
(*output omitted*)

◁

Saved Results

If the individual option is not specified, varirf table saves in r():

Scalars
    r(ncols)   number of columns in table
    r(k_umax) number of distinct keys
    r(k)   number of specific table commands
Macros
    r(key*i*)   *i*th key
    r(tnotes) list of keys applied to each column

If the individual option is specified, then for each irfname, varirf table saves in r():

Scalars
    r(*irfname*_ncols)   number of columns in table for *irfname*
    r(*irfname*_k_umax) number of distinct keys in table for *irfname*
    r(*irfname*_k)   number of specific table commands used to create table for *irfname*
Macros
    r(*irfname*_key*i*)   *i*th key for *irfname* table
    r(*irfname*_tnotes) list of keys applied to each column in table for *irfname*

Methods and Formulas

varirf table is implemented as an ado-file.

Also See

Complementary:	[TS] **var**, [TS] **var svar**, [TS] **varbasic**, [TS] **varirf add**, [TS] **varirf create**, [TS] **varirf describe**, [TS] **varirf dir**, [TS] **varirf drop**, [TS] **varirf erase**, [TS] **varirf rename**, [TS] **varirf set**
Related:	[TS] **arima**, [TS] **varirf cgraph**, [TS] **varirf ctable**, [TS] **varirf graph**, [TS] **varirf ograph**
Background:	[U] **14.4.3 Time-series varlists**, [TS] **var intro**, [TS] **varirf**

Title

> **varlmar** — Obtain LM statistics for residual autocorrelation after var or svar

Syntax

varlmar [, m̲lag(#) e̲stimates(*estname*) s̲eparator(#)]

varlmar can only be used with var or svar e() results; see [TS] **var** and [TS] **var svar**.

varlmar is for use with time-series data; see [TS] **tsset**. You must tsset your data before using varlmar.

Description

Most post-estimation analysis of VAR(p) models and SVAR models is conducted under the assumption that the disturbances are not autocorrelated. varlmar implements a Lagrange multiplier (LM-test) for autocorrelation in the residuals of VAR(p) models presented in Johansen (1995).

Options

m̲lag(#) specifies the maximum order of autocorrelation that should be tested. The integer specified in mlag() must be greater than 0; the default is 2.

e̲stimates(*estname*) specifies that varlmar is to use the previously obtained set of var or svar estimates saved as *estname*. By default, varlmar uses the currently active results. See [R] **estimates** for information on saving and restoring e() results.

s̲eparator(#) specifies how often separator lines should be drawn between rows. By default, separator lines do not appear.

Remarks

One of the assumptions upon which inference and post-estimation analysis after var and var svar are predicated is that the errors are not autocorrelated. varlmar implements the LM-test for autocorrelation in the residuals of a VAR(p) model discussed in Johansen (1995, 21–22). The test is performed at lags $j = 1, \ldots, \text{mlag}()$. For each j, the null hypothesis of the test is that there is no autocorrelation at lag j.

varlmar uses the e() results saved by var or svar. By default, varlmar uses the active e() results. However, varlmar can use any previously saved var or svar e() results that are specified using the estimates() option.

▷ Example

Let's quietly fit the VAR(2) that was fitted in [TS] **var**, and then call varlmar.

```
. use http://www.stata-press.com/data/r8/lutkepohl
(Quarterly SA West German macro data, Bil DM, from Lutkepohl 1993 Table E.1)

. quietly var dlinvestment dlincome dlconsumption if qtr>=q(1961q2) &
> qtr <= q(1978q4), lags(1/2) dfk

. varlmar, mlag(5)

H0: no autocorrelation at lag order j
```

j	chi2	df	p
1	4.0531	9	0.90788
2	4.9264	9	0.84068
3	7.9976	9	0.53439
4	12.0709	9	0.20934
5	5.3671	9	0.80120

Since we cannot reject the null hypothesis that there is no autocorrelation in the residuals for any of the five orders tested, this test does not provide any hint of model misspecification. Although we fit the VAR with the dfk option to be consistent with the example in [TS] **var**, varlmar will always use the ML estimator of Σ. In other words, the results obtained from varlmar are invariant to whether dfk was specified.

◁

▷ Example

When varlmar is applied to e() results produced by svar, the sequence of LM-tests is applied to the underlying var. See [TS] **var svar** for how an SVAR model builds on a VAR(p). In this example, we fit an SVAR that has an underlying VAR(2) that is identical to the one fitted in the previous example.

```
. matrix A = (.,.,0\0,.,0\.,.,.)

. matrix B = I(3)

. quietly svar dlinvestment dlincome dlconsumption if qtr>=q(1961q2) &
> qtr <= q(1978q4), lags(1/2) dfk aeq(A) beq(B)

. varlmar, mlag(5)

H0: no autocorrelation at lag order j
```

j	chi2	df	p
1	4.0531	9	0.90788
2	4.9264	9	0.84068
3	7.9976	9	0.53439
4	12.0709	9	0.20934
5	5.3671	9	0.80120

Since the underlying VAR(2) is the same as the previous example (we tell you that this is true), the output from varlmar is also the same.

◁

Saved Results

varlmar saves in r():

Matrices

r(lm) χ^2, df, and p-values

Methods and Formulas

`varlmar` is implemented as an ado-file.

The formula for the LM test statistic at lag j is

$$
\text{LM}_s = (T - d - .5) \ln \left(\frac{|\widehat{\Sigma}|}{|\widetilde{\Sigma}_s|} \right)
$$

where T is the number of observations in the VAR(p), d is explained below, $\widehat{\Sigma}$ is the maximum likelihood estimate of Σ, the variance–covariance matrix of the disturbances from the VAR(p), and $\widetilde{\Sigma}_s$ is the maximum likelihood estimate of Σ from the following augmented VAR(p).

If there are K equations in the VAR(p), then we can define $\mathbf{e}_t$ to be a $K \times 1$ vector of residuals. After we create the K new variables e1, e2, ..., eK, we can augment the original VAR(p) with lags of these K new variables. For each lag s, we form an augmented regression in which the new residual variables are lagged s times. Following the method of Davidson and MacKinnon (1993), the missing values from these s lags are replaced with zeros. Then, $\widetilde{\Sigma}_s$ is the maximum-likelihood estimate of Σ from this augmented VAR(p), and d is the number of coefficients estimated in the augmented VAR(p). See [TS] **var** for a discussion of the maximum likelihood estimate of Σ in a VAR(p).

The asymptotic distribution of LM_s is χ^2 with K^2 degrees of freedom.

References

Davidson, R. and J. G. MacKinnon. 1993. *Estimation and Inference in Econometrics.* Oxford: Oxford University Press.

Johansen, S. 1995. *Likelihood-Based Inference in Cointegrated Vector Auto-Regressive Models.* Oxford: Oxford University Press.

Also See

Complementary: [TS] **var**, [TS] **var svar**, [TS] **varbasic**

Background: [TS] **var intro**

Title

> **varnorm** — Test for normally distributed disturbances after var or svar

Syntax

varnorm $\left[\right.$, jbera skewness kurtosis cholesky estimates(*estname*) separator(#) $\left. \right]$

varnorm may only be used with var or svar e() results; see [TS] **var** and [TS] **var svar**.

varnorm is for use with time-series data; see [TS] **tsset**. You must tsset your data before using varnorm.

Description

varnorm computes and reports a series of statistics against the null hypothesis that the disturbances in a VAR(p) are normally distributed. For each equation, and for all equations jointly, up to three statistics may be computed: a skewness statistic, a kurtosis statistic, and the Jarque–Bera statistic. By default, all three statistics are reported.

Options

jbera specifies that the Jarque–Bera statistic be reported. By default, the Jarque–Bera, skewness, and kurtosis statistics are all reported. When jbera is specified, only the Jarque–Bera statistic and any other explicitly requested statistics are reported.

skewness specifies that the skewness statistic be reported. By default, the Jarque–Bera, skewness, and kurtosis statistics are all reported. When skewness is specified, only the skewness statistic and any other explicitly requested statistics are reported.

kurtosis specifies that the kurtosis statistic be reported. By default, the Jarque–Bera, skewness, and kurtosis statistics are all reported. When kurtosis is specified, only the kurtosis statistic and any other explicitly requested statistics are reported.

cholesky specifies that varnorm should use the Cholesky decomposition of the estimated variance–covariance matrix of the disturbances, $\widehat{\boldsymbol{\Sigma}}$, to orthogonalize the residuals when varnorm is applied to svar e() results. By default, when varnorm is applied to svar e() results, it uses the estimated structural decomposition, $\widehat{\mathbf{A}}^{-1}\widehat{\mathbf{B}}$, to orthogonalize the residuals. When applied to var e() results, varnorm always uses the Cholesky decomposition of $\widehat{\boldsymbol{\Sigma}}$. For this reason, the cholesky option may not be specified when using var e() results.

estimates(*estname*) specifies that varnorm is to use the previously obtained set of var or svar estimates saved as *estname*. By default, varnorm uses the currently active results. See [R] **estimates** for information on saving and restoring e() results.

separator(#) specifies how often separator lines should be drawn between rows. By default, separator lines do not appear.

Remarks

Some of the post-estimation statistics for VAR and SVAR use the assumption that the K disturbances have a K dimensional multivariate normal distribution. varnorm uses the e() results produced by var or svar to produce a series of statistics against the null hypothesis that the K disturbances in the VAR(p) are normally distributed.

Following the notation in Lütkepohl (1993), call the skewness statistic $\widehat{\lambda}_1$, the kurtosis statistic $\widehat{\lambda}_2$, and the Jarque–Bera statistic $\widehat{\lambda}_3$. The Jarque–Bera statistic is a combination of the other two. The single-equation results are from tests against the null hypothesis that the disturbance for that particular equation is normally distributed. The results for all the equations are from tests against the null hypothesis that the K disturbances follow a K-dimensional multivariate normal distribution. Failure to reject indicates a lack of model misspecification.

▷ Example

This example uses varnorm to test for normality after fitting a VAR(2). Since we fitted this VAR(2) in [TS] **var**, we begin by quietly fitting the VAR(2) and then calling varnorm.

```
. use http://www.stata-press.com/data/r8/lutkepohl
(Quarterly SA West German macro data, Bil DM, from Lutkepohl 1993 Table E.1)

. quietly var dlinvestment dlincome dlconsumption if qtr <= q(1978q4),
> lags(1/2) dfk

. varnorm
```

```
            Jarque-Bera results
```

Equation	Chi2	df	Prob > chi2
dlinvestment	2.8215	2	0.24397
dlincome	3.4500	2	0.17817
dlconsumption	1.5661	2	0.45702
ALL	7.8375	6	0.25025

```
            Skewness results
```

Equation	Skewness	Chi2	df	Prob > chi2
dlinvestment	0.1194	0.1733	1	0.67718
dlincome	-0.3832	1.7862	1	0.18139
dlconsumption	-0.3127	1.1900	1	0.27532
ALL	.	3.1496	3	0.36913

```
            Kurtosis results
```

Equation	Kurtosis	Chi2	df	Prob > chi2
dlinvestment	3.9331	2.6481	1	0.10367
dlincome	3.7396	1.6637	1	0.19710
dlconsumption	2.6484	0.3760	1	0.53973
ALL	.	4.6879	3	0.19613

In this example, neither the single-equation Jarque–Bera statistics nor the joint Jarque–Bera statistic come close to rejecting the null hypothesis.

The skewness and kurtosis results have analogous structures.

The Jarque–Bera results use the sum of the skewness and kurtosis statistics. The skewness and kurtosis results are based on the skewness and kurtosis coefficients, respectively. See *Methods and Formulas*.

◁

▷ Example

The test statistics are computed on the orthogonalized VAR(p) residuals; see *Methods and Formulas*. When `varnorm` is applied to `var e()` results, `varnorm` uses a Cholesky decomposition of the estimated variance–covariance matrix of the disturbances, $\widehat{\Sigma}$, to orthogonalize the residuals.

By default, when `varnorm` is applied to `svar e()` results, it uses the estimated structural decomposition to orthogonalize the residuals of the underlying VAR(p). Alternatively, when `varnorm` is applied to `svar e()` results, and the `cholesky` option is specified, `varnorm` uses the Cholesky decomposition of $\widehat{\Sigma}$ to orthogonalize the residuals of the underlying VAR(p).

Let's fit an SVAR that is based on an underlying VAR(2), which is the same as the one fit in the previous example. We impose a structural decomposition that is the same as a Cholesky decomposition. (See [TS] **var svar** for more examples that illustrate this relationship.)

```
. matrix a = (.,0,0\.,.,0\.,.,.)
. matrix b = I(3)
. quietly svar dlinvestment dlincome dlconsumption if qtr <= q(1978q4),
> lags(1/2) dfk aeq(a) beq(b)
. varnorm
```

                Jarque-Bera results

Equation	Chi2	df	Prob > chi2
dlinvestment	2.8215	2	0.24397
dlincome	3.4500	2	0.17817
dlconsumption	1.5661	2	0.45702
ALL	7.8375	6	0.25025

                Skewness results

Equation	Skewness	Chi2	df	Prob > chi2
dlinvestment	0.1194	0.1733	1	0.67718
dlincome	-0.3832	1.7862	1	0.18139
dlconsumption	-0.3127	1.1900	1	0.27532
ALL	.	3.1496	3	0.36913

                Kurtosis results

Equation	Kurtosis	Chi2	df	Prob > chi2
dlinvestment	3.9331	2.6481	1	0.10367
dlincome	3.7396	1.6637	1	0.19710
dlconsumption	2.6484	0.3760	1	0.53973
ALL	.	4.6879	3	0.19613

Since the estimated structural decomposition is the same as the Cholesky decomposition, the `varnorm` results are the same as those from the previous example.

◁

❏ Technical Note

The statistics computed by varnorm depend on $\overset{\approx}{\Sigma}$, the estimated variance–covariance matrix of the disturbances. var uses the maximum likelihood estimator of this matrix by default, but the dfk option produces an estimator that uses a small-sample correction. Thus, specifying dfk in the call to var or svar will affect the test results produced by varnorm.

❏

Saved Results

varnorm saves in r():

Matrices

r(kurtosis)	kurtosis test, df, and p-values
r(skewness)	skewness test, df, and p-values
r(jb)	Jarque–Bera test, df, and p-values

Methods and Formulas

varnorm is implemented as an ado-file.

varnorm is based on the derivations found in Lütkepohl (1993, 152–158). Let $\widehat{\mathbf{u}}_t$ be the $K \times 1$ vector of residuals from the K equations in a previously fitted VAR(p), or the residuals from the K equations of the VAR(p) underlying a previously fit SVAR. Similarly, let $\widehat{\Sigma}$ be the estimated covariance matrix of the disturbances. (Note that $\widehat{\Sigma}$ depends on whether the dfk option was specified.) The skewness, kurtosis, and the Jarque–Bera statistics must be computed using the orthogonalized residuals.

Since

$$\widehat{\Sigma} = \widehat{\mathbf{P}}\widehat{\mathbf{P}}'$$

implies that

$$\widehat{\mathbf{P}}^{-1}\widehat{\Sigma}\widehat{\mathbf{P}}^{-1\prime} = \mathbf{I}_K$$

Premultiplying $\widehat{\mathbf{u}}_t$ by $\widehat{\mathbf{P}}$ is one way of performing the orthogonalization. When varnorm is applied to var results, $\widehat{\mathbf{P}}$ is defined to be the Cholesky decomposition of $\widehat{\Sigma}$. When varnorm is applied to svar e() results, by default $\widehat{\mathbf{P}}$ is set to the estimated structural decomposition; i.e., $\widehat{\mathbf{P}} = \widehat{\mathbf{A}}^{-1}\widehat{\mathbf{B}}$, where $\widehat{\mathbf{A}}$ and $\widehat{\mathbf{B}}$ are the svar estimates of the $\mathbf{A}$ and $\mathbf{B}$ matrices. (See [TS] **var svar** for more on the origin and estimation of the $\mathbf{A}$ and $\mathbf{B}$ matrices.) When varnorm is applied to svar e() results and the cholesky option is specified, $\widehat{\mathbf{P}}$ is set to the Cholesky decomposition of $\widehat{\Sigma}$.

Define $\widehat{\mathbf{w}}_t$ to be the orthogonalized VAR(p) residuals given by

$$\widehat{\mathbf{w}}_t = (\widehat{w}_{1t}, \ldots, \widehat{w}_{Kt})' = \widehat{\mathbf{P}}^{-1}\widehat{\mathbf{u}}_t$$

The $K \times 1$ vectors of skewness and kurtosis coefficients are then computed using the orthogonalized residuals by

$$\widehat{\mathbf{b}}_1 = (\widehat{b}_{11}, \ldots, \widehat{b}_{K1})'; \qquad \widehat{b}_{k1} = \frac{1}{T} \sum_{i=1}^{T} \widehat{w}_{kt}^3$$

$$\widehat{\mathbf{b}}_2 = (\widehat{b}_{12}, \ldots, \widehat{b}_{K2})'; \qquad \widehat{b}_{k2} = \frac{1}{T} \sum_{i=1}^{T} \widehat{w}_{kt}^4$$

Under the null hypothesis of multivariate Gaussian disturbances,

$$\widehat{\lambda}_1 = \frac{T\widehat{\mathbf{b}}_1'\widehat{\mathbf{b}}_1}{6} \quad \overset{d}{\to} \quad \chi^2(K)$$

$$\widehat{\lambda}_2 = \frac{T(\widehat{\mathbf{b}}_2 - 3)'(\widehat{\mathbf{b}}_2 - 3)}{24} \quad \overset{d}{\to} \quad \chi^2(K)$$

and

$$\widehat{\lambda}_3 = \widehat{\lambda}_1 + \widehat{\lambda}_2 \quad \overset{d}{\to} \quad \chi^2(2K)$$

$\widehat{\lambda}_1$ is the skewness statistic, $\widehat{\lambda}_2$ is the kurtosis statistic, and $\widehat{\lambda}_3$ is the Jarque–Bera statistic.

$\widehat{\lambda}_1$, $\widehat{\lambda}_2$, and $\widehat{\lambda}_3$ are for tests of the null hypothesis that the $K \times 1$ vector of disturbances follows a multivariate normal distribution. The corresponding statistics against the null hypothesis that the disturbances from the kth equation come from a univariate normal distribution are

$$\widehat{\lambda}_{1k} = \frac{T\widehat{b}_{k1}^2}{6} \quad \overset{d}{\to} \quad \chi^2(1)$$

$$\widehat{\lambda}_{2k} = \frac{T(\widehat{b}_{k2}^2 - 3)^2}{24} \quad \overset{d}{\to} \quad \chi^2(1)$$

and

$$\widehat{\lambda}_{3k} = \widehat{\lambda}_1 + \widehat{\lambda}_2 \quad \overset{d}{\to} \quad \chi^2(2)$$

References

Jarque, C. M. and A. K. Bera. 1987. A test for normality of observations and regression residuals. *International Statistical Review* 2: 163–172.

Lütkepohl, H. 1993. *Introduction to Multiple Time Series Analysis*. 2d ed. New York: Springer.

Hamilton, J. D. 1994. *Time Series Analysis*. Princeton: Princeton University Press.

Also See

Complementary:	[TS] **var**, [TS] **var svar**, [TS] **varbasic**
Background:	[TS] **var intro**

Title

> **varsoc** — Obtain lag-order selection statistics for a set of VARs

Syntax

Pre-estimation syntax

varsoc *depvarlist* [if *exp*] [in *range*] [, maxlag(*#*) exog(*varlist*)

    constraints(*numlist*) noconstant lutstats separator(*#*) level(*#*)]

Post-estimation syntax

varsoc [*depvarlist*] [if *exp*] [in *range*], estimates(*estname*) [maxlag(*#*)

    [exog(*varlist*) | rmexog] [constraints(*numlist*) | rmconstraints]

    [noconstant | addconstant] [lutstats | rmlutstats] separator(*#*)

    level(*#*)]

varsoc is for use with time-series data; see [TS] **tsset**. You must tsset your data before using varsoc.

by..: may be used with varsoc; see [R] **by**.

Description

    varsoc reports the final prediction error (FPE), Akaike's information criterion (AIC), the Bayesian information criterion (BIC), and the Hannan and Quinn information criterion (HQIC) lag order selection statistics for a series of vector autoregressions of order 1, ..., maxlag(). These statistics are commonly used to assist researchers in fitting a VAR(*p*) of the correct order. A sequence of likelihood-ratio test statistics for all of the full VARs of order less than or equal to the highest requested lag order is also reported. These LR-tests are used to determine whether the additional parameters in the VAR, with one additional lag, are zero. See Lütkepohl (1993) and Hamilton (1994) for an introduction to these statistics.

Options

maxlag(*#*) specifies the maximum lag order for which the statistics are to be obtained. If the estimates() option is not specified, the default is 4. maxlag(*#*) must specify an integer that is greater than or equal to 2.

exog(*varlist*) specifies any exogenous variables.

rmexog specifies that any exogenous variables from the model associated with *estname* are not to be included in the VARs fit by varsoc.

estimates(*estname*) specifies the name of a previously stored set of var or svar estimates. To request the results currently stored in e(), you must specify estimates(.) since, by default, varsoc uses the *pre-estimation syntax*. When e() results are specified via estimates(*estname*), varsoc uses the *post-estimation syntax*, and it obtains all of its defaults from the e() results specified in estimates(*estname*). See [R] **estimates** for more on saving and restoring e() results.

constraints(*numlist*) specifies a list of constraints on the exogenous variables that are to be applied. Do not specify constraints on the lags of the endogenous variables, since specifying one would mean that at least one of the VAR models considered by varsoc will not contain the lag specified in the constraint. Use [TS] **var** directly to obtain selection order criteria with constraints on lags of the endogenous variables.

rmconstraints specifies that any constraints specified in the model associated with estimates(*estname*) are not to be applied to models considered by varsoc.

noconstant suppresses the constant terms from the model. In the pre-estimation syntax, by default constant terms are included in the model. In the post-estimation syntax, the default is what was in the model associated with estimates(*estname*).

addconstant specifies that constant terms are to be included in the model, even though they may have been suppressed in the model associated with estimates(*estname*).

lutstats specifies that the Lütkepohl (1993) versions of the information criteria should be calculated and reported.

rmlutstats specifies that varsoc should not compute the Lütkepohl (1993) versions of the information criteria even though it was specified in the command that generated the estimates in *estname*.

separator(*#*) specifies how often separator lines should be drawn between rows. By default, separator lines do not appear.

level(*#*) specifies the confidence level, in percent, that is used to identify the first likelihood-ratio test that rejects the null hypothesis that the additional parameters from adding a lag are jointly zero. The default is level(95) or as set by set level; see [U] **23.6 Specifying the width of confidence intervals**.

Remarks

A number of selection order statistics have been developed to assist researchers in fitting a VAR(p) of the correct order. Several of these selection order statistics appear in the [TS] **var** and [TS] **var svar** output. The varsoc command will compute these statistics over a wider range of p while maintaining a common sample and option specification.

varsoc can be used as a pre-estimation or a post-estimation command. When the estimates() option is not specified, varsoc is a pre-estimation command. When used as a pre-estimation command, a *depvarlist* is required, and the default maximum lag is 4.

When the estimates() option is specified, varsoc operates as a post-estimation command, obtaining its defaults from the information contained in *estname* varsoc contains three special "rm" options to negate information saved in the e() results specified in estimates(). The rmexog option specifies that any exogenous variables included in the specified e() results are not to be included in the models fit by varsoc. The rmlutstats option specifies that the standard versions of the information criteria are to be reported, even though the Lütkepohl (1993) versions were requested in the original model. Finally, the noconstraints option specifies that the constraints applied in the original model are not to be applied to the models fit by varsoc. This is especially useful because varsoc does not allow constraints to be placed on lags of the endogenous variables, since any constraint on a lag of an endogenous variable will be invalid for at least one of the models fit by varsoc.

varsoc computes four information criteria, the FPE, AIC, HQIC, and SBIC, as well as a series of likelihood-ratio tests for all of the full VARs of order less than or equal to the highest lag order requested. While the likelihood-ratio tests are standard tests for whether the additional parameters in a VAR with one more lag are zero, Lütkepohl (1993, 125–127) presents a useful top-down lag-order selection strategy using these statistics. One starts by looking at the results of the test for the model with the most lags. Starting at the bottom, the first test that rejects the null hypothesis is the lag order selected by this process. See Lütkepohl (1993, 125–126) for a nice discussion of why the multiple testing problem causes the Type I error of the procedure to differ from the significance level chosen for each test. For each of the four information criteria, the lag with the smallest value is the order selected by that criteria. An '*' appears next to the smallest statistic indicating that the corresponding lag is optimal.

In a sense, the FPE is really not an information criteria. As the name suggests, the FPE has its origins as a measure of forecast inaccuracy. Thus, it is intuitive that the VAR(p) with the lowest FPE is, by some measure, optimal. The FPE is quite similar to AIC. Conceptually, the AIC is designed to measure the discrepancy between the given model and the true model, which, of course, we would want to minimize. Amemiya (1985) provides an intuitive discussion of the arguments in Akaike (1973). The BIC and the HQIC can be interpreted similarly to the AIC, but the BIC and the HQIC have a theoretical advantage over the AIC and the FPE. As Lütkepohl (1993, 130–133) demonstrates, choosing p to minimize the BIC or the HQIC provides consistent estimates of the true lag order p. In contrast, minimizing the AIC or the FPE will overestimate the true lag order with positive probability, even with an infinite sample size.

▷ Example

Here, we use varsoc as a pre-estimation command.

```
. use http://www.stata-press.com/data/r8/lutkepohl
(Quarterly SA West German macro data, Bil DM, from Lutkepohl 1993 Table E.1)
. varsoc dlinvestment dlincome dlconsumption if qtr<=q(1978q4), lutstats
Selection order criteria (lutstats)
endogenous variables:
    dlinvestment dlincome dlconsumption
constant included in models
Sample:  1961q2    1978q4
Obs = 71
```

lag	LL	LR	df	p	FPE	AIC	HQIC	SBIC
0	564.784	.	.	.	2.69e-11	-24.423	-24.423*	-24.423*
1	576.409	23.249	9	0.006	2.50e-11	-24.497	-24.3829	-24.2102
2	588.859	24.901*	9	0.003	2.27e-11*	-24.5942*	-24.3661	-24.0205
3	591.237	4.757	9	0.855	2.75e-11	-24.4076	-24.0655	-23.5472
4	598.457	14.438	9	0.108	2.91e-11	-24.3575	-23.9012	-23.2102

Notice that the sample used begins in 1961q2. This is because all the VARs are fit to the sample defined by any if or in conditions, and the available data, for the maximum lag specified. The output indicates that we accepted the default of a maximum lag of 4. An asterisk next to a statistic indicates that it corresponds to the lag length chosen using the methods described in Lütkepohl (1993, Chapter 4), which were outlined earlier in the *Remarks*.

Since we specified the lutstats option, the table contains the Lütkepohl (1993) versions of the information criteria, which differ from the standard definitions in that they drop the constant term from the log likelihood. In this example, the likelihood-ratio tests selected a model with 2 lags. AIC and FPE have also both chosen a model with 2 lags, whereas BIC and HQIC have both selected a model with 0 lags.

◁

▷ Example

varsoc may be used as a post-estimation command by specifying the estimates() option. If you do not want to use all of these results, specify rmexog, rmconstraints, or rmlutstats.

```
. qui var dlincome dlconsumption if qtr<=q(1978q4), lutstats exog(l.dlinvestment)
. varsoc, est(.)
Selection order criteria (lutstats)
endogenous variables:
    dlincome dlconsumption

exogenous variables:
    L.dlinvestment

constant included in models
Sample:  1960q4    1978q4
Obs = 73
```

lag	LL	LR	df	p	FPE	AIC	HQIC	SBIC
0	460.646	.	.	.	1.26e-08	-18.2962	-18.2962	-18.2962*
1	467.606	13.919	4	0.008	1.17e-08	-18.3773	-18.3273	-18.2518
2	477.087	18.962*	4	0.001	1.00e-08*	-18.5275*	-18.4274*	-18.2764

One lag of dlinvestment was included in each model as an exogenous variable.

Suppose that we are willing to accept all the choices made in the VAR we fit above, except that we want to exclude any lags of dlinvestment as exogenous variables. We need to specify the rmexog option.

```
. qui var dlincome dlconsumption  if qtr<=q(1978q4), exog(l.dlinvestment)
. varsoc, est(.) rmexog
Selection order criteria
endogenous variables:
    dlincome dlconsumption

constant included in models
Sample:  1960q4    1978q4
Obs = 73
```

lag	LL	LR	df	p	FPE	AIC	HQIC	SBIC
0	458.535	.	.	.	1.27e-08	-12.5078	-12.4828	-12.4451
1	466.534	15.998	4	0.003	1.14e-08	-12.6174	-12.5423	-12.4291
2	475.740	18.411*	4	0.001	9.85e-09*	-12.76*	-12.635*	-12.4462*

rmlutstats and rmconstraints work analogously. rmlutstats specifies that the Lütkepohl (1993) versions of the statistics are not to be computed, even though they were specified in the original VAR model. rmconstraints specifies that the constraints applied to the original VAR specified in estimates() are not to be applied to models fit by varsoc.

Suppose that we wanted to include two lags of dlinvestment as exogenous variables instead of one.

```
. varsoc, est(.) exog(l(1/2).dlinvestment)
Selection order criteria
endogenous variables:
    dlincome dlconsumption
exogenous variables:
    L.dlinvestment L2.dlinvestment
constant included in models
Sample:  1960q4    1978q4
Obs = 73
```

lag	LL	LR	df	p	FPE	AIC	HQIC	SBIC
0	463.046	.	.	.	1.25e-08	-12.5218	-12.4468	-12.3336*
1	469.589	13.086	4	0.011	1.17e-08	-12.5915	-12.4664	-12.2777
2	478.593	18.008*	4	0.001	1.02e-08*	-12.7286*	-12.5535*	-12.2893

When `varsoc` is used as a post-estimation command, explicitly specifying an option will override what was saved in the specified `estimates()`.

◁

▷ Example

When `varsoc` is applied to `svar e()` results, `varsoc` obtains its defaults for the VAR(p) that underlies the SVAR that was fitted.

```
. mat A = (.,0\.,.)
. mat B = I(2)
. qui svar dlincome dlconsumption if qtr<=q(1978q4), aeq(A) beq(B)
> exog(l(1/2).dlinvestment)
. varsoc, est(.)
Selection order criteria
endogenous variables:
    dlincome dlconsumption
exogenous variables:
    L.dlinvestment L2.dlinvestment
constant included in models
Sample:  1960q4    1978q4
Obs = 73
```

lag	LL	LR	df	p	FPE	AIC	HQIC	SBIC
0	463.046	.	.	.	1.25e-08	-12.5218	-12.4468	-12.3336*
1	469.589	13.086	4	0.011	1.17e-08	-12.5915	-12.4664	-12.2777
2	478.593	18.008*	4	0.001	1.02e-08*	-12.7286*	-12.5535*	-12.2893

Since the sample and structure placed on the VAR(p)s fit in the two calls to `varsoc` are the same (we tell you that this is true), the output is the same.

◁

(Continued on next page)

Saved Results

varsoc saves in r():

Scalars
r(N)	number of observations	r(tmin)	first time period in sample
r(tmax)	last time period in sample	r(mlag)	maximum lag order

Macros
r(endog)	names of endogenous variables	r(exog)	names of exogenous variables
r(lutstats)	lutstats, if specified	r(rmlutstats)	rmlutstats, if specified
r(cns#)	the #th constraint	r(N_gaps)	the number of gaps in the sample

Matrices
r(stats)	LL, LR, FPE, AIC, HQIC, SBIC, and p-values

Methods and Formulas

varsoc is implemented as an ado-file.

As shown by Hamilton (1994, 295–296), the log likelihood for a VAR(p) is

$$
\text{LL} = \left(\frac{T}{2}\right)\left\{\ln\left(|\widehat{\boldsymbol{\Sigma}}^{-1}|\right) - K\ln(2\pi) - K\right\}
$$

where T is the number of observations, K is the number of equations, and $\widehat{\boldsymbol{\Sigma}}$ is the maximum likelihood estimate of $E[\mathbf{u}_t\mathbf{u}_t']$, where $\mathbf{u}_t$ is the $K \times 1$ vector of disturbances. (See [TS] **var** for more on $\mathbf{u}_t$.) However, since

$$
\ln\left(|\widehat{\boldsymbol{\Sigma}}^{-1}|\right) = -\ln\left(|\widehat{\boldsymbol{\Sigma}}|\right)
$$

the log likelihood can be rewritten as

$$
\text{LL} = -\left(\frac{T}{2}\right)\left\{\ln\left(|\widehat{\boldsymbol{\Sigma}}|\right) + K\ln(2\pi) + K\right\}
$$

Letting LL(j) be the value of the log likelihood with j lags yields LR statistic for lag order j as

$$
\text{LR}(j) = 2\left\{\text{LL}(j) - \text{LL}(j-1)\right\}
$$

Model order statistics

The formula for the FPE given in Lütkepohl (1993, 128) is

$$
\text{FPE} = |\boldsymbol{\Sigma}_u|\left(\frac{T + Kp + 1}{T - Kp - 1}\right)^K
$$

This formula, however, assumes that there is a constant in the model and that none of the variables are dropped due to collinearity. To deal with these problems, the FPE is implemented as

$$\text{FPE} = |\mathbf{\Sigma}_u| \left(\frac{T + \overline{m}}{T - \overline{m}} \right)^K$$

where $\overline{m}$ is the average number of parameters over the K equations. This implementation will account for variables dropped due to collinearity.

By default, the AIC, SBIC, and HQIC are computed according to their standard definitions, which include the constant term from the log likelihood. That is,

$$\text{AIC} = -2 \left(\frac{\text{LL}}{T} \right) + \frac{2t_p}{T}$$

$$\text{SBIC} = -2 \left(\frac{\text{LL}}{T} \right) + \frac{\ln(T)}{T} t_p$$

$$\text{HQIC} = -2 \left(\frac{\text{LL}}{T} \right) + \frac{2\ln\{\ln(T)\}}{T} t_p$$

where t_p is the total number of parameters in the model and LL is the log likelihood.

Lutstats

Lütkepohl (1993) advocated dropping the constant term from the log likelihood since it does not affect inference. The Lütkepohl versions of the information criteria are

$$\text{AIC} = \ln(|\mathbf{\Sigma}_u|) + \frac{2pK^2}{T}$$

$$\text{BIC} = \ln(|\mathbf{\Sigma}_u|) + \frac{\ln(T)}{T} pK^2$$

$$\text{HQIC} = \ln(|\mathbf{\Sigma}_u|) + \frac{2\ln\{\ln(T)\}}{T} pK^2$$

References

Akaike, H. 1973. Information theory and an extension of the maximum likelihood principle. In *Second International Symposium on Information Theory*, eds B. Petrov and F. Csaki, 267–281. Budapest: Académiai Kiadó.

Amemiya, T. 1985. *Advanced Econometrics*. Cambridge MA: Harvard University Press.

Hamilton, J. D. 1994. *Time Series Analysis*. Princeton: Princeton University Press.

Lütkepohl, H. 1993. *Introduction to Multiple Time Series Analysis*. 2d ed. New York: Springer.

Also See

Complementary:	[TS] **var**, [TS] **var svar**, [TS] **varbasic**
Background:	[TS] **var intro**

Title

> **varstable** — Check stability condition of var or svar estimates

Syntax

> varstable [, <u>a</u>mat(*matrix_name*) <u>est</u>imates(*estname*)]

varstable may only be used with var or svar e() results; see [TS] **var** and [TS] **var svar**.

Description

varstable checks the eigenvalue stability condition after estimating the parameters of a vector autoregression using var or svar.

Options

amat(*matrix_name*) specifies a valid Stata matrix name in which the **A** matrix can be saved (see *Methods and Formulas* for the definition of the matrix **A**). The default is not to save the **A** matrix.

estimates(*estname*) specifies that varstable should check the stability condition using the VAR estimates stored in *estname*. By default, varstable uses the currently active e() results. See [R] **estimates** for information on saving and restoring previously obtained estimates.

Remarks

Inference after var and svar requires that variables be covariance stationary. The variables in $\mathbf{y}_t$ are covariance stationary if their first two moments exist and are independent of time. More explicitly, a variable y_t is covariance stationary if

1. $E[y_t]$ is finite and independent of t.

2. $Var[y_t]$ is finite and independent of t

3. $Cov[y_t, y_s]$ is a finite function of $|t - s|$, but neither t nor s.

Interpretation of VAR(p) models, however, requires that an even stricter stability condition be met. If a VAR(p) is stable, then it is invertible and has an infinite-order vector moving-average representation. If the VAR(p) is stable, impulse–response functions and forecast-error variance decompositions have known interpretations.

Lütkepohl (1993) and Hamilton (1994) both show that if the modulus of each eigenvalue of the matrix **A** is strictly less than one, then the estimated VAR(p) is stable (see *Methods and Formulas* for the definition of the matrix **A**). We begin with some examples of how to use varstable. Later, some intuition for results that Lütkepohl (1993) and Hamilton (1994) develop in detail will be provided.

▷ Example

After fitting a VAR with var, you can use varstable to check the stability condition. Using the same VAR(2) model that was used in [TS] **var**, we demonstrate the use of varstable.

```
. use http://www.stata-press.com/data/r8/lutkepohl
(Quarterly SA West German macro data, Bil DM, from Lutkepohl 1993 Table E.1)
. quietly var dlinvestment dlincome dlconsumption if qtr>=q(1961q2) &
> qtr <= q(1978q4), lags(1/2)
. varstable
        Eigenvalue stability condition
```

Eigenvalue	Modulus
.5456253	.54562531
-.0643276 + .45959443	.46407444
-.0643276 - .45959443	.46407444
-.3785754 + .38539823	.54023248
-.3785754 - .38539823	.54023248
-.3698058	.36980578

```
All the eigenvalues lie inside the unit circle
VAR satisfies stability condition
```

Since the modulus of each of the eigenvalues is strictly less than 1, the estimates satisfy the eigenvalue stability condition.

◁

▷ Example

This example illustrates two other features of the `varstable` command. First, `varstable` can check the stability of the estimates of the VAR underlying an SVAR fitted by `var svar`. Second, `varstable` can be used to check the stability of any previously stored `var` or `var svar` estimates.

Let's begin by `quietly` refitting the previous VAR and storing the results as `var1`. Since this is the same VAR that was fitted in the previous example, the stability results should be identical.

```
. quietly var dlinvestment dlincome dlconsumption if qtr>=q(1961q2) &
> qtr <= q(1978q4), lags(1/2)
. estimates store var1
```

Now, let's use `svar` to fit an SVAR with a different underlying VAR, and then check the estimates of that underlying VAR for stability.

```
. matrix A = (.,0\.,.)
. matrix B = I(2)
. quietly svar d.lincome d.lconsumption, aeq(A) beq(B)
. varstable
        Eigenvalue stability condition
```

Eigenvalue	Modulus
.548711	.54871097
-.2979493 + .43280132	.52544339
-.2979493 - .43280132	.52544339
-.3570825	.35708246

```
All the eigenvalues lie inside the unit circle
VAR satisfies stability condition
```

The `estimates()` option allows us to check the stability of the `var` results stored as `var1`.

```
. varstable, est(var1)
        Eigenvalue stability condition
```

Eigenvalue	Modulus
.5450253	.54562531
-.0643276 + .45959443i	.46407444
-.0643276 - .45959443i	.46407444
-.3785754 + .38539823i	.54023248
-.3785754 - .38539823i	.54023248
-.3698058	.36980578

```
All the eigenvalues lie inside the unit circle
VAR satisfies stability condition
```

Since the results are identical to those obtained in the previous example, this confirms that we were checking the results in `var1`.

◁

Now, let's develop some intuition for the results discussed in Lütkepohl (1993) and Hamilton (1994).

Consider a VAR(1):

$$\mathbf{y}_t = \mathbf{v} + \mathbf{A}_1 \mathbf{y}_{t-1} + \mathbf{u}_t \tag{1}$$

By repeated substitution, this equation can be rewritten as

$$\mathbf{y}_t = (\mathbf{I}_K + \mathbf{A}_1 + \mathbf{A}_1^2 + \cdots + \mathbf{A}_1^j)\mathbf{v} + \mathbf{A}_1^{j+1}\mathbf{y}_{t-j-1} + \sum_{i=0}^{j} \mathbf{A}_1^i \mathbf{u}_{t-i} \tag{2}$$

Much of the interest in these models focuses on how *innovations* affect the dependent variables. An infinite-order vector moving-average representation of this model (VMA(∞)), if it exists, would therefore be of interest, since it would explicitly express the functional relationship between the dependent variable and the entire history of the innovations. Thus, we are interested in the subset of VAR(1) models that can be written as

$$\mathbf{y}_t = \boldsymbol{\mu} + \sum_{i=0}^{\infty} \mathbf{A}_1^i \mathbf{u}_{t-i} \tag{3}$$

Equation (2) tells us that we can rewrite the VAR(1) as a VMA(∞) if

1. $(\mathbf{I}_K + \mathbf{A}_1 + \mathbf{A}_1^2 + \cdots + \mathbf{A}_1^j)\mathbf{v} \xrightarrow[j \to \infty]{} \boldsymbol{\mu} = (\mathbf{I}_K - \mathbf{A}_1)^{-1}\mathbf{v}$,

2. $\mathbf{A}_1^{j+1}\mathbf{y}_{t-j-1} \xrightarrow[j \to \infty]{} \mathbf{0}$ sufficiently rapidly, and

3. $\sum_{i=0}^{j} \mathbf{A}_1^i \mathbf{u}_{t-i} \xrightarrow[j \to \infty]{} \mathbf{w}_t$, where $\mathbf{w}_t$ is a well-defined random vector for all t.

A sufficient condition for all three of these results is that the eigenvalues of $\mathbf{A}_1$ have modulus strictly less than 1.

To obtain the equivalent condition for the VAR(p), rewrite (1) as

$$\mathbf{Y}_t = \mathbf{v} + \mathbf{A}\mathbf{Y}_{t-1} + \mathbf{U}_t \tag{4}$$

where

$$\mathbf{Y}_t = \begin{pmatrix} \mathbf{y}_t \\ \mathbf{y}_{t-1} \\ \vdots \\ \mathbf{y}_{t-p+1} \end{pmatrix}, \quad \mathbf{v} = \begin{pmatrix} \mathbf{v} \\ \mathbf{0} \\ \vdots \\ \mathbf{0} \end{pmatrix}, \quad \mathbf{A} = \begin{pmatrix} \mathbf{A}_1 & \mathbf{A}_2 & \cdots & \mathbf{A}_{p-1} & \mathbf{A}_p \\ \mathbf{I}_K & \mathbf{0} & \cdots & \mathbf{0} & \mathbf{0} \\ \mathbf{0} & \mathbf{I}_K & & \mathbf{0} & \mathbf{0} \\ \vdots & & \ddots & \vdots & \vdots \\ \mathbf{0} & \mathbf{0} & \cdots & \mathbf{I}_K & \mathbf{0} \end{pmatrix}, \quad \mathbf{U}_t = \begin{pmatrix} \mathbf{u}_t \\ \mathbf{0} \\ \vdots \\ \mathbf{0} \end{pmatrix}$$

The same logic used on the VAR(1) can be used to show that if the modulus of all the eigenvalues of $\mathbf{A}$ are strictly less than 1, the VMA(∞) representation of (4),

$$\mathbf{Y}_t = \boldsymbol{\mu} + \sum_{i=0}^{\infty} \mathbf{A}^i \mathbf{U}_{t-i} \tag{5}$$

is well-defined.

Although (5) provides the result that we need, we can obtain even more intuition for the usefulness of these results by a simple algebraic manipulation of (5). If the VAR(p) is stable, it can be inverted and rewritten as an infinite-order vector moving-average process. Furthermore, invertibility of a VAR(p) is a sufficient condition for computing impulse–response functions that have a known and meaningful interpretation. To illustrate, we will manipulate (5) to obtain the simple impulse–response functions. First, define the $K \times Kp$ matrix $\mathbf{J} = [\mathbf{I}_K, \mathbf{0}, \ldots, \mathbf{0}]$. Since $\mathbf{y}_t = \mathbf{J}\mathbf{Y}_t$, this matrix allows us to easily obtain $\mathbf{y}_t$ from $\mathbf{Y}_t$. While $\mathbf{J}\mathbf{U}_t = \mathbf{u}_t$, the white-noise structure of $\mathbf{U}_t$ implies that $\mathbf{U}_t = \mathbf{J}'\mathbf{J}\mathbf{U}_t$. These simple relationships yield

$$\mathbf{J}\mathbf{Y}_t = \mathbf{J}\boldsymbol{\mu} + \mathbf{J}\sum_{i=0}^{\infty} \mathbf{A}^i \mathbf{U}_{t-i}$$

$$\mathbf{y}_t = \mathbf{J}\boldsymbol{\mu} + \sum_{i=0}^{\infty} \mathbf{J}\mathbf{A}^i \mathbf{J}'\mathbf{J}\mathbf{U}_{t-i}$$

$$\mathbf{y}_t = \boldsymbol{\mu} + \sum_{i=0}^{\infty} \boldsymbol{\Phi}_i \mathbf{u}_{t-i}$$

where $\boldsymbol{\mu} = \mathbf{J}\boldsymbol{\mu}$ and $\boldsymbol{\Phi}_i = \mathbf{J}\mathbf{A}^i\mathbf{J}'$. The moving-average coefficients, $\boldsymbol{\Phi}_i$, are also known as the impulse–response functions at horizon i. See [TS] **varirf** for information on obtaining VAR impulse–response functions in Stata. See Lütkepohl (1993) and Hamilton (1994) for general introductions to impulse–response functions.

The above discussion ignores the fact that var and var svar fit VAR(p) models that may include exogenous variables, but if these exogenous variables are covariance stationary, the stability of the VAR(p) depends only on the endogenous variables. However, for the impulse–response functions and the forecast-error variance decompositions to have their causal interpretations, the exogenous variables must also be strictly exogenous.

Saved Results

varstable saves in r():

Matrices
r(Re)	real part of the eigenvalues of A
r(Im)	imaginary part of the eigenvalues of A
r(Modulus)	modulus of the eigenvalues of A

Methods and Formulas

varstable is implemented as an ado-file.

varstable forms the matrix $\mathbf{A}$ in (4) and obtains its eigenvalues using matrix eigenvalues. The modulus of the complex eigenvalue $r + ci$ is $\sqrt{r^2 + c^2}$. As shown by Lütkepohl (1993) and Hamilton (1994), the VAR(p) is stable if the modulus of each eigenvalue of $\mathbf{A}$ is strictly less than 1.

References

Hamilton, J. D. 1994. *Time Series Analysis*. Princeton: Princeton University Press.

Lütkepohl, H. 1993. *Introduction to Multiple Time Series Analysis*. 2d ed. New York: Springer.

Also See

Complementary:	[TS] **var**, [TS] **var svar**, [TS] **varbasic**
Background:	[TS] **var intro**

Title

> **varwle** — Obtain Wald lag exclusion statistics after var or svar

Syntax

<code>varwle [, <u>est</u>imates(estname) <u>sep</u>arator(#)]</code>

varwle may only be used with var or svar e() results; see [TS] **var** and [TS] **var svar**.

Description

After fitting a VAR(p), one hypothesis of interest is that all the endogenous variables at a given lag are jointly zero. varwle reports Wald tests of this hypothesis for each equation and for all equations jointly.

See [TS] **var** and [TS] **var svar** for an introduction to fitting VARs and SVARs in Stata. See Lütkepohl (1993), Hamilton (1994), and Amisano and Giannini (1997) for an introduction to VARs and SVARs.

Options

<u>est</u>imates(estname) specifies that varwle is to use the previously obtained set of var or svar estimates saved as estname. By default, varwle uses the currently active e() results. See [R] **estimates** for information on saving and restoring e() results.

<u>sep</u>arator(#) specifies how often separator lines should be drawn between rows. By default, separator lines do not appear.

Remarks

varwle uses the e() results from a previously fitted var or var svar. By default, varwle uses the active e() results, but one may also use a stored set of estimates by specifying the estimates() option.

If the VAR(p) was fitted with the small option, then varwle also presents small-sample F statistics; otherwise, varwle presents large-sample chi-squared statistics.

▷ Example

Let's begin with an example using the Lütkepohl (1993) data. We quietly fit a VAR previously considered in [TS] **var**, and then run varwle.

```
. use http://www.stata-press.com/data/r8/lutkepohl
(Quarterly SA West German macro data, Bil DM, from Lutkepohl 1993 Table E.1)
. quietly var dlinvestment dlincome dlconsumption if qtr <= q(1978q4),
> lags(1/2) dfk small
. varwle
```

Equation: dlinvestment

Lag	F	df	df_r	Prob > F
1	2.649	3	66	0.056
2	1.258	3	66	0.296

Equation: dlincome

Lag	F	df	df_r	Prob > F
1	2.193	3	66	0.097
2	0.907	3	66	0.442

Equation: dlconsumption

Lag	F	df	df_r	Prob > F
1	1.808	3	66	0.154
2	5.576	3	66	0.002

Equation: All

Lag	F	df	df_r	Prob > F
1	3.789	9	66	0.001
2	2.968	9	66	0.005

Since the VAR(2) was fitted with the **dfk** and the **small** options, **varwle** used the small-sample estimator of $\widehat{\Sigma}$ in constructing the VCE, and it produced an F statistic. It appears that the first two equations may have a different lag structure than the third. In the first two equations, it is not possible to reject the null hypothesis that all three endogenous variables have zero coefficients at the second lag. The hypothesis that all three endogenous variables have zero coefficients at the first lag can be rejected at the 10% level for each of the first two equations. In contrast, in the third equation, the coefficients on the second lag of the endogenous variables are jointly significant, but not those on the first lag. However, the hypothesis that the coefficients on the first lag of the endogenous variables are zero in all three equations jointly is strongly rejected. Similarly, the hypothesis that the coefficients on the second lag of the endogenous variables are zero in all three equations jointly is strongly rejected.

If we believe these results strongly enough, we might want to refit the original VAR(2), placing some constraints on the coefficients. See [TS] **var** for the details on how to fit VAR(p) models with constraints.

◁

(*Continued on next page*)

▷ Example

Here we fit a simple SVAR, and then run `varwle`:

```
. matrix a = (.,0\.,.)
. matrix b = I(2)
. svar dlincome dlconsumption, aeq(a) beq(b)
Iteration 0:    log likelihood = -163.59202
Iteration 1:    log likelihood =  490.93031
Iteration 2:    log likelihood =  529.02422
Iteration 3:    log likelihood =  573.77591
Iteration 4:    log likelihood =  578.03111
Iteration 5:    log likelihood =  578.27621
Iteration 6:    log likelihood =  578.27699
Iteration 7:    log likelihood =  578.27699

Structural vector autoregression

Constraints:
 ( 1)   [a_1_2]_cons = 0
 ( 2)   [b_1_1]_cons = 1
 ( 3)   [b_1_2]_cons = 0
 ( 4)   [b_2_1]_cons = 0
 ( 5)   [b_2_2]_cons = 1
```

| Sample: 1960q4 1982q4 | Number of obs | = | 89 |
| | Log likelihood | = | 578.27699 |

Just-identified model

Equation	Obs	Parms	RMSE	R-sq	chi2	P
dlincome	89	5	.011472	0.0918	9.000554	0.0611
dlconsumpt~n	89	5	.01005	0.2037	22.76921	0.0001

VAR Model lag order selection statistics

FPE	AIC	HQIC	SBIC	LL	Det(Sigma_ml)
9.752e-09	-12.770269	-12.657562	-12.490647	578.27699	7.788e-09

	Coef.	Std. Err.	z	P>|z|	[95% Conf. Interval]
a_1_1						
_cons	89.72411	6.725107	13.34	0.000	76.54315	102.9051
a_2_1						
_cons	-64.73622	10.67698	-6.06	0.000	-85.66271	-43.80973
a_2_2						
_cons	126.2964	9.466318	13.34	0.000	107.7428	144.8501

The output table from `var svar` provides information about the estimates of the parameters in the **A** and **B** matrices in the structural VAR. But, as discussed in [TS] **var svar**, an SVAR model builds on an underlying VAR(p). When `varwle` uses the e() results produced by `var svar`, it performs Wald lag exclusion tests on the underlying VAR(p) model. Now, we run `varwle` on these svar e() results.

(Continued on next page)

```
. varwle, separator(2)
```

Equation: dlincome

Lag	chi2	df	Prob > chi2
1	6.888	2	0.032
2	1.874	2	0.392

Equation: dlconsumption

Lag	chi2	df	Prob > chi2
1	9.939	2	0.007
2	13.900	2	0.001

Equation: All

Lag	chi2	df	Prob > chi2
1	34.543	4	0.000
2	19.441	4	0.001

Now, let's quietly fit the underlying VAR(2), and apply varwle to these e() results.

```
. quietly var dlincome dlconsumption
. varwle, separator(2)
```

Equation: dlincome

Lag	chi2	df	Prob > chi2
1	6.888	2	0.032
2	1.874	2	0.392

Equation: dlconsumption

Lag	chi2	df	Prob > chi2
1	9.939	2	0.007
2	13.900	2	0.001

Equation: All

Lag	chi2	df	Prob > chi2
1	34.543	4	0.000
2	19.441	4	0.001

Since varwle produces the same results in these two cases, we can conclude that when varwle is applied to svar e() results, it performs Wald lag exclusion tests on the underlying VAR(p).

◁

(Continued on next page)

Saved Results

varwle saves in r():

Matrices

if e(small)==""

r(chi2)	χ^2 test statistics
r(df)	degrees of freedom
r(p)	p-values

if e(small)!=""

r(F)	F test statistics
r(df_r)	numerator degrees of freedom
r(df)	denominator degree of freedom
r(p)	p-values

Methods and Formulas

varwle is implemented as an ado-file.

varwle uses test to obtain Wald statistics of the hypotheses that all the endogenous variables at a given lag are jointly zero, for each equation and for all equations jointly. Like test, varwle will use estimation results saved by var or var svar to determine whether to calculate and report small-sample F statistics or large-sample chi-squared statistics.

References

Amisano, G. and C. Giannini. 1997. *Topics in Structural VAR Econometrics*. 2d ed. Heidelberg: Springer.

Hamilton, J. D. 1994. *Time Series Analysis*. Princeton: Princeton University Press.

Lütkepohl, H. 1993. *Introduction to Multiple Time Series Analysis*. 2d ed. New York: Springer.

Also See

Complementary:	[TS] **var**, [TS] **var svar**, [TS] **varbasic**
Background:	[TS] **var intro**

Title

> **wntestb** — Bartlett's periodogram-based test for white noise

Syntax

> wntestb *varname* [if *exp*] [in *range*] [, table level(*#*) plot(*plot*)
>
> *scatter_options* *twoway_options*]

wntestb is for use with time-series data; see [TS] **tsset**. You must tsset your data before using wntestb. In addition, the time series must be dense (nonmissing and no gaps in the time variable) in the specified sample.

varname may contain time-series operators; see [U] **14.4.3 Time-series varlists**.

Description

wntestb performs Bartlett's periodogram-based test for white noise. The result is presented graphically by default, but may be optionally presented as text (table output).

Options

table specifies that the test results should be printed as a table instead of as the default graph.

level(*#*) specifies the confidence level, in percent, for the confidence bands included on the graph. The default is level(95) or as set by set level; see [U] **23.6 Specifying the width of confidence intervals**.

plot(*plot*) provides a way to add other plots to the generated graph; see [G] *plot_option*.

scatter_options affect the rendition of the plotted points; see [G] **graph twoway scatter**.

twoway_options are any of the options documented in [G] *twoway_options* excluding by(). These include options for titling the graph (see [G] *title_options*) and options for saving the graph to disk (see [G] *saving_option*).

Remarks

Bartlett's test is a test of the null hypothesis that the data come from a white-noise process of uncorrelated random variables having a constant mean and a constant variance.

For a discussion of this test, see Bartlett (1955, 92–94), Newton (1988, 172), or Newton (1996).

▷ Example

In this example, we generate two time series and show the graphical and statistical tests that can be obtained from this command. The first time series is a white-noise process, and the second is a white-noise process with an embedded deterministic cosine curve.

```
. drop _all
. set seed 12393
. set obs 100
obs was 0, now 100
. gen x1 = invnorm(uniform())
. gen x2 = invnorm(uniform()) + cos(2*_pi*(_n-1)/10)
. gen time = _n
. tsset time
        time variable:  time, 1 to 100
```

We can then submit the white-noise data to the `wntestb` command by typing

```
. wntestb x1
```

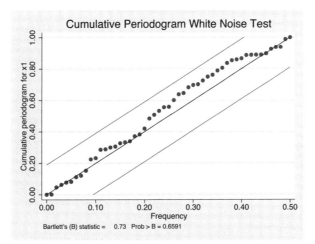

We can see in the graph that the values never appear outside of the confidence bands. We also note that the test statistic has a p-value of .66, so we would conclude that the process is not different from white noise. If we had only wanted the statistic without the plot, we could have used the `table` option.

Turning our attention to the other series (x2), we type

```
. wntestb x2
```

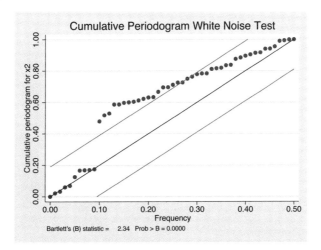

Here the process does appear outside of the bands. In fact, it steps out of the bands at a frequency of .1 (exactly as we synthesized this process). We also have confirmation from the test statistic, at a p-value less than .0001, that the process is significantly different from white noise.

◁

Saved Results

wntestb saves in r():

Scalars

r(stat)	Bartlett's statistic	r(p)	Probability value

Methods and Formulas

wntestb is implemented as an ado-file.

If $x(1), \ldots, x(T)$ is a realization from a white-noise process with variance σ^2, the spectral distribution would be given by $F(\omega) = \sigma^2 \omega$ for $\omega \in [0, 1]$, and we would expect the cumulative periodogram (see [TS] **cumsp**) of the data to be close to the points $S_k = k/q$ for $q = [n/2] + 1, k = 1, \ldots, q$. $[n/2]$ is the maximum integer less than or equal to $n/2$.

Except for $\omega = 0$ and $\omega = .5$, the random variables $2\widehat{f}(\omega_k)/\sigma^2$ are asymptotically independent and identically distributed as χ_2^2. Since χ_2^2 is the same as twice a random variable distributed exponentially with mean 1, the cumulative periodogram has approximately the same distribution as the ordered values from a uniform (on the unit interval) distribution. Feller (1948) shows that this results in (where U_k is the ordered uniform quantile)

$$\lim_{q \to \infty} \Pr\left(\max_{1 \le k \le q} \sqrt{q}\left|U_k - \frac{k}{q}\right| \le a\right) = \sum_{j=-\infty}^{\infty} (-1)^j e^{-2a^2 j^2} = G(a)$$

The Bartlett statistic is computed as

$$B = \max_{1 \le k \le q} \sqrt{\frac{n}{2}}\left|\widehat{F}_k - \frac{k}{q}\right|$$

where $\widehat{F}_k$ is the cumulative periodogram defined in terms of the sample spectral density $\widehat{f}$ (see [TS] **pergram**) as

$$\widehat{F}_k = \frac{\sum_{j=1}^{k} \widehat{f}(\omega_j)}{\sum_{j=1}^{q} \widehat{f}(\omega_j)}$$

The associated p-value for the Bartlett statistic and the confidence bands on the graph are computed as $1 - G(B)$ using Feller's result.

Maurice S. Bartlett (1910-2002) was a British statistician. Apart from a short period in industry, he spent his career teaching and researching at the universities of Cambridge, Manchester, London (University College) and Oxford. His many contributions include work on the statistical analysis of multivariate data (especially factor analysis) and time series and on stochastic models of population growth, epidemics, and spatial processes.

Acknowledgment

`wntestb` is based on the `wntestf` command by H. Joseph Newton (1996), Department of Statistics, Texas A&M University.

References

Bartlett, M. S. 1955. *An Introduction to Stochastic Processes with Special Reference to Methods and Applications*. Cambridge: Cambridge University Press.

Feller, W. 1948. On the Kolmogorov–Smirnov theorems for empirical distributions. *Annals of Mathematical Statistics* 19: 177–189.

Newton, H. J. 1988. *TIMESLAB: A Time Series Laboratory*. Pacific Grove, CA: Wadsworth & Brooks/Cole.

——. 1996. sts12: A periodogram-based test for white noise. *Stata Technical Bulletin* 34: 36–39. Reprinted in *Stata Technical Bulletin Reprints*, vol. 6, pp. 203–207.

Olkin, I. 1989. A conversation with Maurice Bartlett. *Statistical Science* 4: 151–163.

Also See

Complementary:	[TS] **tsset**
Related:	[TS] **corrgram**, [TS] **cumsp**, [TS] **pergram**, [TS] **wntestq**
Background:	*Stata Graphics Reference Manual*

Title

> **wntestq** — Portmanteau (Q) test for white noise

Syntax

> wntestq *varname* [if *exp*] [in *range*] [, lags(#)]

wntestq is for use with time-series data; see [TS] **tsset**. You must tsset your data before using wntestq. In addition, the time series must be dense (nonmissing and no gaps in the time variable) in the specified sample.

varname may contain time-series operators; see [U] **14.4.3 Time-series varlists**.

Description

wntestq performs the portmanteau (or Q) test for white noise.

Options

lags(#) specifies the number of autocorrelations to calculate. The default is to use $\min([n/2]-2, 40)$ where $[n/2]$ is the greatest integer less than or equal to $n/2$.

Remarks

Box and Pierce (1970) developed a portmanteau test of white noise that was refined by Ljung and Box (1978). Also see Diggle (1990, section 2.5).

▷ Example

In the example shown in [TS] **wntestb**, we generated two time series. One (x1) was a white-noise process, and the other (x2) was a white noise process with an embedded cosine curve. Here we compare the output of the two tests.

```
. set seed 12393
. set obs 100
obs was 0, now 100
. gen x1 = invnorm(uniform())
. gen x2 = invnorm(uniform()) + cos(2*_pi*(_n-1)/10)
. gen time = _n
. tsset time
        time variable:  time, 1 to 100
. wntestb x1, table
Cumulative periodogram white-noise test
```

Bartlett's (B) statistic	=	0.7311
Prob > B	=	0.6591

```
. wntestq x1

Portmanteau test for white noise
```

Portmanteau (Q) statistic	=	27.2378
Prob > chi2(40)	=	0.9380

```
. wntestb x2, table

Cumulative periodogram white-noise test
```

Bartlett's (B) statistic	=	2.3364
Prob > B	=	0.0000

```
. wntestq x2

Portmanteau test for white noise
```

Portmanteau (Q) statistic	=	182.2446
Prob > chi2(40)	=	0.0000

This example shows that both tests agree. For the first process, the Bartlett and portmanteau result in nonsignificant test statistics: a p-value of 0.6591 for wntestb, and one of 0.9380 for wntestq.

For the second process, each of the tests has a significant result to less than 0.0001.

◁

Saved Results

wntestq saves in r():

Scalars

r(stat)	Q statistic	r(p)	Probability value
r(df)	degrees of freedom		

Methods and Formulas

wntestq is implemented as an ado-file.

The portmanteau test relies on the fact that if $x(1), \ldots, x(n)$ is a realization from a white-noise process, then

$$Q = n(n+2) \sum_{j=1}^{m} \frac{1}{n-j} \, \widehat{\rho}^2(j) \longrightarrow \chi_m^2$$

where m is the number of autocorrelations calculated (equal to the number of lags specified) and $\longrightarrow$ indicates convergence in distribution to a χ^2 distribution with m degrees of freedom. $\widehat{\rho}_j$ is the estimated autocorrelation for lag j; see [TS] **corrgram** for details.

References

Box, G. E. P. and D. A. Pierce. 1970. Distribution of residual autocorrelations in autoregressive-integrated moving average time series models. *Journal of the American Statistical Association* 65: 1509–1526.

Diggle, P. J. 1990. *Time Series: A Biostatistical Introduction.* Oxford: Oxford University Press.

Ljung, G. M. and G. E. P. Box. 1978. On a measure of lack of fit in time series models. *Biometrika* 65: 297–303.

Sperling, R. and C. F. Baum. 2001. sts19: Multivariate portmanteau (Q) test for white noise. *Stata Technical Bulletin* 60: 39–41. Reprinted in *Stata Technical Bulletin Reprints*, vol. 10, pp. 373–375.

Also See

Complementary: [TS] **tsset**

Related: [TS] **corrgram**, [TS] **cumsp**, [TS] **wntestb**

Title

xcorr — Cross-correlogram for bivariate time series

Syntax

xcorr *varname*$_1$ *varname*$_2$ $\left[\text{if } exp\right]$ $\left[\text{in } range\right]$ $\left[\text{, } \underline{\text{gen}}\text{erate}(newvarname) \ \underline{\text{lags}}(\#)\right.$

$\underline{\text{tab}}\text{le noplot plot}(plot)$ *dropline_options twoway_options* $\left.\right]$

xcorr is for use with time-series data; see [TS] **tsset**. You must tsset your data before using xcorr.

varname$_1$ and *varname*$_2$ may contain time-series operators; see [U] **14.4.3 Time-series varlists**.

Description

xcorr plots the sample cross-correlation function.

Options

$\underline{\text{generate}}(newvarname)$ specifies a new variable to contain the cross-correlation values.

lags(#) indicates the number of lags and leads to include in the graph. The default is to use $\min(\lfloor n/2 \rfloor - 2, 20)$.

table requests that the results be presented as a table rather than the default graph.

noplot requests that the tabular output not include the character-based plot of the cross-correlations.

plot(*plot*) provides a way to add other plots to the generated graph. See [G] *plot_option*.

dropline_options affect the rendition of the plotted cross-correlations; see [G] **graph twoway dropline**.

twoway_options are any of the options documented in [G] *twoway_options* excluding by(). These include options for titling the graph (see [G] *title_options*) and options for saving the graph to disk (see [G] *saving_option*).

Remarks

▷ Example

We have a bivariate time series (Series J, Box, Jenkins, and Reinsel 1994) on the input and output of a gas furnace, where 296 paired observations on the input gas rate and output percent CO_2 were recorded every 9 seconds. The cross-correlation function is given by

(Continued on next page)

341

```
. use http://www.stata-press.com/data/r8/furnace
(TIMESLAB: Gas furnace)

. xcorr input output, xline(5) lags(40)
```

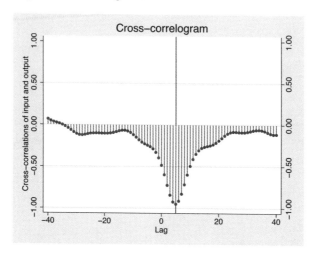

Note that we included a vertical line at lag 5, as there is a well-defined peak at this value. This peak indicates that the output lags the input by 5 time periods. Further, the fact that the correlations are negative indicates that as input (coded gas rate) is increased, output ($\%CO_2$) decreases.

We may obtain the table of autocorrelations and the character-based plot of the cross-correlations (analogous to the univariate time-series command `corrgram`) by specifying the `table` option.

```
. xcorr input output, table lags(20)
                          -1       0       1
      LAG      CORR      [Cross-correlation]

      -20    -0.1033              |
      -19    -0.1027              |
      -18    -0.0998              |
      -17    -0.0932              |
      -16    -0.0832              |
      -15    -0.0727              |
      -14    -0.0660              |
      -13    -0.0662              |
      -12    -0.0751              |
      -11    -0.0927              |
      -10    -0.1180              |
       -9    -0.1484            --|
       -8    -0.1793            --|
       -7    -0.2059            --|
       -6    -0.2266            --|
       -5    -0.2429            ---|
       -4    -0.2604            ---|
       -3    -0.2865            ---|
       -2    -0.3287            ---|
       -1    -0.3936           ----|
        0    -0.4845           ----|
        1    -0.5985          -----|
        2    -0.7251         ------|
        3    -0.8429        -------|
        4    -0.9246       --------|
        5    -0.9503       --------|
```

```
 6      -0.9146    ——————
 7      -0.8294    —————
 8      -0.7166    ————
 9      -0.5998    ———
10      -0.4952    ——
11      -0.4107    ——
12      -0.3479    —
13      -0.3049    —
14      -0.2779    —
15      -0.2632    —
16      -0.2548    —
17      -0.2463    —
18      -0.2332    —
19      -0.2135    —
20      -0.1869    —
```

Once again, the well-defined peak is apparent in the plot.

◁

Methods and Formulas

The cross-covariance function of lag k for time series x_1 and x_2 is given by

$$\text{Cov}\left\{x_1(t), x_2(t+k)\right\} = R_{12}(k)$$

Note that this function is not symmetric about lag zero; that is,

$$R_{12}(k) \neq R_{12}(-k)$$

Define the cross-correlation function as

$$\rho_{ij}(k) = \text{Corr}\left\{x_i(t), x_j(t+k)\right\} = \frac{R_{ij}(k)}{\sqrt{R_{ii}(0)R_{jj}(0)}}$$

where ρ_{11} and ρ_{22} are the autocorrelation functions for x_1 and x_2, respectively. The sequence $\rho_{12}(k)$ is the cross-correlation function, and is drawn for lags $k \in (-Q, -Q+1, \ldots, -1, 0, 1, \ldots, Q-1, Q)$.

Note that if $\rho_{12}(k) = 0$ for all lags, then we say that x_1 and x_2 are not cross-correlated.

References

Box, G. E. P., G. M. Jenkins, and G. C. Reinsel. 1994. *Time Series Analysis: Forecasting and Control*. 3d ed. Englewood Cliffs, NJ: Prentice–Hall.

Hamilton, J. D. 1994. *Time Series Analysis*. Princeton: Princeton University Press.

Newton, H. J. 1988. *TIMESLAB: A Time Series Laboratory*. Pacific Grove, CA: Wadsworth & Brooks/Cole.

Also See

Complementary: [TS] **tsset**

Related: [TS] **corrgram**, [TS] **pergram**

Subject and author index

This is the subject and author index for the *Stata Time-Series Reference Manual*. Readers interested in topics other than time series or graphics should see the combined subject index at the end of Volume 4 of the *Stata Base Reference Manual*, which indexes the *Stata Base Reference Manual*, the *Stata User's Guide*, the *Stata Programming Reference Manual*, the *Stata Cluster Analysis Reference Manual*, the *Stata Cross-Sectional Time-Series Reference Manual*, the *Stata Survey Data Reference Manual*, the *Stata Survival Analysis & Epidemiological Tables Reference Manual*, and this manual.

Readers interested in graphics topics should see the index at the end of the *Stata Graphics Reference Manual*.

Semicolons set off the most important entries from the rest. Sometimes no entry will be set off with semicolons; this means all entries are equally important.

A

Abraham, B., [TS] **tssmooth**, [TS] **tssmooth dexponential**, [TS] **tssmooth exponential**, [TS] **tssmooth hwinters**, [TS] **tssmooth shwinters**
ac command, [TS] **corrgram**
Akaike, H., [TS] **varsoc**
Amemiya, T., [TS] **varsoc**
Amisano, G., [TS] **var intro**, [TS] **var svar**, [TS] **vargranger**, [TS] **varirf create**, [TS] **varwle**
Ansley, C. F., [TS] **arima**
ARCH, [TS] **arch**
ARCH effects,
    testing for, [TS] **regression diagnostics**
    estimation, [TS] **arch**
arch command, [TS] **arch**
archlm command, [TS] **regression diagnostics**
ARIMA, [TS] **arch**, [TS] **arima**
arima command, [TS] **arima**
ARMA, [TS] **arch**, [TS] **arima**
autocorrelation, [TS] **arch**, [TS] **arima**, [TS] **corrgram**, [TS] **prais**, [TS] **newey**, [TS] **regression diagnostics**, [TS] **var**, [TS] **varlmar**
autoregression, [TS] **var** or *see* vector autoregression
autoregressive conditional heteroskedasticity, [TS] **arch**, [TS] **regression diagnostics**
autoregressive integrated moving average, [TS] **arch**, [TS] **arima**
autoregressive moving average, [TS] **arch**, [TS] **arima**

B

Bartlett, M. S., [TS] **wntestb**
Bartlett's bands, [TS] **corrgram**
Bartlett's periodogram test, [TS] **wntestb**

Baum, C. F., [TS] **arch**, [TS] **arima**, [TS] **dfgls**, [TS] **regression diagnostics**, [TS] **tsset**, [TS] **wntestq**
Becketti, S., [TS] **corrgram**
Bera, A. K., [TS] **arch**, [TS] **varnorm**
Berndt, E. K., [TS] **arch**, [TS] **arima**
bgodfrey command, [TS] **regression diagnostics**
Black, F., [TS] **arch**
Bollerslev, T., [TS] **arch**, [TS] **arima**
Bowerman, B., [TS] **tssmooth**, [TS] **tssmooth dexponential**, [TS] **tssmooth exponential**, [TS] **tssmooth hwinters**, [TS] **tssmooth shwinters**
Box, G. E. P., [TS] **arch**, [TS] **arima**, [TS] **corrgram**, [TS] **cumsp**, [TS] **dfuller**, [TS] **pergram**, [TS] **pperron**, [TS] **wntestq**, [TS] **xcorr**
Breusch–Godfrey test, [TS] **regression diagnostics**

C

Chatfield, C., [TS] **corrgram**, [TS] **pergram**, [TS] **tssmooth**, [TS] **tssmooth dexponential**, [TS] **tssmooth exponential**, [TS] **tssmooth hwinters**, [TS] **tssmooth ma**, [TS] **tssmooth shwinters**
Chatterjee, S., [TS] **prais**
Cheung, Y., [TS] **dfgls**
Chou, R. Y., [TS] **arch**
Christiano, L. J., [TS] **varirf create**
Cochrane, D., [TS] **prais**
Cochrane–Orcutt regression, [TS] **prais**
conditional variance, [TS] **arch**
correlogram, [TS] **corrgram**
corrgram command, [TS] **corrgram**
cross-correlogram, [TS] **xcorr**
Csaki, F., [TS] **varsoc**
cumsp command, [TS] **cumsp**
cumulative spectral distribution, empirical, [TS] **cumsp**

D

data manipulation, [TS] **tsappend**, [TS] **tsfill**, [TS] **tsreport**, [TS] **tsset**
David, J. S., [TS] **arima**
Davidson, R., [TS] **arch**, [TS] **arima**, [TS] **regression diagnostics**, [TS] **varlmar**
dfgls command, [TS] **dfgls**
dfuller command, [TS] **dfuller**
Dickens, R., [TS] **prais**
Dickey, D. A., [TS] **dfgls**, [TS] **dfuller**, [TS] **pperron**
Dickey–Fuller test, [TS] **dfuller**, [TS] **dfgls**
Diggle, P. J., [TS] **arima**, [TS] **wntestq**
DiNardo, J., [TS] **arch**, [TS] **prais**, [TS] **regression diagnostics**
Ding, Z., [TS] **arch**
double exponential smoothing, [TS] **tssmooth dexponential**
Durbin, J., [TS] **prais**, [TS] **regression diagnostics**

W

X

Y

Z